Lecture Notes in Logistics

More information about this series at http://www.springer.com/series/11220

Uwe Clausen · Sven Langkau ·
Felix Kreuz
Editors

Advances in Production, Logistics and Traffic

Proceedings of the 4th Interdisciplinary Conference on Production Logistics and Traffic 2019

Springer

Editors
Uwe Clausen
Institute of Transport Logistics
TU Dortmund University
Dortmund, Germany

Sven Langkau
Institute of Transport Logistics
TU Dortmund University
Dortmund, Germany

Felix Kreuz
Institute of Transport Logistics
TU Dortmund University
Dortmund, Germany

ISSN 2194-8917 ISSN 2194-8925 (electronic)
Lecture Notes in Logistics
ISBN 978-3-030-13534-8 ISBN 978-3-030-13535-5 (eBook)
https://doi.org/10.1007/978-3-030-13535-5

Library of Congress Control Number: 2019932165

This Springer imprint is published by the registered company Springer Nature Switzerland AG
The registered company address is: Gewerbestrasse 11, 6330 Cham, Switzerland

Foreword

The series of Interdisciplinary Conferences on Production, Logistics and Traffic (ICPLT) address the research community as well as practitioners in these fields with special attention to links and interfaces between the three disciplines. The fourth ICPLT in particular deals with technology from intralogistics to automated trucking driving as well as the societal aspects of commercial transport.

It took place on March 27–28, 2019, at TU Dortmund University, representing a joint effort by TU Dortmund University and TU Darmstadt University. Transport demand is driving by production and consumption demand, affected by manifold trends such as new production technologies, maintenance, or control methods.

Urbanization and e-commerce are changing our cities, offering tremendous opportunities as well as difficult challenges. High-frequency supply with small delivery quantities meets a heavy-loaded infrastructure. Urban freight management may have to adapt more rapidly. With consumers getting more and more digitally fluent and socially conscious, redesign and optimization of logistics systems becomes a constant effort.

International networks follow the intensified division of labor and exploit the capabilities of all transport modes, recently facing rising trade tensions and uncertainties regarding the economic and legal framework. We observe all these challenges in a time with fast progress in computing technology and unseen capabilities from simulation to artificial intelligence at our disposal. With commercial transport being a major prerequisite for prosperity, it is at the same time driving negative effects on air quality, climate, noise, or traffic safety.

Against this background, the 4th ICPLT closely cooperates with the Dortmund-based Center of Excellence Logistics and IT in order to strengthen research excellence in the scientific fields of production, logistics, and traffic. The Center of Excellence Logistics and IT—within a larger initiative of federal state and Fraunhofer society—wants to positioning logistics as the central science of the industrial future and strengthening excellent logistics research at the interface between logistics and IT. The research group will draw up a comprehensive road map with strategic topics that may serve as a plan of action for logistics research in the coming years and will form the basis for new research and transfer programmes in logistics and information logistics, the systematic combination of logistics and information technology.

Center of Excellence Logistics and IT Partners are TU Dortmund University (with five chairs), the Fraunhofer Institute for Material Flow and Logistics (IML) and for Software and Systems Engineering (ISST), the Leibniz Research Centre for Working Environment and Human Factors (IfADo), companies like Boehringer Ingelheim Pharma, Deutsche Telekom AG, EffizienzCluster Management GmbH, and numerous initiatives like International Data Spaces Association, Digital.Hub Logistics, Digital in NRW—Kompetenzzentrum Mittelstand 4.0, Graduate School of Logistics, Innovationlab Hybrid Services in Logistics, and more. It is great to be part of this effort, and I am thankful that the Center of Excellence Logistics and IT is supporting the 4th ICPLT at TU Dortmund University.

To contribute to a high-level and beneficial exchange between authorities in politics and municipalities with researchers and practitioners in production and logistics management, the ICPLT has asked for contributions from the three disciplines to better understand innovative technologies, best practises, and latest results. These contributions have been evaluated and selected based on a double-blind review process to become part of this book. It comprises 21 contributions examining trends and challenges for commercial transport as the essential link for production, logistics, and society. Therefore, innovative technologies and strategies are presented and discussed to better understand the interdependencies and conflicts of interest and to develop feasible solutions.

The focus of this book is on the following core topics:

- Simulation and Optimization in Production and Logistics
- Freight Transport Demand Modeling
- Intralogistics and Logistics Facilities
- Policy and Human Factors
- Production and Maintenance
- Supply Chain Management
- Sustainable Logistics and Energy

As the Chair of the 4th ICPLT, I would like to thank all authors for their contributions, my co-chairs and all members of the Scientific Committee for their advice, everyone in our Organization Committee, especially Dr. Sven Langkau and Felix Kreuz for their efforts and commitment to the success of the conference and this book.

January 2019 Uwe Clausen

Acknowledgements

The editors would like to thank all involved members of the Conference Chair and the Scientific Committee of 4th ICPLT, who reviewed the contributions to assess their scientific and practical relevance, quality, and originality. Without the untiring and voluntary effort of the Scientific Committee, this publication would not have been possible. In particular, the editors also would like to thank all authors for sharing their work and the Organization Committee for their effort for making 4th ICPLT possible.

Scientific Committee of 4th ICPLT 2019

Alexander Eisenkopf	Lehrstuhl für Wirtschafts- und Verkehrspolitik, Zeppelin University, Friedrichshafen, Germany
Pietro Evangelista	Research Institute on Innovation and Services for Development, National Research Council Naples, Italy
Nathalie Fabbe-Costes	Transportation and Logistics, Aix-Marseille Université, France
Heike Flämig	Institut für Verkehrsplanung und Logistik, Technische Universität Hamburg-Harburg, Germany
Hanno Friedrich	Freight Transportation-Modelling and Policy, Kühne Logistics University, Germany
George A. Giannopoulos	Hellenic Institute of Transport, National Centre for Research & Development, Greece
Christoph Glock	Production and Supply Chain Management, Technische Universität Darmstadt, Germany
Genevieve Giuliano	Ferraro Chair in Effective Government, University of Southern California, USA
Carlos Jahn	Institut für Maritime Logistik, CML - Fraunhofer-Center für Maritime Logistik und Dienstleistungen, Germany
Danuta Kisperska-Moron	Department of Business Logistics, University of Economics in Katowice, Poland
Bernd Kuhlenkötter	Lehrstuhl für Produktionssysteme, Ruhr-University Bochum, Germany
Sebastian Kummer	Transportwirtschaft und Logistik, Wirtschaftsuniversität Wien, Austria
Bert Leerkamp	Lehr- und Forschungsgebiet für Güterverkehrsplanung und Transportlogistik, Bergische Universität Wuppertal, Germany
Tone Lerher	Department of Technical Logistics, University of Maribor, Slovenia
Barbara Lenz	Institut für Verkehrsforschung, German Aerospace Center (DLR), Germany
Gernot Liedtke	Institut für Verkehrsforschung, German Aerospace Center (DLR), Germany
Maria Rosário Macário	Civil Engineering Research and Innovation for Sustainability, University of Lisbon, Portugal
Alan McKinnon	Logistics & Supply Chain Management, Kühne Logistics University, Germany
Joachim Metternich	Institut für Produktionsmanagement, Technologie und Werkzeugmaschinen, Technische Universität Darmstadt, Germany

Herbert Meyr	Supply Chain Management, University of Hohenheim, Germany
Stefan Minner	Logistics and Supply Chain Management, Technische Universität München, Germany
Andreas Norrman	Department of Industrial Management and Logistics, Lund University, Sweden
Hans-Christian Pfohl	Chair of Supply Chain and Network Management, Technische Universität Darmstadt, Germany
Markus Rabe	Chair of IT in Production and Logistics, TU Dortmund University, Germany
Günther Schuh	Lehrstuhl für Produktionssystematik, RWTH Aachen, Germany
Stefan Seuring	Supply Chain Management, University of Kassel, Germany
Wolfgang Stölzle	Logistics Management, University St. Gallen, Switzerland
Frank Straube	Logistics, Technische Universität Berlin, Germany
Lóránt Tavasszy	Faculty of Technology, Policy and Management, Delft University of Technology, Netherlands
Eddy Van De Voorde	Department of Transport an Regional Economics, Universiteit Antwerpen, Belgium
Boris Varlamov	Economics and Strategic Management, ITMO University Saint Petersburg, Russia
Gyula Vastag	Operations and Supply Chain Management, Corvinus University of Budapest, Hungary
Inge Vierth	Transport Economics, VTI Stockholm, Sweden
Carl Marcus Wallenburg	Logistics and Services Management, WHU Vallendar, Germany
Jaroslaw Witkowski	Strategic Management and Logistics, Wroclaw University of Economics, Poland
Johan Woxenius	Industrial and Financial Management & Logistics, University of Gothenburg, Sweden

Organization Committee of 4th ICPLT 2019

Uwe Clausen	Institute of Transport Logistics, TU Dortmund University, Germany
Sven Langkau	Institute of Transport Logistics, TU Dortmund University, Germany
Felix Kreuz	Institute of Transport Logistics, TU Dortmund University, Germany
Christoph Besenfelder	Chair in Enterprise Logistics, TU Dortmund University, Germany
Christoph Pott	Department of Intralogistics and IT Planning, Fraunhofer Institute for Material Flow and Logistics, Germany
Michael Schmidt	Department of Intralogistics and IT Planning, Fraunhofer Institute for Material Flow and Logistics, Germany
André Terharen	Chair of Materials Handling and Warehousing, TU Dortmund University, Germany
Karolin Wywial	Institute of Transport Logistics, TU Dortmund University, Germany

Chair of 4th ICPLT 2019

Prof. Dr.-Ing. Uwe Clausen born in 1964 in Düsseldorf, studied computer sciences at the University of Karlsruhe (TH), and graduated at TU Dortmund University.

After having worked as scientific employee, he became Head of the Traffic Logistics department at Fraunhofer IML. Subsequently, he worked for Deutsche Post AG as Project Manager for logistics and later on as Managing Director of the subsidiary IPP Paketbeförderung GmbH in Austria. In July 1999, he joined Amazon.de in Bad Hersfeld and, finally, he became European Operations Director at Amazon.com before he returned to Fraunhofer IML on February 1, 2001, as one of the directors of the institute. At the same time, he started his work as Director of the Institute of Transport Logistics. From July 2002 until July 2005, he was Dean of the Faculty of Mechanical Engineering at TU Dortmund University, and from July 2005 until August 2008, he was Vice Dean of the very same faculty.

Today, he is Director of the Institute of Transport Logistics at TU Dortmund University and Director of Fraunhofer Institute for Material Flow and Logistics IML. He is also representing Fraunhofer in ECTRI European Conference of Transport Research Institutes and Member of the Advisory Council of the Association of German Transportation Companies (VDV) and the Scientific Advisory Board of the Bundesvereinigung Logistik (BVL) e.V. His research areas include green logistics, commercial traffic modeling, intermodal transportation, mathematical optimization, network optimization, and distribution systems.

Prof. Dr.-Ing. Manfred Boltze is Civil Engineer by profession, and he became Research Associate and obtained his doctoral degree for a study on optimal cycle times in traffic signal control for road networks from TU Darmstadt in 1988. After working as Head of the Department of Transport Planning and Traffic Engineering for Albert Speer & Partner in Frankfurt, he was appointed Chair of the Institute for Transport Planning and Traffic Engineering at TU Darmstadt in 1997. Since his appointment, the institute's activities cover, among others, interdisciplinary research fields such as environmentally induced traffic signal control, freight transport demand management, traffic, and health or sustainable road freight traffic. He was one of the initiators of the interdisciplinary research project "Dynamo PLV" which is the nucleus of the ICPLT conference series.

Prof. Dr. Ralf Ebert is Professor and Chair of Management and Logistics at TU Darmstadt since 2011. From 2009 to 2011, he held an assistant professorship at TU Berlin for Logistics Services and Transportation. His research focuses on the management and planning of transportation networks (especially intermodal transportation networks), specifically on the analysis of freight mode choice decisions, efficiency improvements by information sharing, and measures for increasing utilization of transport capacities. Further research fields are warehouse management and the integration of human factors in intralogistics systems as well as the management of logistics and production networks. Simulation modeling is the preferred research method throughout most of his work.

Prof. Dr. Michael Henke is Director of Fraunhofer Institute for Material Flow and Logistics IML in Dortmund and holds the Chair in Enterprise Logistics (LFO) at the Faculty of Mechanical Engineering at TU Dortmund University. Furthermore, he is Adjunct Professor for supply chain management at the School of Business and Management of Lappeenranta Univcrsity of Technology in Finland.

His research focuses inter alia on management of Industrie 4.0 and platform economy, blockchain and smart contracts, financial supply chain management, supply chain risk management, procurement, logistics, and supply chain management. Doing this, he is combining his practical experience from entrepreneurial practice and his extensive knowledge from research.

He studied brewing and beverage technology (Dipl.-Ing.) and gained his doctorate and habilitation in business and economics at Technical University of Munich, Germany. During and after his habilitation, he worked for the Supply Management Group (SMG) in St. Gallen, Switzerland. From 2007 until 2013, he was active in teaching and research as Professor at European Business School (EBS).

Prof. Dr. Dr. h. c. Michael ten Hompel is Full Professor and holds the Chair of Materials Handling and Warehousing at TU Dortmund University. At the same time, he is Managing Director of Fraunhofer Institute for Material Flow and Logistics and Director of Fraunhofer Institute for Software and Systems Engineering, both in Dortmund. He is also Board Member of BITKOM, Member of acatech (National Academy of Science and Engineering), BVL, and Logistics Hall of Fame, and Founding Member of WGTL (Scientific Association of Technical Logistics). He published more than 400 papers and more than a dozen books.

Contents

Simulation of an Optimized Last-Mile Parcel Delivery Network Involving Delivery Robots

Moritz Poeting(✉), Stefan Schaudt, and Uwe Clausen

Institute of Transport Logistics, Dortmund, Germany
{moritz.poeting,stefan.schaudt}@tu-dortmund.de,
clausen@itl.tu-dortmund.de

Abstract. Delivery robots represent a rather new technology, which has been developed in recent years. These robots drive with walking speed on sidewalks and have unit capacity. The advantage is that these robots are able to drive autonomous and a single operator can manage a fleet of robots. This enables deliveries to be made within a time slot chosen by the customer. Delivery robots are suitable for deliveries of small goods such as groceries, medicine, food, or parcels.

In this contribution, a simulation model for a parcel delivery network on the last-mile is presented. The model represents deliveries from a hub, which is close to an urban area, to customers. In addition to conventional delivery vehicles, the delivery with parcel robots is examined. Decision processes of the simulation are supported by mathematical optimization. In the optimization, two extension of the Traveling Salesman Problem are solved heuristically. To evaluate the simulation we created a case study based on real world data from parcel delivery company Hermes.

Keywords: Delivery robots · Last-mile · Simulation · TSP

1 Introduction

Concentration tendencies of population and economic activities are observable in cities world-wide. Besides, e-commerce demands in the B2C segment as well as offered e-commerce activities within most business sectors (B2B) are increasing [18]. As a result, the parcel delivery industry is seeking new ways to increase efficiency and investigates new technologies for last-mile delivery. Innovative concepts such as drones, autonomous vehicles, sharing economy, deliveries to the trunk of private cars, and parcel robots are tested [7].

Field tests for parcel delivery robots have been carried out by manufacturer Starship Technologies in cooperation with parcel delivery company Hermes. They tested between mid-October 2016 and mid-March 2017 and successfully realized the delivery of parcels including the collection of returns in three districts of Hamburg, Germany. Customers were able to choose an arrival time slot

U. Clausen et al. (Eds.): ICPLT 2019, LNLO, pp. 1–19, 2019.
https://doi.org/10.1007/978-3-030-13535-5_1

of 15 min. In total, the robots completed 600 tours and traveled 3,500 km. After this test in Hamburg, Hermes and Starship started another pilot test in Southwark, London. They tested on-demand collections of parcels from customers' homes. Similar to the pilot test in Hamburg, the customers were able to select a due window of 30 min in which the robot should arrive [3].

In this respect, the application field of autonomous delivery robots on the last-mile represents a promising field of research. Not only for logistics, but also for mathematical optimization and simulation.

This paper is structured as follows: Sect. 2 provides an literature overview. In Sect. 3, the problem of simulating and optimizing a parcel delivery network for the last-mile with delivery robots is described. In Sect. 4, two heuristics to solve extensions of the Traveling Salesman Problem are presented. The developed simulation framework is presented in Sect. 5. In Sect. 6, computational experiments are explained and results are examined. The last Sect. 7 concludes the findings and gives an outlook of future work.

2 Literature Review

In the following chapter, the state of the art of simulation and optimization in the context of last-mile parcel delivery is presented. With respect to the challenges of parcel delivery using autonomous robots the research gap is derived for this approach.

2.1 Simulation of Urban Delivery

In the context of urban logistics, simulation is an important tool for decision support and widely used by several works that focus on urban distribution (e.g. the distribution of food and fast-moving consumer goods (FMCG), or waste collection) using conventional trucks, electric vehicles, or cargo bikes. Most of the existing approaches follow the Agent-Based Simulation (ABS) paradigm (e.g. [2,6,10]).

ABS is a variant of Discrete-Event Simulation (DES) in which entities can actively interact with each other by consistently applying object-oriented programming. Interacting entities, represented by agents, can be for example people, companies, or other target objects. An agent is an autonomous unit that is able to perceive its environment including other agents. Perceived information can be used to make decisions. The reactive behavior of the agents is determined by a series of attributes and decision rules. A proactive behavior is to be represented by the implementation of objectives of the individual agents. In addition, agents have a memory that allows them to learn and adapt. Therefore, they are capable of learning from the behavior of other agents and independently adapt their own behavior by modifying their decision rules. According to Law (2014), ABS can be used when the system has entities that naturally interact with each other and with the environment, when entities learn or adapt behavior, and when entities movements take place according to situational awareness and not to prescribed patterns [11].

Rabe et al. [16] used supply-chain simulation in order to determine the performance of freight delivery in an urban environment. In this context, the application of urban consolidation centers (UCC) and collaboration between different actors are evaluated within a supply-chain. The simulation approach uses real application data from companies in Athens. Decision making in their model is limited to the planning of daily tours, which are optimized according to the load factor of the vehicles. The customer assignment is re-determined by the input data. However, the model uses approximated distances without consideration of the actual urban infrastructure.

A decision support system for urban last-mile distribution was presented by Fikar et al. [5]. Orders are delivered from micro-hubs using both conventional vehicles and cargo-bikes. They use ABS implemented with the simulation software *AnyLogic* to generate and select vehicle routes and transhipment points. The model is based on an architecture that integrates geographic information system (GIS) in ABS with dynamic vehicle routing and was previously used for a case-study in disaster management [6].

Asish et al. [2] have used the simulation software Netlogo to create an ABS model that integrates Particle Swarm Optimization to solve a Dynamic Vehicle Routing Problem (DVRP) for urban freight transport. The model evaluates urban freight transport of sixty-two retailers in Yogyakarta city (Indonesia) and also includes GIS data with dynamism and stochastics. During the simulation experiments, the effects of dynamism are evaluated on logistics performance.

Hofmann et al. [10] combined simulation with tour planning algorithms for cargo-bikes in a GIS-based environment using *AnyLogic*. The tool is applied to evaluate the potential use of cargo bikes for B2B e-commerce deliveries in Grenoble, France. Orders are consolidated in a two-level urban distribution system using a depot outside the city and an UCC at the city-edge for consolidation of freight to and from the depots.

2.2 Optimization

In the simulation, two tour planning problems are solved; both are extensions of the well known Traveling Salesman Problem (TSP). The Traveling Salesman Problem is to find an hamiltonian cycle of minimum cost in a complete graph $G = (V, E)$, starting and returning to the same node.

The first problem that is used in the simulation, is the Traveling Salesman Problem with Precedence Constraints (TSPPC). These precedence constraints require that certain nodes must precede certain other nodes in any feasible tour. An exact solution for the TSPPC using a dynamic program was proposed by Mingozzi et al. [15]. Ascheuer et al. [1] proposed a branch-and-cut algorithm for the asymmetric TSPPC.

Since the TSPPC belongs to the class of NP-hard problem, optimal solutions cannot be obtained within a reasonable computational time for a large instances. In recent years, many heuristics have been proposed for the TSPPC and the Precedence Constrained Sequencing Problem (PCSP). Sung and Jeong [17] introduced an evolutionary algorithm for the TSPPC. The various parameters within the algorithm are adapted (changed) with the number of iterations based on the performance over the iterations. Yun and Moon [22] proposed a GA approach with a topological sort-based representation procedure for solving various types of PCSP. Yun et al. [23] presented a hybrid genetic algorithm to solve the PCSP, which uses an adaptive local search scheme. Maadi et al. [13] proposed a modified Cuckoo-Optimization Algorithm to solve several sequencing problems. The Cuckoo Algorithm is a nature inspired meta-heuristic based on a evolutionary algorithm inspired from the obligate brood parasitic behavior of certain cuckoo species.

The second optimization problem is a Orienteering Problem with Multiple Time Windows (OPMTW), which is an extension of the class of orienteering problems. In a orienteering problem, a set of customers is given, each customer has a profit that is collected, if the tour is included in a tour. The goal is to maximize the collected profit. Tour constraint can be for example maximal duration or length of a tour. Vansteenwegen et al. [21] provided an excellent review of OP, TOP, and TOPTW. However, only a small number of publications considers the OPMTW. A simulated annealing heuristic for the Team Orienteering Problem with Multiple Time Windows is proposed by Lin and Yu [12]. In a team orienteering problems the target is to find a set of tours that maximize the collected profit. Tricoire et al. [20] proposed a variable neighborhood search heuristic for the Multi-Period Orienteering Problem with Multiple Time Windows. In this problem, a OP over a period of three days is considered and time windows of customers can change from day to day.

Considering the various works in this field it is evident that simulation in combination with optimization algorithms for route- and tour planning is the most suitable method for decision support in the context of urban logistics. The topic of parcel robots is a new field in the context of city logistics for which no related works were found during the literature review.

3 Problem Description

In this section, the observed problem is stated. The problem can be described as a simulation of a parcel delivery network for the last-mile involving parcel delivery robots. First, the problem without delivery robots is described. In this case, the parcels get delivered by trucking. The delivery usually starts at a hub location where a set of parcels for a delivery region is consolidated. The delivery region is subdivided into smaller patches where each patch represents one delivery tour. Each patch contains one or more micro-depots, which can be for example a kiosk or a supermarket. Assigned to each micro-depot is a set of customers and in case of a failed delivery, the parcel gets transported to the closest corresponding micro-depot later in the delivery tour. If the vehicle arrives at the micro-depot outside the opening hours, the parcels gets transported back to the hub and a second delivery attempt is made the next day.

In the extension of this model, delivery robots are used. Before the extension is described, a small introduction into parcel delivery with robots is given. A picture of a delivery robot is given in Fig. 1. The main challenge in the parcel delivery using robots arises from the limited driving range, driving speed, and the given infrastructure. Delivery robots have a range of 6 to 10 Km with a single charge. They use sidewalks and therefore, drive with walking speed (in average 3.6 km/h) [7]. The starting point of a parcel delivery with a robot has to be close to the customer and an additional handling step is needed to transfer the parcels to this starting point. During the pilot of Hermes in Hamburg 2016, parcel shops were used as starting points for robot deliveries [4]. Furthermore, parcel robots require that customers pick up their order directly from the robot on the sidewalk. Hermes offered a time window for the delivery, so customers are able to plan the arrival time of their order.

In the extension of the model, we assume that a small percentage of the parcels is delivered with delivery robots. In the following, the delivery of these parcels is described. All other parcels will be delivered as described previously. First, a due window is assigned to all parcels delivered by a robot. The micro-depots represent starting points for a robot delivery and each micro-depot has a single robot assigned. The same assignment between micro-depots and customer from the original model is used to decide from which micro-depot a robot starts. The simulation model calculates, based on a given probability, if a parcel is delivered by a robot or a delivery van. All parcels that are delivered by a robot are consolidated at the hub and loaded on a separate vehicle for a robot delivery. For this set of parcels a tour is calculated, visiting all micro-depots that are used to deliver the set. If the delivery vehicle arrives at a micro-depot, the parcels get unloaded and the starting times for the robot are calculated in a way that the deviation to the due windows is minimized.

In the following two subsections, the calculation and optimization of the truck delivery tour and the robot parcel delivery tour are explained.

Fig. 1. Delivery robot manufactured by ©Starship Technologies

4 Optimization

In this section, two tour planning heuristics are described. The first heuristic solves the tour planning for parcel delivery with trucks and the second algorithm solves the tour planning for a delivery tour that only visits micro-depots.

4.1 Truck Delivery Tour

The task is to calculate an order for a given set of parcels to minimize the total travel distance of the resulting tour. As mentioned in Sect. 3, each customer is assigned to a micro-depot. At the beginning, each parcel should be delivered to a customer. In the event of a failed delivery the parcel is reassigned and then transported to the assigned micro-depot. If the delivery vehicle arrives at a micro-depot outside the opening hours, the parcel is reassigned and transported to the hub. A second delivery attempt for those parcels will be made the next day. The probability of a failed delivery, p_fd, is an input parameter and assumed to be equal for all customers. It is assumed that a micro-depot is only visited once after all assigned customer have been visited.

The described problem can be transformed into a Traveling Salesman Problem with Precedence Constraints (TSPPC). The TSPPC is to find an Hamiltonian cycle of minimum length in a complete graph G = (V, E), starting and returning to the same node, where precedence constraints require that certain nodes must precede certain other nodes in any feasible tour. In our case, all customer nodes, assigned to a micro-depot node, have to precede the micro-depot node. A few assumptions need to be made; first, opening hours of micro-depots are not included in the optimization. Second, uncertainty of a micro-depot visit is not included in the MIP formulation for the TSPPC. These simplifications are needed to find a solution in reasonable time, since TSPPC belongs to NP-hard problems. Let $G = (V, A)$ be a complete graph, where, $V = \{0, 1, \ldots, d, d+1, \ldots, d+c\}$ is the node set and $A = \{(i, j) : i, j \in V, i \neq j\}$ is the arc set. Node 0 represents the hub node, nodes 1 to d the micro-depots, and nodes $d+1$ to $d+c$ the customer nodes. Graph G is partitioned into $d+1$ clusters, where each cluster V_p is identified by p from within a set K of indices, with p = 0 showing the (singleton) cluster that the hub belongs to and V_p, with $p \in K \setminus 0$, is the cluster of the p'th depot that the p'th depot and all its assigned customers belongs to. The travel time between two nodes $i, j \in V$ is represented by the arc weight of arc$(i, j) \in E$. A heuristic solution for the stated problem is described in the following. A pseudocode for this heuristic is provided in the Appendix in Algorithm 1.

A micro-depot visit can be very unlikely, depending on the number of assigned customer nodes and the probability p_fd of a failed delivery. We remove micro-depots nodes with a visiting probability of less than 50% and calculate based on the reduced node set a TSPPC tour. Later the removed nodes are reinserted with a simple insertion heuristic that respects the precedence. Let *removed* be the set of micro-depots with a probability of less than 50% and let V' be the reduced node set, with $V = V' \cup removed$. The probability of micro-depot

$i \in \{1, 2, \ldots, d\}$ can be calculated as the converse probability that all customers in cluster V_i are visited, with the following formula:

$$P(X) = 1 - (1 - p\_fd)^{|V_i|-1}$$

Next, based on node set V' the TSPPC is solved with a Miller-Tucker-Zemlin (MTZ) formulation presented at the end of this subsection (cf. [14]). The solution is calculated using the mixed integer programming solver *Gurobi 7.0.2* and the run time is set to 15 s. To improve the running time, a starting solution based on a nearest neighbor heuristic is calculated. Afterwards, all micro-depots from set *removed* are reinserted with an insertion heuristic.

This heuristic considers the first node of the set *removed* and places it at every possible position of the tour, so that the increase of tour length is minimal and precedence constraints are observed. After a position is found, the node is inserted in the tour and the next node from the set *removed* is taken. This procedure is done until the set *removed* is empty. Note that after the insertion procedure the tour includes all nodes of V. In the event of a failed delivery at a customer, the optimization procedure is called and returns the position of the assigned micro-depot in the tour. In the next paragraph, an example of the described problem is given.

In Fig. 2 a solution for a problem instance with two micro-depots and five customers is displayed. For this instance the probability of a failed delivery is assumed to be equal to 25%. The bold edges display the route calculated without micro-depots that have a probability of less than 50%. The dashed edges display the tour found after reinserting these micro-depots. The first returned tour is $(c_1, c_2, c_3, c_4, c_5)$ and the complete tour would be $(0, c_1, c_2, c_3, c_4, d_1, c_5, d_2, 0)$. If a delivery fails, for example, at the customer c_1, the position of micro-depot d_1 would be returned.

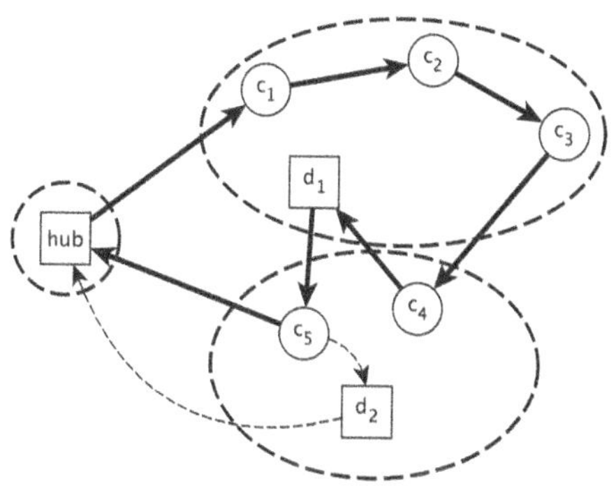

Fig. 2. An example solution to a problem instance with two micro-depots and five customers

The MTZ formulation uses binary variable x_{ij} for each arc $(i,j) \in A$ that is equal to one, if arc (i, j) is traversed. The continuous variables u_i for each node $i \in V$ correspond to the position of node i in the tour. Each tour starts and ends at node 0.

Mixed Integer Program for the TSP with Precedence Constraints

$$\min \quad \sum_{\substack{i,j \in V \\ i \neq j}} t_{ij} x_{ij} \tag{1}$$

$$\text{s. t.} \quad \sum_{\substack{i \in V \\ i \neq j}} x_{ij} = 1 \qquad \forall\, j \in V \tag{2}$$

$$\sum_{\substack{i \in V \\ i \neq j}} x_{ji} = 1 \qquad \forall\, j \in V \tag{3}$$

$$u_i + x_{ij} \leq u_j \qquad \forall\, i \in V, j \in V \setminus \{0\},\ i \neq j \tag{4}$$

$$u_i < u_j \qquad \forall\, j \in \{1, \ldots, d\}, i \in V_j \tag{5}$$

$$0 \leq u_i \leq d + c \qquad \forall\, i \in V \tag{6}$$

$$x_{ij} \in \{0, 1\} \qquad \forall\, i, j \in V, i \neq j \tag{7}$$

The objective (1) is to minimizes the tour length. Constraints (2) and (3) ensure that each node is visited exactly once. MTZ constraint (4) eliminates sub tours and constraint (5) ensures that precedence is observed. Constraints (6)–(7) impose the restrictions on the variables relating to the order and tour decisions.

4.2 Robot Parcel Tour

In this section, the optimization of a truck delivery tour for robot deliveries is considered. The truck has only loaded parcels, which are intended to be delivered by a delivery robot. Since delivery robots always start from a micro-depot, the truck tour only visits those.

Given a single one delivery truck, a set of parcels, micro-depots and their opening hours, and a travel time matrix. The assignment of parcels to micro-depots and of parcels to the truck is assumed to be given. Optimization is used to decide in which order the micro-depots are visited. A micro-depot can only be visited within the opening hours. If the vehicle arrives early, it waits until the micro-depot opens. The objective is to maximize number of micro-depots visited. If two tours found that visit the same number of micro-depots, the tour with a smaller duration is taken. If a micro-depot can not be visited within its opening hours, a second delivery attempt for the assigned parcels is made the next day.

The described problem can be transformed into an Orienteering Problem with Multiple Time Windows. Given is a complete undirected weighted graph $G = (V, E)$, with node set $V = \{0, 1, \ldots, d\}$, whereas node 0 corresponds to the hub and nodes $1, \ldots, d$ to the micro-depot and edge set E. The edge weight,

$c(E) \mapsto \mathbb{R}^+$, corresponds to the travel time between to micro-depots or the hub node and a micro-depots. All edge weights are stored in a matrix, $M \in \mathbb{R}^{n+1} \times \mathbb{R}^{n+1}$. Each node i has a service time $service_i$ and all service times are stored in a list, $service \in \mathbb{R}^{n+1}$. The service time of hub node 0, $service_0$, corresponds to the loading time of the vehicle and for all other nodes this time correspond to the unloading time. Each node $i \in V$ may have several time windows, which corresponds to the opening hours of the micro-depot. All opening hours are stored in a matrix $open \in \mathbb{R}^{n+1} \times \mathbb{R}^K$. Entry $open_{ij} = [l_{ij}, u_{ij}]$ is equal to the $j'th$ time window of node $i \in V$, whereas index j belongs to a set of K_i indices. A vehicle that arrives at time t_i at a node i, is able to serve it if and only if there exists an index $j \in K_i$ and a value $w_i \geq 0$, such that $t_i + w_i \in [l_{ij}, u_{ij}]$. The value w corresponds to the waiting time at node i. If the vehicle is traversing edge $\{i, j\} \in E$, the arrival time t_j at node j is calculated as the sum of $service_i$, $M_{i,j}$, w_i, w_j, and t_i.

In the following, a heuristic to solve the TSPMTW is described. The heuristic is displayed as a pseudocode in the Appendix in Algorithm 2. First, a traveling salesman tour without time windows on graph G is calculated. For this calculation an MTZ formulation is used, which is similar to the formulation given in Subsect. 4.1, but without precedence constraints. The model is solved with MIP solver *Gurobi 7.0.2*. After 30 s the optimization is stopped and the best known integer solution is returned. However, this returned solution may not led to a feasible solution, because time windows are not taken into account. For this reason, arrival times of the returned sequence are calculate as mentioned earlier. The nodes in the tour are checked in ascending order, if the arrival time at a node is not within one of its time windows. In this event, the node is removed from the sequence and arrival times are updated. All removed nodes are stored in the set, *removed*. Next, nodes from the set *removed* are tried to be reinserted such that the resulting tour duration in minimized. First, the set *removed* is shuffled. The first node is taken from the set and reinserted in the tour at all possible positions. The position which results in a minimal tour duration is chosen. If it is not possible to place a node, the next node is considered until all nodes of the set *removed* have been considered. The reinsertion procedure is called $|removed|^2$ times with a different order of the set *removed*. After calling the insertion heuristic $|removed|^2$ times, the solution with a maximal number of nodes is taken. If two tours have the same number of nodes, the tour with the smaller duration is taken. In the end, a sequence of micro-depots and thus, a sequence of parcels to be delivered, can be returned. If a micro-depot is not contained in the tour, all assigned parcels return to the hub node.

5 Simulation Model

In this section, our conception model to simulate the last-mile of parcel delivery using conventional trucks and parcel robots is presented. First, the functional requirements on modeling the last-mile distribution are described. Then, the implementation in an ABS model is presented. We used *AnyLogic*, which is one of the most common and flexible tools for modeling agents in a GIS environment.

5.1 Functional Requirements

As the main focus of the simulation lies in the last-mile of delivery, we set the system boundaries starting with the sorting process in the hub location near the target area. The main run of parcels is not considered in the model. After sorting, we focus on tour planning and delivery using the actual infrastructure for trucks and delivery robots. Therefore, we use the GIS environment which is provided by *AnyLogic* and uses *OpenStreetMaps* (*OSM*). To display the actual distribution of parcels within the urban area correctly, we need to model separate shipment volumes for each district. Therefore, we chose a data driven modeling approach where parcels and customers are generated automatically based on parcel volume information from a database. This includes the name of the district in OSM and the relative share of parcels that are allocated to the area. Furthermore, the addresses of hub and micro-depot locations are stored in a database table and generated when the simulation model is initialized. Thus, we are able to generate simulation models for different urban environments based on the provided data. For the assessment of the tours and deliveries it is required to measure a set of economic and environmental KPIs (e.g. km per tour, duration of a tour, and CO_2 consumption).

5.2 Implementation

Based on the functional requirements, several agent types are created and implemented in *AnyLogic* in a data-driven simulation approach. This includes agents for the hub, loading gates, micro-depots, parcels, customers, regions, trucks, parcel robots, tours, and an evaluation module. Furthermore, there is a main window in which the graphical user interface (GUI) is implemented, see Fig. 3.

In order to generate the model, the simulation tool uses an underlying data model that includes information about the GIS Regions, the share of shipping volumes, a list of micro-depots with their GIS position, and opening hours. The data model also includes the input data for the simulation experiment. Input data are:

- Model parameters (number of parcels p.a., number of delivery trucks, the average rate of failed deliveries, and the average number of parcels per stop),
- truck parameters (average loading time, average stop time, average speed and capacity),
- robot parameters (share of orders for robot delivery, average loading time, average stop time, and average speed),
- process parameters (start of sorting, end of sorting, start of the first tour, start of the last tour).

Generation of Orders

Customer orders are the basis of deliveries in the simulation and permanently generated by means of a fixed rate. This rate is calculated by the input parameter of parcels p.a. and converted to a generation rate. When an order is generated, the model randomly chooses the region of the customer, according to the share

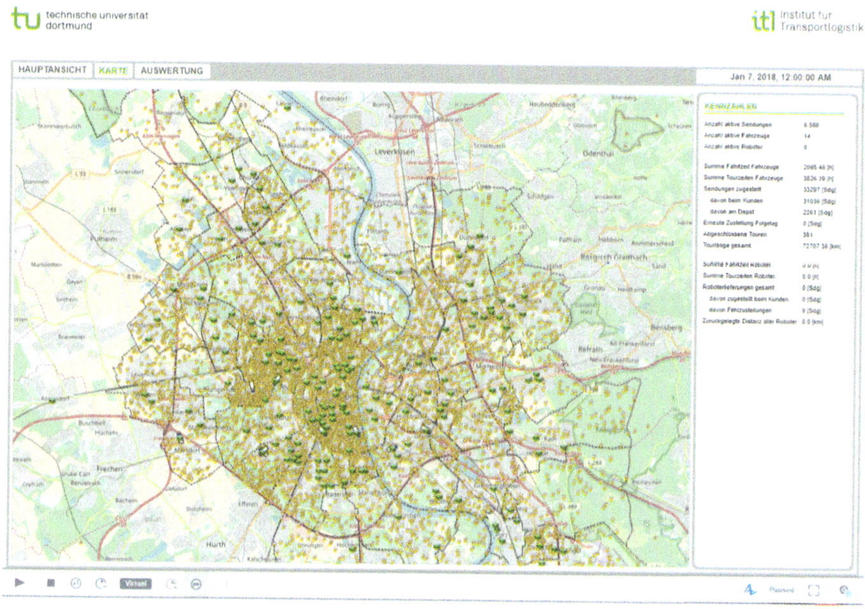

Fig. 3. The GUI with the main view of the simulation model

of deliveries from the region data set. The customer is randomly placed within the designated area and the parcel is generated in the sorting queue of the hub.

Sorting of Parcels

The hub is the starting point for delivery tours on the last-mile. Sorting activities start and end every day at a preset time (e.g. from around 5:00am to 1:00pm). The parcels pass through a sorting process and are moved to the loading gate of the corresponding target area (cluster). The number of loading gates is predetermined by the number of delivery regions and can be set individually. The first delivery tours also start within a preset start time window (e.g. from 8:00am to 1:00pm) with the assignment of an available delivery truck.

Tour Planning and Delivery

After sorting, the parcels are queued within the loading gate agents until the start of the delivery tours. The loading process is triggered by a loading event which is parametrized for the simulation experiment. For each tour, a vehicle is assigned from a designated resource pool. The assigned vehicle receives the loaded parcels and individually plans his delivery tour. For tour planning, we implemented the tour planning algorithms that are described in Sect. 4. During the delivery tour, each vehicle processes a circle of states that are defined in the vehicle's state chart (Fig. 4).

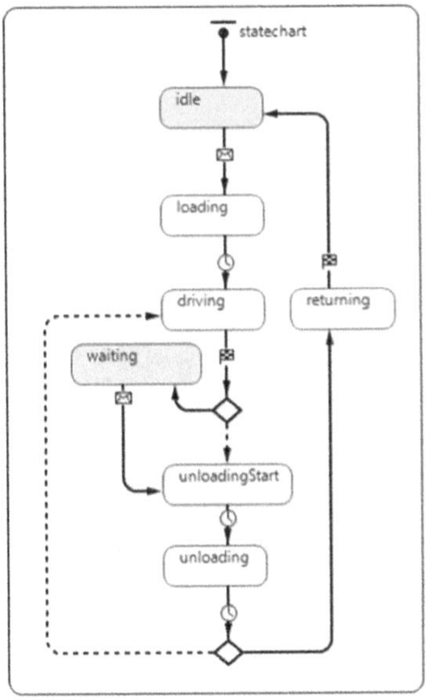

Fig. 4. The state chart of a delivery truck

The default state of a vehicle is “idle”. After the loading gate calls the beginning of the tour, the vehicle changes its status to“loading” for the loading duration, which is defined in the starting parameters. After loading, the vehicle changes in the “driving” state, until it has reached the location of the parcels destination on the GIS map. The loading and unloading process duration can be preset in the model parameters.

The average driving speed is defined in the starting parameters and preset to 36 km/h. At the target destination, the agent checks, if the parcel is successfully delivered. We implemented a stochastic rate for failed deliveries (i.e. the customer is not at home). When the delivery is successful, the vehicle changes in the “unloading” state. After unloading the parcel, the agent checks, if there are additional parcels and changes back in the “driving” state, until every parcel on the truck is delivered. If the parcel could not be delivered, the vehicle agent replans its tour by resorting the parcels with respect to the nearest micro-depots. Parcels from failed deliveries are transported to the nearest micro-depot after visiting the last customer within the micro-depots target area (cluster). The new position for these parcels in the tor is given by the optimization algorithm. If this micro-depot is closed on arrival, the parcels get transported to the hub. Therefore, we defined parameters in order to distinguish whether a parcel is intended for a customer, a micro-depot, or the starting hub. After a failed delivery, the parameters are changed accordingly.

After delivering the last parcel, the vehicle changes in the "returning" state and drives back to the starting hub. At the starting hub, parcels that could not be delivered are unloaded for delivery on the following day. After unloading, the vehicle changes back in the"idle" state, where it is released from the tour and available for the assignment on a new tour.

Robot Deliveries
The delivery of parcels using parcel robots is implemented similar to the delivery of regular parcels with a few major differences. First, the parcels that are ordered by customers for robot delivery are sorted in a separate loading gate. The tour planning at this point is more complex compared to regular parcels, because on one hand the customers can choose a desired time window for delivery and on the other hand the micro-depots that are used as starting points for the robots have their own opening hours. The tour planning algorithm for robot delivery parcels is described in Sect. 4.2. After tour planning, the vehicle agent begins its delivery tour that follows the same logic as regular tours. At the micro-depots, the parcels for robot delivery are queued until the beginning of the first delivery time window. Every micro-depot has exactly one delivery robot assigned that can be used for delivery. The queued orders are then processed one after another by the robot agent. Parcel robots are using the infrastructure of the same GIS environment as the vehicles. They are parametrized with a default average speed of 3.6 km/h. The loading and unloading times of a robot are assumed to be 10 min.

6 Experimentation and Results

The model was developed using *AnyLogic 8.3.3* and coded in Java. To solve the route planning problems we integrated with *Gurobi 7.0.2* into the simulation model. The routes are calculated on the basis of *OpenStreetMaps (2018)* network data. The simulation experiments are run on an Intel Xenon E5-1620 MS-Windows 10 machine with 3.5 GHz Quadcore and 64 GB Ram.

For the distribution of parcels we used public primary data from Hermes for the city of Cologne, Germany [9]. The simulation period for the experiments is 3 days. Four experiments were made with a robot parcel share of 0% to 3%. The model was validated and verificated (V&V) using tracing, observing the animation of the simulation output and structured walk-through of the program (cf. [11]). First, we assessed the efficiency of the tour planning algorithm in e_1, where every parcel was delivered by truck. Thereby, we collected data for every tour and determined the objective function value for the starting solution using the nearest neighbor heuristic and the objective function value after the 15 s of exact optimization. Overall, an improvement could be achieved for each of the tours up to a 23% reduction of the tour length. The mean reduction over all tours is 11% compared with the heuristic starting solution. In view of the short computational time of 15 s, the procedure turns out to be very rewarding.

The data-driven approach has proven to be very suitable for the application of our case-study. Furthermore, the approach makes it easy to adopt other case studies by changing the underlying information for hubs, parcels, micro-depots, and regions in the database. Another main advantage is the use of GIS information to include the actual infrastructure of the studied region. However, with focus on Table 1, the tour distance and duration shows that the tours in the current simulation model are too long. For example, the average length of a tour for the city of Hamburg is, depending on the area, between 40 Km and 130 km [7]. Therefore, we presume that the amount of delivery areas in the current model is insufficient and the covered delivery regions are oversized for one vehicle.

Table 1. Results for truck deliveries

Exp.	Vehicle stops	Tours	Aver. length	Aver. dur.	Total stops	Failed deliveries	CO_2
			[km]	[h]		[stops]	[t]
e_1	14,508	169	189.48	10.03	15,475	967	6.98
e_2	14,496	169	176.80	9.58	15,484	979	6.51
e_3	14,486	167	190.38	10.05	15,364	989	6.93
e_4	14,551	171	183.55	9.72	15,391	924	6.84

In the experiments $e2$ to $e4$, which have a robot parcel share of 1%, 2% and 3%, we assessed the overall performance of last-mile delivery. Also the driven distance for each tour is collected. The framework integrates successfully the delivery of parcels with robots using the implemented algorithms throughout the observed period. Most of the time windows are met or only have minor delays. With regard to the findings of our secondary data analysis, the tours that are planned in our simulation are too long. The average tour length of 10 h exceeds the regular shift duration of a driver by 2 h.

The results indicate that the use of parcel robots has no significant impact on the total number of kilometers traveled and CO_2 emissions. On the one hand additional tours are needed to supply the micro-depots with parcels for the robots. On the other hand the delivery with robots relieves the burden on the regular delivery. Table 2 shows the total number of deliveries using robots in each experiment and the average length and duration of the tours.

Table 2. Results for robot deliveries

Exp.	Tours	Aver. length	Aver. dur.
		[km]	[min]
e_2	136	363.2	69.65
e_3	240	395.1	85.46
e_4	386	379.6	78.26

It is particularly noticeable that the length and duration exceeds what we expected for the tours, as there are plenty of micro-depots in close proximity to the customers. During the V&V process we noticed two reasons for these results. The first reason might be, that we created the customers randomly inside the areas of Cologne which can lead to unrealistic customer locations, such as inside the river Rhine or inside a forest. This results in the robots taking unrealistic routes by crossing bridges or using motorways, which leads to extraordinary detours. The other reason might be the geographical circumstances of the river Rhine crossing the city. We allocate customers to the nearest micro-depot without calculating the actual route, which would lead to severe delays in the computation due to the high number of parcels per day. Therefore, some customers are assigned to micro-depots on the other side of the river, which also results in extensive detours by using a bridge.

7 Conclusion and Outlook

This contribution presents an ABS freight transport simulation that simulates the last-mile of delivery for parcels using trucks and the innovative technology of parcel robots. First, the state of the art in simulation and optimization on the last-mile was discussed to identify the strength and weaknesses of the respective approaches. On this basis we identified a research gap in the field of last-mile delivery, where we found no papers that address parcel robots in the context of simulation. Then, our framework for combining ABS with optimization of route planning was presented and the implemented optimization algorithms described. This methodological approach was used to investigate trends in urban freight transport on the example of parcel robot delivery in the city of cologne, using primary data from the parcel company Hermes. The evaluation of four simulation experiments showed, that our framework is capable to simulate several days of delivery, including regular trucks and parcel robots. Our tour-planning algorithm that uses two heuristics gave near-optimal tours in short computational time, which is essential concerning the large number of tours per day. The evaluation of the experiments showed, that the use of parcel robots for time window delivery does not affect the driven distance and total CO_2 consumption in a significant way. On the one hand additional trucks are used to deliver parcels with time window restriction to the micro-depots in a timely manner. On the other hand this reduces the workload of the conventional delivery tours. From a customer's point of view, the selection of individual time windows improves the service level and is very convenient.

However, the results showed that the underlying input data should be improved to adjust tour length and duration. This leads to a need for further developments, such as a more realistic attempt to generate customers in the urban environment. Further steps will foresee to apply this approach to a suitable examination area, to parametrize and calibrate the model with additional secondary data, and to verify the model. This includes the location of hubs that serve as a starting point for delivery and the optimal number of delivery areas to achieve more realistic delivery tours. Further investigations will also focus on the implementation of other means of transport for the last-mile, such as electric vehicles or cargo bikes.

Acknowledgements. This paper is a research result in context of the Leistungszentrum (Center of Excellence) for Logistics & IT, an initiative of the Fraunhofer-Gesellschaft and the state of North Rhine-Westphalia.

Appendix

Algorithm 1. Delivery tour

Input: Graph $G = (V, E)$ with $V = (0, 1, \ldots, d, d+1, \ldots, d+c)$, cost fct. $\tilde{c}(e)$, *integer* timeOut
list $V' \leftarrow V$,
list removed $\leftarrow \emptyset$
for i = 1 to d **do**
    $P(X_i) \leftarrow (1 - (1 - p\_fd)^{|V_i|-1})$
    if $P(X_i) < 0.5$ **then**
        $V' \leftarrow V' \setminus \{i\}$
        removed.*add*(i)
    end if
end for
// Calculate tour with Nearest Neighbor with precedence constraints
NN_tour $\leftarrow$ NN_heuristic (V',$\tilde{c}$)
Gurobi calculate TSPPC tour, **Input:** V', NN_tour, timeOut
list return_sequence $\leftarrow$ TSP_tour $\setminus\{0, \ldots, d\}$
//reinsert nodes from list removed
insertion heuristic(TSP_tour, removed)
return *return_sequence*

Algorithm 2. Robot parcel delivery tour

Input: Graph $G = (V, E)$ with $V = \{0, 1, \ldots, n\}$, cost fct. $\tilde{c}(E)$, *list* sT,*matrix* open
//calculate TSP tour, without opening hours
Gurobi calculate TSP tour, **Input:** $V, \tilde{c}(E)$, timeOut
integer $t \leftarrow 0$
list timed_tour $\leftarrow \emptyset$
for $\{i, j\} \in$ TSP_tour **do**
  $t_j \leftarrow t + sT_i + \tilde{c}(\{i, j\})$
  if $t_j \in open_{j,k}$, with $k \in K_j$ **then**
    timed_tour.*add*(j)
    $t \leftarrow t_j$
  else
    removed.*add*(j)
  end if
end for
best_dur $\leftarrow t_0$
best_tour $\leftarrow$ timed_tour
for i = 1 to $|removed|^2$ **do**
  shuffle removed
  boolean feas
  list temp_tour $\leftarrow$ timed_tour
  for $j \in$ removed **do**
    insertion_heuristic(j, temp_tour, feas)
    if feas **then**
      temp_tour.*add*(j)
    end if
  end for
  if | temp_tour | > | best_tour | **or** (| temp_tour | == | best_tour | **and** $t_{temp\_tour_0}$ < best_dur) **then**
    best_tour $\leftarrow$ *temp_tour*
    best_dur $\leftarrow t_{temp\_tour_0}$
  end if
end for
return best_tour

References

1. Ascheuer, N., Juenger, M., Reinelt, G.: A branch and cut algorithm for the asymmetric traveling salesman problem with precedence constraints. Comput. Optim. Appl. **17**(1), 61–84 (2000). https://doi.org/10.1023/a:1008779125567
2. Asih, A.M.S., Jatiningrum, W.S., Sopha, B.M.: Collaborative distribution - application to the city of yogyakarta, indonesia. In: 2016 IEEE International Conference on Industrial Engineering and Engineering Management (IEEM), IEEE (2016). https://doi.org/10.1109/ieem.2016.7798056
3. Bertram, I.: Hermes testet Starship-Roboter – Die wichtigsten Fragen (2016). https://newsroom.hermesworld.com/hermes-testet-starship-roboter-die-wichtigsten-fragen-10720/. Accessed 19 Oct 2018

4. Bertram, I.: Hermes startet Test mit Starship-Robotern in London (2017). https://newsroom.hermesworld.com/hermes-startet-test-mit-starship-robotern-in-london-12325/. Accessed 19 Oct 2018
5. Fikar, C., Gronalt, M., Hirsch, P.: A decision support system for coordinated disaster relief distribution. Expert Syst. Appl. **57**, 104–116 (2016). https://doi.org/10.1016/j.eswa.2016.03.039
6. Fikar, C., Hirsch, P., Gronalt, M.: A decision support system to investigate dynamic last-mile distribution facilitating cargo-bikes. Int. J. Logistics Res. Appl. **21**(3), 300–317 (2017). https://doi.org/10.1080/13675567.2017.1395830
7. Gester, F., Bogdanski, R.: Innovationen auf der letzten Meile - Bewertung von Chancen für die nachhaltige Stadtlogistik von morgen. Technical report, Bundesverband Paket & Expresslogistik (BIEK) (2017). https://www.biek.de/download.html?getfile=508
8. Govindan, K., Jafarian, A., Khodaverdi, R., Devika, K.: Two-echelon multiple-vehicle location–routing problem with time windows for optimization of sustainable supply chain network of perishable food. Int. J. Prod. Econ. **152**, 9–28 (2014). https://doi.org/10.1016/j.ijpe.2013.12.028
9. Hermes: Hermes PaketRadar. Website (2017). https://newsroom.hermesworld.com/wp-content/uploads/heatmaps2017/koeln.html. Accessed 21 Oct 2018
10. Hofmann, W., Assmann, T., Neghabadi, P.D., Tolujew, J.: A simulation tool to assess the integration of cargo bikes into an urban distribution system. In: The 5th International Workshop on Simulation for Energy, Sustainable Development & Environment, Barcelona, Spain (2017)
11. Law, A.M.: Simulation Modeling and Analysis, 5th edn. McGraw-Hill Education, New York (2015)
12. Lin, S.W., Yu, V.F.: A simulated annealing heuristic for the multiconstraint team orienteering problem with multiple time windows. Appl. Soft Comput. **37**, 632–642 (2015). https://doi.org/10.1016/j.asoc.2015.08.058
13. Maadi, M., Javidnia, M., Ramezani, R.: Modified cuckoo optimization algorithm (MCOA) to solve precedence constrained sequencing problem (PCSP). Appl. Intell. **48**(6), 1407–1422 (2017). https://doi.org/10.1007/s10489-017-1022-0
14. Miller, C.E., Tucker, A.W., Zemlin, R.A.: Integer programming formulation of traveling salesman problems. J. ACM **7**(4), 326–329 (1960). https://doi.org/10.1145/321043.321046
15. Mingozzi, A., Bianco, L., Ricciardelli, S.: Dynamic programming strategies for the traveling salesman problem with time window and precedence constraints. Oper. Res. **45**(3), 365–377 (1997). https://doi.org/10.1287/opre.45.3.365
16. Rabe, M., Klueter, A., Clausen, U., Poeting, M.: An approach for modeling collaborative route planning in supply chain simulation. In: Roeder, T.M.K., Frazier, P.I., Szechtman, R., Zhou, E., Huschka, T., Chick, S.E. (eds.) Proceedings of the 2016 Winter Simulation Conference, pp. 2228–2238. IEEE, Piscataway (2016)
17. Sung, J., Jeong, B.: An adaptive evolutionary algorithm for traveling salesman problem with precedence constraints. Sci. World J. **2014**, 1–11 (2014). https://doi.org/10.1155/2014/313767
18. Thaller, C., Clausen, U., Kampmann, R.: System dynamics based, microscopic freight transport simulation for urban areas. In: Commercial Transport, Springer International Publishing, pp. 55–72 (2015). https://doi.org/10.1007/978-3-319-21266-1_4
19. Tricoire, F., Romauch, M., Doerner, K.F., Hartl, R.F.: Algorithms for the multi-period orienteering problem with multiple time windows. EU Meeting 2008 - Troyes, France, October 23–24 (2008)

20. Tricoire, F., Romauch, M., Doerner, K.F., Hartl, R.F.: Heuristics for the multi-period orienteering problem with multiple time windows. Comput. Oper. Res. **37**(2), 351–367 (2010). https://doi.org/10.1016/j.cor.2009.05.012
21. Vansteenwegen, P., Souffriau, W., Oudheusden, D.V.: The orienteering problem: a survey. Eur. J. Oper. Res. **209**(1), 1–10 (2011). https://doi.org/10.1016/j.ejor.2010.03.045
22. Yun, Y., Moon, C.: Genetic algorithm approach for precedence-constrained sequencing problems. J. Intell. Manuf. **22**(3), 379–388 (2009). https://doi.org/10.1007/s10845-009-0296-4
23. Yun, Y., Chung, H., Moon, C.: Hybrid genetic algorithm approach for precedence-constrained sequencing problem. Comput. Ind. Eng. **65**(1), 137–147 (2013). https://doi.org/10.1016/j.cie.2011.11.019

Stack Shuffling Optimization of Steel Bars by Using Genetic Algorithms

Jakob Marolt(✉), Bojan Rupnik, and Tone Lerher

Faculty of Logistics, University of Maribor, Celje, Slovenia
{jakob.marolt,bojan.rupnik,tone.lerher}@um.si

Abstract. A steel plant company producing steel bars has large assortment of the end products, with similar appearance and attributes. The steel bars are stored on the floor in a stacking frame. For the order picking of steel bars, an overhead crane is used for reshuffling all the necessary steel bars to get access to the required product. While the production schedule allows for anticipating the storage occupancy, a stochastic transport arrival prevents optimal product stacking for efficient order-picking operation. Due to this, any order-picking sequence may require reshuffling of the stacked material, which increases working cost, order-picking times, and complicates material tracking. This paper presents a method for minimizing the order-picking times by overhead crane movements through proper reshuffling of the steel bars. Similar research was done on container yard pre-marshalling and reshuffling problem, while the presented approach handles with the special situation in the steel plant. Various optimization approaches including linear programming, simulated annealing, taboo search, branch and bound and genetic algorithms were used by researchers to solve comparable problems. The proposed method for solving the specific problem of reshuffling steel bars uses genetic algorithms to find a feasible solution in real-time. The proposed solution reduces intralogistics cost and increases order-picking efficiency.

Keywords: Steel bars · Genetic algorithms · Stack shuffling

1 Introduction

The steel plant company is storing sets of steel bars (SSB) ranging from 3 to 6 meters in stacking frames (Fig. 1). There are up to six tiers of SSB in each stacking frame. This is partly because of safety reasons and partly because of possible damage of product due to mechanical deformation. In each tier of a stack there are three SSBs. A full stacking frame contains eighteen elements. When warehouse workers try to pick a SSB with highest picking priority, they must clear all SSBs that are blocking targeted one.

The problem of reshuffling SSBs in the steel plant is comparable to reshuffling operations in container yards [3]. The difference is that vertically stacked containers have only one column in a stack. There are more than one columns in a stacking frame, meaning that in order to retrieve the targeted SSB, warehouse workers need to reshuffle all the elements in the tiers above the targeted one. Reshuffling operations are estimated over a half of total warehouse expenditure [1].

U. Clausen et al. (Eds.): ICPLT 2019, LNLO, pp. 20–31, 2019.
https://doi.org/10.1007/978-3-030-13535-5_2

Fig. 1. Stacking frame for SSB

An example of the problem is displayed in Fig. 2 (first stage), where the element with the highest priority (green) is blocked by other SSBs (orange) and the solution in last stage.

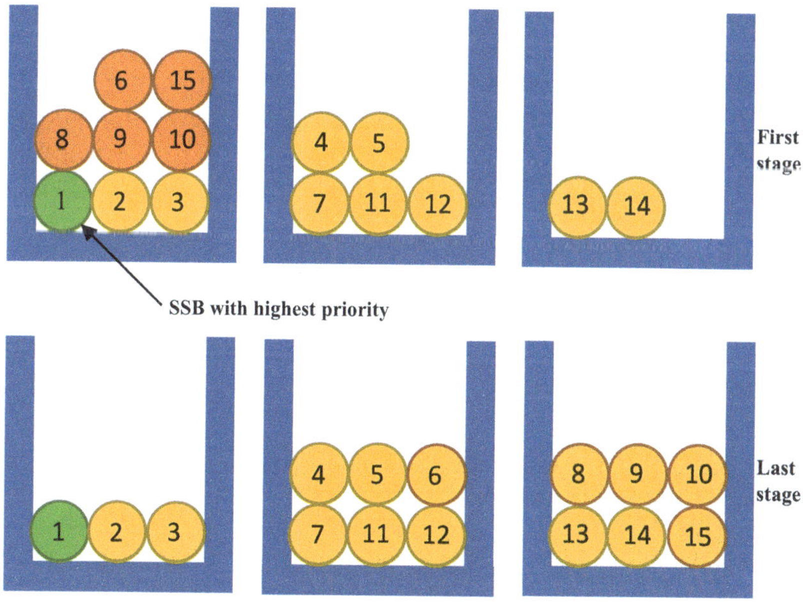

Fig. 2. An example of optimal reshuffling

Mainly two approaches are considered when solving similar problems, the first being pre-marshalling, and the second reshuffling [3, 6]. By pre-marshalling, the elements reposition is done in correct priority order for later retrieval [9]. In this paper we focus on reshuffling, where the elements with highest priority can be retrieved, if no other element is blocking it.

Related problems are well studied in scientific literature. Kim et al. [2] presented a branch and bound algorithm and a decision rule heuristic based on probability with an estimator for an expected number of additional relocations for block stacking problem. Tang et al. [6] studied the item shuffling problem of coils and plates in a steel plant industry. The authors formulated a linear program with some special cases where polynomial time of an algorithm was obtained. Greedy heuristics made real-time calculation possible and was later improved by tabu search optimization. Storage assignment of containers in a container yard was studied by Wan et al. [7]. They solved both static and dynamic problem, with new container arrivals.

The optimal solution for the presented problem requires finding the minimum number of necessary reshuffles for item dispatchment. With larger warehouse and consequently more elements the complexity of the problem grows exponentially and so the optimal solutions become numerically unfeasible. Similar problems have been proven to be NP complete [8].

For the purpose of clarity, the following requirements for the presented problem are specified:

1. The problem at hand is assumed to be static, meaning that no new SSBs are being produced while reshuffling takes place.
2. Stacks consist of 6×3 columns and tiers respectively.
3. All stored SSBs are assigned a priority number, which defines the order picking schedule of retrieval
4. To solve the problem all SSB inside a warehouse must be dispatched.

2 Method for Reducing the Number of Necessary Reshuffles Using Genetic Algorithm

The general algorithm for retrieving elements for dispatchment consists of six cyclical steps:

1. Try to retrieve the SSB with the highest priority
2. Return positions of all reachable SSBs
3. Select an SSB
4. Return positions of all suitable empty slots
5. Select an empty slot
6. Move selected SSB to an empty slot

First, the crane operator retrieves an SSB with highest priority if possible. Otherwise, the operator needs to rearrange the configuration of SSBs in order to make an SSB with highest priority reachable. Steps 2 and 4 return the information about currently reachable SSBs and suitable empty slots. The most significant steps are step 3, where an element is selected from a list of reachable elements, and step 5, where an empty slot is selected from a list of suitable empty slots. By executing these two steps perfectly we get an optimal solution to any given configuration. However, a perfect execution is hard to achieve, because it is difficult to predict how one relocation of an element will impact the complexity of the problem in future iteration. Instead, we apply heuristics for selection of SSBs and empty slots.

2.1 Heuristics for Selection of Reachable SSBs and Suitable Empty Slots

We propose a series of different heuristics for selection of SSBs and empty slots. The operations employed by the heuristics are described using configuration variables in Table 1.

Table 1. Configuration variables

$i \in \mathbb{N}$	Current stack
$j \in \mathbb{N}$	Current row
$k \in \mathbb{N}$	Current column
$I \in \mathbb{N}$	Number of stacks
$J \in \mathbb{N}$	Number of rows
$K \in \mathbb{N}$	Number of columns
$N_{i,j,k} \in \{0,1\}$	Binary representation of empty slots $N_{i,j,k} = \begin{cases} 0, \text{ at full slots} \\ 1, \text{ at empty slots} \end{cases}$
$P_{i,j,k} \in \mathbb{N}$	Priority number of SSB
$S_i \subset \mathbb{N}$	Stack summation variable
$T_{x,i} \in \mathbb{N}$	Element variable
$M \in \mathbb{N}$	Number of elements

2.1.1 Heuristics for SSB Selection

In the second step of the algorithm, positions of all reachable elements are retrieved from a given configuration and are stored in a list. The positions contain the information about the stack *(i)*, row *(j)* and column *(k)*. Selection of SSB heuristics then reorders this list based on a summation variable (S_i) and stack index *(i)* of elements. The first SSB in the list after rearrangement of the list elements is the selected SSB.

Three Highest Priorities Heuristic. The heuristic sums three highest priorities of each stack. The stack list is ordered by sum in ascending order. Reachable SSBs are reordered in the same manner as rearranged list of stack summations (S_i). The first reachable SSB on the list becomes the selected SSB.

$$T_{i,1} = \min\left(P_{i,j,k}\right) \tag{1}$$

$$T_{i,2} = \min\left(P_{i,j,k}/\left\{T_{i,1}\right\}\right) \tag{2}$$

$$T_{i,3} = \min\left(P_{i,j,k}/\left\{T_{i,1}, T_{i,2}\right\}\right) \tag{3}$$

$$S_i = T_{i,1} + T_{i,2} + T_{i,3} \tag{4}$$

Total Priority Heuristic for SSB Selection. The heuristic sums up all priorities of each stack. The stack list is ordered by sum in ascending order. Reachable SSBs are reordered in the same manner as rearranged list of stack summations (S_i). The first reachable SSB on the list becomes the selected SSB.

$$S_i = \sum_{k=1}^{K} \sum_{j=1}^{J} P_{i,j,k} \tag{5}$$

Highest-slot Heuristic. The heuristic sums up all the empty slots of each stack. The stack list is ordered by sum in ascending order. Reachable SSBs are reordered in the same manner as rearranged list of stack summations (S_i). The first reachable SSB on the list becomes the selected SSB.

$$S_i = \sum_{k=1}^{K} \sum_{j=1}^{J} N_{i,j,k} \tag{6}$$

Highest Priority First Heuristic. The heuristic finds the element with the highest priority of each stack. The stack list is ordered by sum in ascending order. Reachable SSBs are reordered in the same manner as rearranged list of stack summations (S_i). The first reachable SSB on the list becomes the selected SSB.

$$S_i = \min\left(P_{i,j,k}\right) \tag{7}$$

2.1.2 Selection of an Empty Slot Heuristics

In the fourth step of the general algorithm, positions of all suitable empty slots are retrieved from a given configuration and are stored in a list. The positions contain stack *(i)*, row *(j)* and column *(k)* information. Selection of an empty slot heuristics then reorder this list based on a summation variable (S_i) and stack index (i) of elements. The suitable empty slot that is the first in the list after rearrangement of the list elements is selected empty slot. Empty slots that have the same stack index (i) as selected SSB would be discarded from the list, in order to prevent relocations of the same stack.

Three Lowest Priorities Heuristic. The heuristic sums three lowest priorities of each stack. The stack list is ordered by sum in descending order. Suitable empty slots are reordered in the same manner as rearranged list of stack summations (S_i). The first suitable empty slot on the list becomes the selected empty slot.

$$T_{i,1} = max\left(P_{i,j,k}\right) \tag{8}$$

$$T_{i,2} = max\left(P_{i,j,k}/\left\{T_{i,1}\right\}\right) \tag{9}$$

$$T_{i,3} = max\left(P_{i,j,k}/\left\{T_{i,1}, T_{i,2}\right\}\right) \tag{10}$$

$$S_i = T_{i,1} + T_{i,2} + T_{i,3} \tag{11}$$

Total Priority Heuristic for Empty Slot Selection. The heuristic sums up all priorities of each stack. The stack list is ordered by sum in descending order. Suitable empty slots are reordered in the same manner as rearranged list of stack summations *(S_i)*. The first suitable empty slot on the list becomes the selected empty slot.

$$S_i = \sum_{k=1}^{K} \sum_{j=1}^{J} P_{i,j,k} \tag{12}$$

Lowest-slot Heuristic. The heuristic sums up all the empty slots of each stack. The stack list is ordered by sum in descending order. Suitable empty slots are reordered in the same manner as rearranged list of stack summations *(S_i)*. The first suitable empty slot on the list becomes the selected empty slot.

$$S_i = \sum_{k=1}^{K} \sum_{j=1}^{J} N_{i,j,k} \tag{13}$$

Lowest Priority First Heuristic. The heuristic finds the element with the lowest priority of each stack. The stack list is ordered by sum in descending order. Suitable empty slots are reordered in the same manner as rearranged list of stack summations *(S_i)*. The first suitable empty slot on the list becomes the selected empty slot.

$$S_i = \max\left(P_{i,j,k}\right) \tag{14}$$

2.2 The Genetic Algorithm

In this paper we aimed for a good solution by optimizing the sequence of heuristics by using genetic algorithms (GA). The GA is a stochastic process in which the random solutions to the problem are presented by chromosomes. The chromosomes in our GA consists of genes where each gene represent a heuristic. The genes at odd positions represent SSB selection heuristics, while genes at even positions represent empty slot heuristics.

The chromosomes are evaluated by a fitness function to rate them according to their performance. Chromosomes that perform poorly are discarded by selection. They are replaced by next generation of chromosomes that are created by crossing the genes of chromosomes that perform better. There is a small chance that the offspring would be subjected to some random mutation in order to avoid local minima or maxima [4, 7].

The proposed GA contains 7 steps (Fig. 3). First, we define parameters that include the number of tiers of each stack, number of elements in each tier, number of stacks, number of empty slots, number of chromosomes and mutation coefficients. Each configuration is represented by a random 3D matrix. After matrix initialization the program creates the initial population consisting of 100 chromosomes. Out of those 84 are created randomly and 16 created artificially. Artificial chromosomes contain repeating sequences of the same genes for empty slots and SSB selection heuristics. We used artificial chromosomes to test, whether the sequence of different heuristics will outperform the sequence of the same two heuristics and for the faster convergence of the optimization. The length of the chromosomes is then set by a function that iteratively increases the size of one artificial chromosome until the matrix is solved. Because the chromosomes do not change the length throughout the optimization, the solution of a given matrix can be reached at some gene that is prior to the last one in a chromosome. Meaning, that a part of chromosome becomes non-significant. The larger the non-significant part of a chromosome is, the least necessary reshuffles are required to solve the matrix.

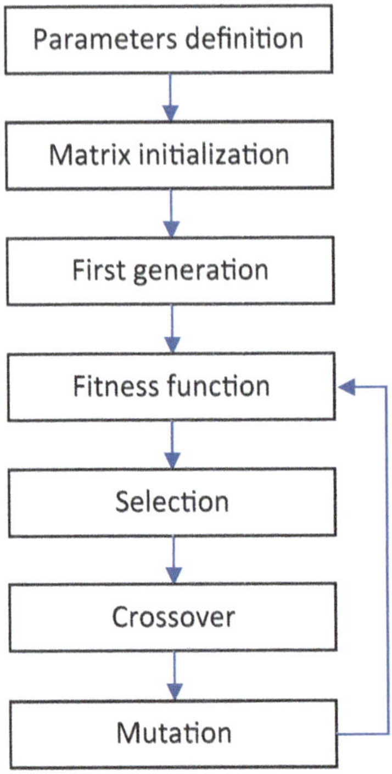

Fig. 3. Steps of GA optimization

The chromosomes are then evaluated by the fitness function. Fitness function is composed of two parts. The first part evaluates if the matrix was solved, meaning that all SSB were retrieved. The second part of the fitness function evaluates the number of reshuffles. The smaller the number, the better is the chromosome rated. The best evaluated half of the population is selected for the next generation, while the rest is discarded. The other half of the population is created by a 10% of the best performed chromosomes together with two newly created random chromosomes by a random selection and two-point crossover. The offspring has 10% chance for one of their genes to mutate to other number. There were 100 iterations of this process for each optimization.

3 Results

The Genetic algorithm optimization was tested on two different cases, with first containing 10, and the second 20 stacks of SSBs. For both instances the reduction of necessary reshuffles in each generation is observed, as well as the gene sequence of the wining chromosome. All of the computer code was written in Mathworks 2017a and ran on a PC with Intel Xeon CPU 3.5 GHz and 16 GB RAM. The occupancy level of elements is 83% for both instances. In the first generation the solution pool was dominated by random chromosomes. These random chromosomes are outperformed by the artificial ones. Selection in the first generation alters the domination of random to artificial. Crossover and mutation distort the sequence of heuristics in later generations and assist the optimization for the speedier convergence. In the Figs. 5 and 8 the sequence of SSB selection heuristics of the best performed chromosome in a solution

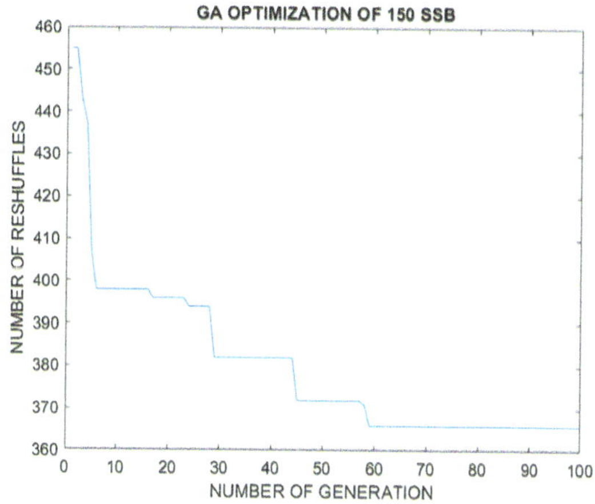

Fig. 4. Number of reshuffles reduction of the best chromosome in a solution pool for the case with 150 SSBs

pool is displayed. It is evident that the final solutions are stemming from one of the artificial chromosomes, because Highest priority first heuristics outperforms the rest. In the Figs. 6 and 9 the sequence of empty slot selection heuristics of the best performed chromosome in a solution pool is displayed. In these two cases the final solutions are significantly different than any artificial chromosome. Also, these two solutions differ from one another, suggesting that no sequence of heuristics could be generalized as a solution.

The first case contained 150 SSBs arranged in 10 stacks with 30 empty slots. After 100 generations the number of necessary reshuffles was reduced from 455 to 366 (Fig. 4). That is approximately 20% fewer relocations for the retrieval of all 150 stored elements. The calculation required 23 min for 100 iterations.

The second case contains 300 SSB arranged in 20 stacks with 60 empty slots. After 100 generations the number of necessary reshuffles was reduced from 1227 to 898

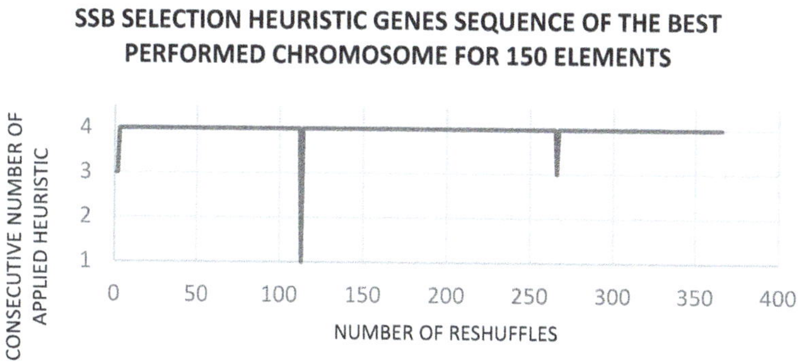

Fig. 5. Genes sequence for the case of 150 SSBs

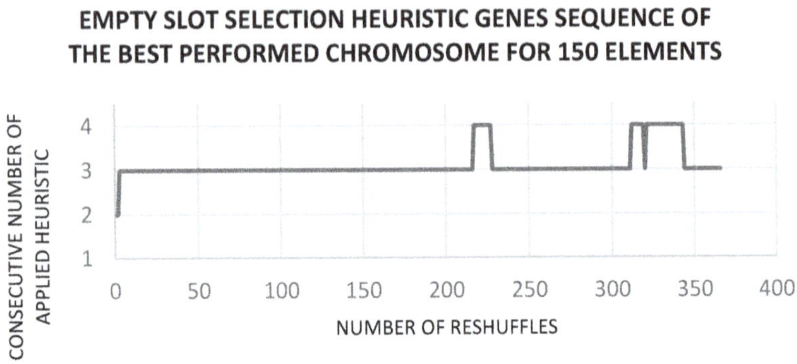

Fig. 6. Genes sequence for the case of 150 SSBs

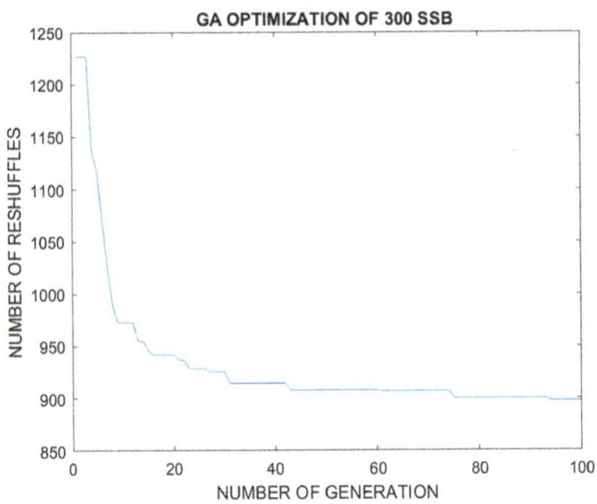

Fig. 7. Number of reshuffles reduction of the best chromosome in a solution pool for the case with 300 SSBs

(Fig. 7). That is approximately 27% fewer relocations for the retrieval of all the 300 elements. The calculation required 84 min for 100 iterations.

The genes inside the best performed chromosome have different sequence of heuristics. Implying that every case must be dealt with separately. From Tables 2 and 3 it is evident that it is not clear which is the best heuristic for empty slot selection while, Highest priority first heuristic dominates SSB selection heuristics.

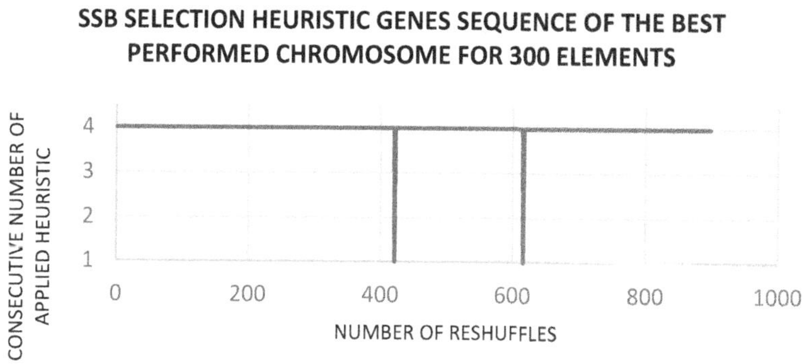

Fig. 8. **Genes sequence for the case of 300 SSBs**

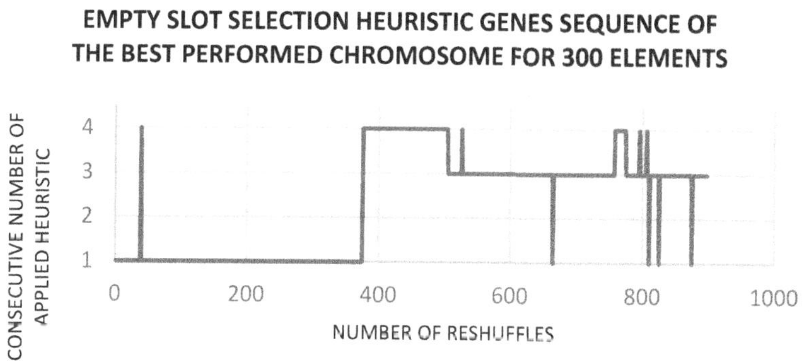

Fig. 9. Genes sequence for the case of 300 SSBs

Table 2. Genes frequency for the case of 150 SSBs

SSB selection			Empty slot selection		
Consecutive number of heuristics	The name of heuristic	Number of occurrences	Consecutive number of heuristics	The name of heuristic	Number of occurrences
4	Highest priority first heuristics	362	3	Lowest-slot heuristics	322
3	Highest-slot heuristics	3	4	Lowest priority first heuristic	42
1	Three highest priorities heuristic	1	2	Total priority heuristic for an empty slot selection	2
2	Total priority heuristic for SSB selection	0	1	Three lowest priority heuristic	0

Table 3. Genes frequency for the case of 300 SSBs

SSB selection			Empty slot selection		
Consecutive number of heuristics	The name of heuristic	Number of occurrences	Consecutive number of heuristics	The name of heuristic	Number of occurrences
4	Highest priority first heuristics	896	1	Three lowest priority heuristic	378
1	Three highest priorities heuristic	2	3	Lowest-slot heuristics	370
2	Three highest priorities heuristic	0	4	Lowest priority first heuristic	150
3	Total priority heuristic for SSB selection	0	2	Total priority heuristic for an empty slot selection	0

4 Conclusions

In this paper we presented the problem of reshuffling SSBs. We solved the problem with genetic algorithm optimization. We proposed four different heuristics for SSB selection and another four different heuristics for empty slot selection. We used the GA to find the best sequence of heuristics to minimize the number of necessary reshuffles. We used artificial chromosomes to test, whether the sequence of different heuristics will outperform the sequence of the same two heuristics and for the faster convergence of the optimization. Two instances were presented, one containing 150 and the second containing 300 elements. We achieved a 20% reduction of necessary reshuffles for the first case and 27% reduction for the second one. We showed that the combination of heuristics outperformed the sequence of the same heuristic.

In the future research we intend to expand the list of proposed heuristics for SSB selection and empty slot selection to also include more complex heuristics proposed by other authors. Similar reshuffling and remarshalling problems could also be tackled with our approach, to search for a mixed sequence of heuristics, instead of repeating the same heuristic throughout the process of elements relocation.

References

1. De Koster, R., Le-Duc, T., Roodbergen, K.J.: Design and control of warehouse order picking: A literature review. Eur. J. Oper. Res. **182**, 481–501 (2007)
2. Kim, K.H., Hong, G.: A heuristic rule for relocating blocks. Comput. Oper. Res. **33**, 940–954 (2006)
3. Lee, Y., Hsu, N.Y.: An optimization model for the container pre-marshalling problem. Comput. Operat. Res. **34**, 3295–3313 (2007)

4. Rao, S.S.: Engineering optimization: theory and practice. New Age International, New Delhi (2002)
5. Tang, L.X., Liu, J.Y., Rong, A.Y., Yang, Z.H.: Modelling and a genetic algorithm solution for the slab stack shuffling problem when implementing steel rolling schedules. Int. J. Prod. Res. **40**, 1583–1595 (2002)
6. Tang, L., Zhao, R., Liu, J.: Models and algorithms for shuffling problems in steel plants. Nav. Res. Logist. **59**(7), 502–524 (2012)
7. Melanie, M.: An Introduction to Genetic Algorithms. MIT Press, Massachusetts (1999)
8. Avriel, M., Penn, M., Shpirer, N.: Container ship stowage problem: complexity and connection to the coloring of circle graphs. Discret. Appl. Math. **103**(1–3), 271–279 (2000)
9. König, F.G., Lübbecke, M., Möhring, R., Schäfer, G., Spenke, I.: Solutions to real-world instances of PSPACE-complete stacking. In: Proceedings 15th European Symposium on Algorithms, Eilat, Israel, pp. 729–740 (2007)

Decoupling the Modeling of Actions in Logistics Networks from the Underlying Simulation Data Model

Markus Rabe[1], Dominik Schmitt[2](✉), Astrid Klüter[1], and Joachim Hunker[1]

[1] TU Dortmund, ITPL, Dortmund, Germany
{markus.rabe, astrid.klueter, joachim.hunker}@tu-dortmund.de
[2] TU Dortmund, GSofLog, Dortmund, Germany
dominik.schmitt@tu-dortmund.de

Abstract. Actions in logistics networks are often determined and evaluated by logistics assistance systems. Typically, these actions are predefined by the logistics assistance system. In case of a data-driven logistics assistance system, these actions are depending on the underlying data structure. This paper describes an approach of decoupling an actions implementation from the underlying data model and, therefore, independent of a given data structure.

Keywords: Logistics assistance system · Decision Support System

1 Introduction

Maintaining a logistics network (LNW) in good conditions, under continuously changing internal and external demands, is a challenging task for a decision maker (DM). In order to adapt to the LNW's changing state, the DM needs to identify promising actions. An action may describe any modification or change to the logistics network, e.g., increasing the stock level of a stock keeping unit (SKU) or changing the transport structure of the logistics network (Werner 2013). These actions are considered to increase the network's performance. However, actions may have contrary effects on the network's performance, e.g., increasing the stock may improve the service level but it may also increase the stock costs. Thus, identifying the most promising actions to improve the overall performance of the LNW, e.g., defined by total costs or service level, is a complex task.

In order to address this issue, the authors have developed a logistics assistance system (LAS) that identifies the most promising actions and that suggests these actions to the decision makers (Rabe et al. 2017). The LAS is using a simheuristic approach (Juan and Rabe 2013), combining a data-driven discrete-event simulation with a metaheuristic. A simulation model, representing the LNW's state, is build, based on data from a database. The simulation is used for evaluating the LNW's performance. The metaheuristic identifies promising actions that are executed on the underlying database in order to alter the LNW's state.

U. Clausen et al. (Eds.): ICPLT 2019, LNLO, pp. 32–44, 2019.
https://doi.org/10.1007/978-3-030-13535-5_3

A basic set of actions is predefined by the LAS. However, enterprises have enterprise-specific requirements for each action and, therefore, enterprises may need an individual set of actions. Thus, the authors developed an approach for integrating user-modeled actions into the LAS (Rabe et al. 2017). However, actions are closely related to the underlying database resulting in multiple issues, e.g., required knowledge of the database's structure when modeling actions or the necessity of adjusting actions in case of the database's structure is changing. To deal with these problems, the authors are investigating an approach for decoupling actions from the underlying simulation data model (SDM).

This paper is organized as follows: First, an overview about related work is given, including logistics networks in general, relevant data of logistics networks, data modeling, logistics assistance systems in general and a description of the author's LAS. After that, the approach for decoupling actions from the underlying data model is described, followed by an overview of the corresponding benefits that come along with this approach. Afterwards, an example of modeling and applying an action using the presented approach is given. The paper closes with a conclusion and an outlook.

2 Related Work

2.1 Logistics Networks

When looking at common definitions of logistics networks, it is noticeable that in the literature there is no clear distinction between logistics networks and supply chains and, therefore, between supply chain management (SCM) and logistics management. (Cordeau et al. 2006; Wang et al. 2016).

Larson and Halldorsson (2004) describe four perspectives on the relationship between logistics and SCM. One phenomenon is the one just described; the relabeling perspective that simply renames logistics; what was logistics is now SCM. Thus, the terms supply chain and logistics network might be used synonymously.

According to Rushton et al. (2010), there is a logistics network between the supply chain's supplier and the customers. Ma and Suo (2006) define a logistics network as the distribution part of a supply chain. These authors describe the distribution network as a critical part of managing a supply chain.

Our application case concerns a special type of logistics network, namely wholesale networks. A wholesaler is a legally and economically independent business which procures goods and sells them mainly to commercial users, commercial resellers, or bulk consumers unchanged or after customary manipulation (Seÿffert 1972). By choosing the trade-level, wholesalers determine their position, within the trade chain, between original production and end customer (Barth et al. 2015).

In the following, we refer to wholesale logistics networks, as it is the internal company's terminology for the given use case. Regarding the exact technical terminology, since we handle data, it ultimately does not matter whether we formally collect data from logistics networks or supply chains, as they obviously do not differ significantly from each other. In the following, we will talk about data in logistics networks in general, knowing that this can also mean data from supply chains.

2.2 Data in Logistics Networks

In logistic networks, data are processed in various IT systems within a company as well as across companies. According to Folinas et al. (2006) a distinction is made between structured content (e.g., relational databases), unstructured content (e.g., plain text), or unstructured content with structured representation (e.g., HTML files) in many different forms (e.g., texts, diagrams, illustrations or tables).

Another distinguishing feature is the time reference. Data in logistics networks can be divided into master data and transaction data (Reuter and Rohde 2015). Master data are data that change only slightly over time and that are used to identify, classify and characterize entities of the logistics network. A classic example is address data as it is stored in logistics networks for suppliers and customers. Transaction data are data that undergo major changes over time. Information, required for planning the logistics network, is obtained from the transaction data that are created and changed by operational performance processes. Examples for transaction data are current inventories, current orders, availability of resources, or planned production quantities and stock levels (Reuter and Rohde 2015).

2.3 Database Systems

Since data occur in different kinds of formats, from different sources and in different relations, especially in the context of a logistics network, they need to be handled adequately. Therefore, data are typically stored in information systems, e.g., a database system. A database system consists of a database and a Database Management System (DBMS) (Elmasri and Navathe 2011).

A database is defined as a coherent, not random, collection of related data which represent an excerpt of the real world (Elmasri and Navathe 2011). Databases follow a database schema, describing the logical structure of a database. To describe such a schema, data models are used. A data model is a formal representation and description of the structure and the relations between data (Connolly and Begg 2015).

In this paper, the authors make use of an enterprise-specific data model that describes any data of the corresponding enterprise in a persistent way. The reader is kindly referred to West (2011), who identifies and distinguishes between different data models discussed in the literature and who highlights that in practice the enterprise-specific data model covers just most of the enterprise's data.

The process of creating and changing a data model is called data modeling. Different techniques for formally describing a data model can be distinguished in the literature, e.g., the entity-relationship-model or the unified modeling language.

For creating and maintaining a database and for manipulating its data, DBMSs are used. A DBMS is a software program that allows users to interact with the associated database. Connolly and Begg (2015) characterize the tasks of a DBMS by

1. handling access restrictions and providing multi-user access to the database,
2. offering a language to specify the database, called Data Definition Language and
3. a Data Manipulation Language to allow specific operations on the data like insert, delete, update, or retrieve.

One of the most-used languages regarding manipulation and querying of data, in the context of relational database systems, is the Structured Query Language (SQL) (Chamberlin 2017). SQL works on the data stored in the database through so-called statements. The data is, in case of SQL and due to the underlying relational data model, stored in the form of multiple tables (Chamberlin 2017).

2.4 Logistics Assistance Systems

To support the decision-making process of decision makers in logistics networks, so-called logistics assistance systems are used (Blutner et al. 2009). LASs are software systems that help the user to monitor a logistics network. Therefore, a LAS might be used to evaluate a specific logistics network's configuration or scenario, e.g. by calculating corresponding Key Performance Indicators. However, a LAS may also combine an optimization approach with an evaluation mechanism, e.g. heuristics and simulation (Juan and Rabe 2013), in order to determine and evaluated multiple logistics network's state configurations. The best solutions can be provided to the user. Each logistics network's configuration can be derived by applying one or more predefined actions to the current logistics network's state.

In the literature, the term Decision Support Systems (DSS) is often used to describe such a system (Kengpol 2008). In the context of logistics, DSS and LAS are used as synonyms. Publications, considering LASs or DSSs in the context of logistics, have been presented for example in the area of automobile production (Bockholt et al. 2011) or in combination with other methods like simulation (Heilala et al. 2010). Throughout this paper, the authors decided to use the term LAS in order to highlight the system's application domain.

2.5 Logistics Assistance System for Wholesale Logistics Networks

The authors have developed a logistics assistance system for managing wholesale logistics networks. This section provides a brief overview of the system's architecture (Fig. 1) and working principles. For a more detailed description of the LAS, the reader is kindly referred to previous publications of the authors (Rabe et al. 2017; Rabe et al. 2018a; Rabe et al. 2018b).

LASs are utilizing master data and transaction data from the real network. Companies use some kind of software in order to store this information, such as SAP R/3. This data is automatically extracted, transformed and loaded into the simulation's database. The LAS uses a data-driven discrete-event simulation combined with a heuristic, also called simheuristic (Juan and Rabe 2013). Therefore, the data from the database is used to instantiate the simulation model that represents the logistics network's state, e.g. the data base table "SiteHaveSKUs" combines the tables "SKU" and "Site" storing the stock for each combination of site and SKU. By running the simulation, the network's performance is evaluated. After each simulation run, the simulation results are fed into the heuristic unit for further investigation.

The logistics network's performance can be improved by altering the network's state, e.g. changing the stock or increasing the transport frequency. For each of these manipulations, a corresponding action needs to be applied to the logistics network.

The heuristic unit (HU) examines the search space, constructed by a collection of all possible actions in the LNW, searching for the most promising actions. Therefore, multiple algorithms are implemented, e.g. Evolutionary Algorithm or Reinforcement Learning. For a detailed description of the HU's working principles, the reader is kindly referred to Rabe et al (2018a) and Rabe et al (2018c) regarding the Evolutionary Algorithm and to Rabe and Dross (2015) for the implementation of the Reinforcement Learning algorithm. Actions, selected by the heuristic unit are forwarded to the execution engine for further processing.

The execution engine transforms these actions into changes to the underlying database, in form of SQL-statements. Before applying these statements to the underlying database, additional predefined metadata will be evaluated. For instance, attributes' default values are used to initialize of new entities. Additionally, attributes' ranges and specific conditions or constraints are used to identify valid changes to the LNW, e.g., capacity constraints when shifting stocks. The actions' effects are evaluated by instantiating a new simulation model, based on the changed data, followed by running the simulation.

This process is run iteratively until a predefined termination criterion is reached, e.g., a specific amount of iterations, a minimum quality of the logistics network's performance or a continuous stagnation in the optimization process. The most promising actions are provided to the decision maker for further investigation.

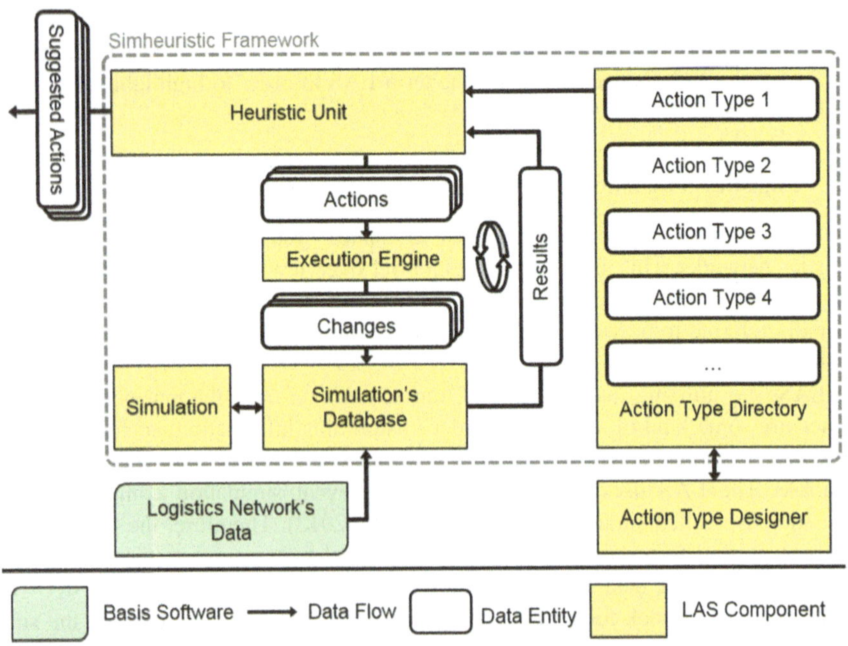

Fig. 1. Simplified architecture of the logistics assistance system, based on Rabe et al. (2018c).

In order to increase the LAS's flexibility, the authors developed a method that allows the user to model action types and integrate them into the system (Rabe et al. 2017).

An action type is a generalization of multiple similar actions, e.g., an action type may describe the general instructions for changing the stock level of any SKUs in any site by any value, whereas a corresponding action describes the increase of stock level for a specific SKU in a specific site by a specific value. Thus, actions can be derived from an action type by specifying its parameter values.

For modeling user-created action types, the user has access to the action type designer, a graphical or textual interface. The action type designer has access to a domain-specific language that is specifically designed to ease the formal definition of action types in logistics networks (Rabe et al. 2017). Thus, the user is able to use predefined language constructs for this modeling process. From the action type designer the user can access all existing action types. Action types are structured as a module. Thus, the user can integrate any existing action type into the modeling process of new action types.

Action types are stored in the action type directory. The heuristic unit is connected with the action type designer to access all action types and, therefore, all corresponding actions in order to create the search space. In contrast to typical approaches for LASs the author's approach for a logistics assistance system offers the possibility to extend the set of predefined actions by user-generated actions.

3 Decoupling Actions from the Underlying Data Model

An action defines specific changes to the logistics network and, therefore, to the underlying SDM (Rabe et al. 2017). Thus, when applying an action, the corresponding entities of the LNW need to be adapted accordingly. Considering a database as the underlying data source of the data-driven simulation, an entity class is defined as a table. A corresponding entry in the table is called entity. For identifying affected entities for a given action, the specific table and attribute names of the underlying simulation's database are specified within the action's definition (Fig. 2). For instance, an action that changes the stock level of an SKU in a site needs to address the specific area of the underlying database, where the corresponding changes are applied. Thus, the attribute "Initial_Stock" in the corresponding table "SiteHaveSKUs" in addition to the change of stock level, e.g., an increase of 20, must be defined.

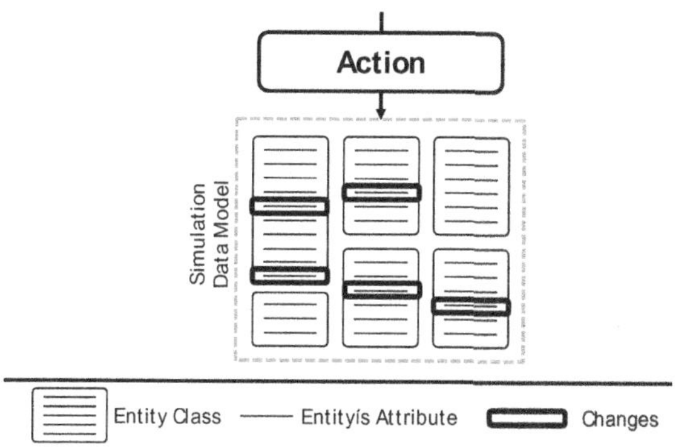

Fig. 2. Applying an action directly to the simulation's database.

However, the SDM is typically predefined by the simulation tool and, thus, not easily adjustable. This aspect may result in multiple issues:

1. The modeling of actions is done in direct relation to the SDM's structure. Thus, users need deep knowledge about the underlying data structure, e.g., the entities' names and attributes, the relations between the entities, and the data that is stored in each entity class. However, users might not be familiar with the given data structure, because it may differ from the company's data structure. Therefore, training the users might become very time-consuming and costly.
2. Updating the simulation tool may change the simulation data model and, thus, actions may need to be adapted accordingly resulting in additional costs and potential downtime of the LAS. In addition, users may need to be trained to the changed data structure in order to model new actions.
3. Predefined metadata is specified against the EDM, e.g., entity classes or its attributes. Thus, changes to the EDM may require adaptions of the meta information leading to additional costs and potential downtime of the LAS.
4. Simulation tools typically have their own individual data structure. Thus, changing the current simulation tool used by the LAS may increase the problems mentioned in 1, 2, and 3.

To address these issues, the authors present an approach for decoupling actions from the simulation data model by adding an additional layer in between. The main part of the additional layer is a so-called enterprise-specific data model. The EDM represents the specific structure of the enterprise's LNW and, therefore, the EDM is typically determined by the company itself.

Each attribute of the SDM is distinctly linked to the corresponding attribute of the EDM. Each connection is defined in the so-called mapping model. The mapping model consists of a number of Json files, one for each entity of the EDM. Such a Json file contains the entity's name of the EDM and a number of tuples for the connections between each of the entity's attributes and the linked attributes of the underlying SDM. An example of such a Json file is pictured in Fig. 3.

Thus, actions will no longer be modelled against the SDM, but exclusively against the EDM instead. In addition, the predefined metadata will be specified against the EDM as well. Therefore, the EDM constitutes an abstract interface to the underlying SDM. When applying an action, the corresponding changes to the EDM are mapped to the SDM utilizing the mapping model and, therefore, the consisting Json files (Fig. 4).

The process of applying an action and transforming the action into changes to the SDM is pictured in Fig. 5. First, a compiler transforms an action into changes to the EDM, e.g., when increasing the stock level of an SKU in a site, the attribute "StockLevel" in the entity class "SitesStoreSku" of the EDM needs to be adjusted for all affected entities accordingly. A mapper transforms these changes to the SDM's structure utilizing the mapping model, e.g., the attribute "StockLevel" in "SitesStoreSku" is linked to the attribute "Initial_Stock" in "SiteHaveSKUs" of the underlying SDM. The utilization and mapping of any metadata is done in the same way. The mapped changes are applied to the SDM and, therefore, altering the LNW's state.

```
{
   "Entity EDM":"SitesStoreSku",
   "Attributes":[
      {
         "Attribute EDM":"SiteId",
         "Entity SDM":"SiteHaveSKUs",
         "Attribute SDM":"Site"
      },
      {
         "Attribute EDM":"SkuId",
         "Entity SDM":"SiteHaveSKUs",
         "Attribute SDM":"SKU"
      },
      {
         "Attribute EDM":"StockLevel",
         "Entity SDM":"SiteHaveSKUs",
         "Attribute SDM":"Initial_Stock"
      }
   ]
}
```

Fig. 3. Example of a Json file for mapping the attributes of entity “SitesStoreSku” from the EDM to the corresponding attributes of the SDM.

Fig. 4. Applying an action to the enterprise-specific data model and mapping the resulting changes to the simulation data model.

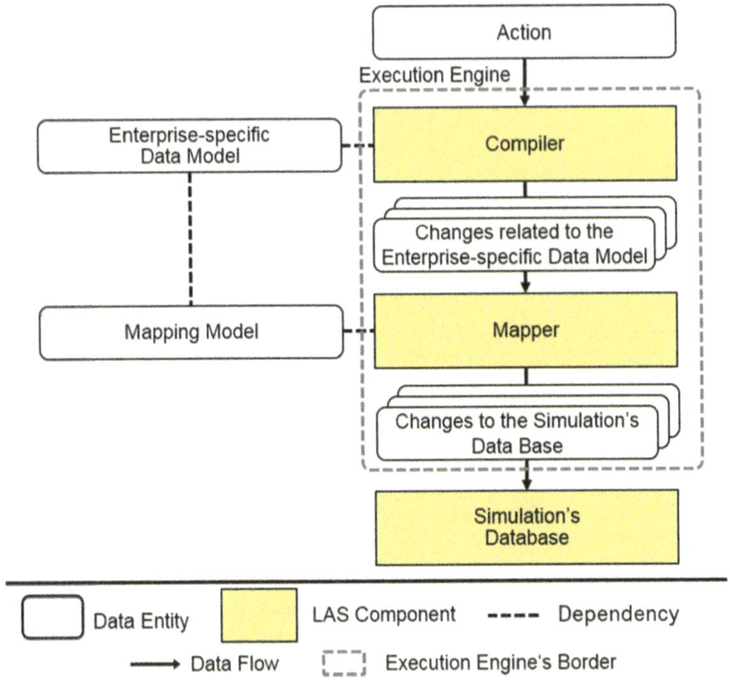

Fig. 5. Process of applying an action using an integrated enterprise-specific data model based on Rabe et al. (2017).

4 Implementation and Results

4.1 Nomenclature and Structure of the Enterprise-Specific Data Model

Typically, an EDM's nomenclature and structure is individually defined by the enterprise itself. However, once specified, an EDM's definition stays persistent. Thus, each of the SDM's attributes need to be linked distinctly to a corresponding attribute of the EDM only once.

For example, an attribute "Initial_Stock" in "SitesHaveSKUs" of the SDM might be linked to an attribute "StockLevel" in "SitesStoreSku" of the EDM. However, the EDM's attribute could also be named "Stock". The only difference between both cases is that the corresponding Json file for the entity class "SitesStoreSku" defines either the attribute "StockLevel" or "Stock" of the EDM as being linked to the attribute "Initial_Stock" in "SitesHaveSKUs" of the SDM.

Information of an EDM's entity class can also be split among several entity classes of the SDM. This information is specified by the attribute's value of "Entity SDM" in the corresponding Json file. For instance, an EDM's entity class may have two attributes "attribute1" and "attribute2", whereas "attribute1" is linked to "attributeA" of an entity class "e1" in the SDM and "attribute2" is linked to "attributeB" of an entity class "e2" accordingly. An extract of the corresponding Json file is displayed in Fig. 6. Consequently, the structure of the EDM is fully independent of the underlying SDM and, therefore, the modeling of action types as well.

```
{
    ...
    "Attributes":[
        {
            "Attribute EDM":"attribute1",
            "Entity SDM":"e1",
            "Attribute SDM":"attributeA"
        },
        {
            "Attribute EDM":"attribute1",
            "Entity SDM":"e2",
            "Attribute SDM":"attributeB"
        },
    ...
}
```

Fig. 6. Example of a Json file for the mapping between the EDM and the SDM.

4.2 Extensibility and Interchangeability of the Underlying Simulation Data Model

The presented approach opens up the possibility of extending the underlying SDM, e.g., by adding new entity classes to the SDM or by extending existing entity classes with new attributes. This supports and preserves the flexibility of the LAS and can be done intentionally and as needed. The use of an EDM allows for simply mapping an entity's attribute of the EDM to the newly added ones in the SDM, as described before. The EDM remains as it is since it covers the whole representation of the enterprise's LNW. For example, the logistics network might be divided into regions and, thus, the information could be stored in a corresponding attribute, e.g., "region" in the master data of the LNW for each of the network's sites. However, the underlying SDM might not be able to represent this information in any way. Adding a new attribute to the SDM and linking it to the corresponding attribute "region" of the EDM opens up the possibility of using this attribute in an action type definition, e.g., for specifying which entities of the LNW are affected by corresponding actions.

Another advantage of this approach is that there is an independency from the underlying SDM and, therefore, from the used simulation tool. In general, simulation tools may have different data model characteristics with respect to their entity classes, attributes, and the relations between those entity classes. If another simulation tool is to be used, the already implemented action types and the EDM are retained in their entirety. Only a new mapping model needs to be created according to the data structure of the new simulation model. Thus, switching the underlying simulation tool can be addressed by changing the mapping model that connects the EDM's structure with the structure of the underlying SDM. The newly created mapping model will be used in the execution engine for further processing of any action. Figure 7 illustrates the procedure of switching the mapping model and, therefore, adapting to a new SDM.

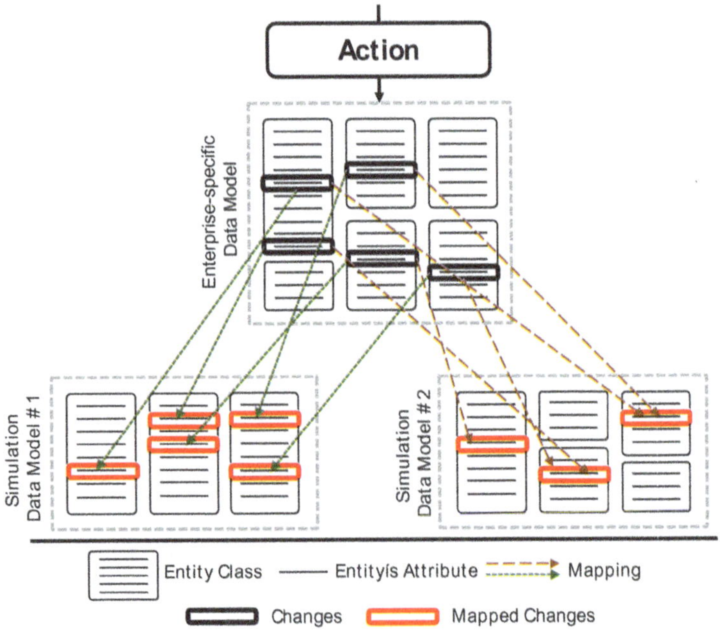

Fig. 7. Applying an action to the enterprise-specific data model and mapping the resulting changes to another simulation data model

5 Conclusion and Outlook

This paper demonstrates a concept for decoupling actions from the underlying SDM by adding an EDM as an additional layer in between the modeling of action types and the application of corresponding actions. The utilization of an EDM results in multiple advantages, e.g., using the EDM's nomenclature and structure for the formal definition of action types in order to improve the LAS's usability. Additionally, the possibility of composing information from multiple entities of the SDM into one EDM's entity and vice versa is shown. Furthermore, an approach for integrating new entities or attributes to the EDM in order to improve the system's quality is presented.

Currently, the authors are working on an object-oriented reference model (RM) for wholesale logistics networks. This RM could be used to derive specific EDMs for companies in that domain. Such EDMs would be linked to the corresponding RM. Following the object-oriented approach could lead to an additional increase of the LAS's usability. Furthermore, the authors are investigating the approach of modeling action types against the RM. These action types could be used in any EDM derived from the corresponding RM, reducing the enterprise's costs of modeling their own action types.

References

Barth, K., Hartmann, M., Schröder, H.: Betriebswirtschaftslehre des Handels. Springer Gabler, Wiesbaden (2015). https://doi.org/10.1007/978-3-8349-7184-5

Blutner, D., Cramer, S., Krause, S., Mönks, T., Nagel, L., Reinholz, A., Witthaut, M.: Assistenzsysteme für die Entscheidungsunterstützung. In: Buchholz, P., Clausen, U. (eds.) Große Netze der Logistik, pp. 241–270. Springer, Berlin, Heidelberg (2009)

Bockholt, F., Raabe, W., Toth, M.: Logistic assistance systems for collaborative supply chain planning. IJSPM **6**(4), 297–307 (2011). https://doi.org/10.1504/IJSPM.2011.048010

Chamberlin, D.: SQL. In: Liu, L., Tamer Özsu, M. (eds.) Encyclopedia of Database Systems, vol. 13, pp. 1–9. Springer, New York (2017)

Connolly, T.M., Begg, C.E.: Database Systems. A Practical Approach to Design, Implementation, and Management. Pearson, Boston (2015). Always Learning

Cordeau, J.-F., Pasin, F., Solomon, M.M.: An integrated model for logistics network design. Ann. Oper. Res. **144**(1), 59–82 (2006). https://doi.org/10.1007/s10479-006-0001-3

Elmasri, R., Navathe, S.: Fundamentals of Database Systems. Addison-Wesley, Boston (2011)

Folinas, D., Manikas, I., Manos, B.: Traceability data management for food chains. Br. Food J. **108**(8), 622–633 (2006). https://doi.org/10.1108/00070700610682319

Heilala, J., Montonen, J., Jarvinen, P., Kivikunnas, S., Maantila, M., Sillanpaa, J., Jokinen, T.: Developing simulation-based decision support systems for customer-driven manufacturing operation planning. In: Dunkin, A., Ingalls, R.G., Yücesan, E., Rossetti, M.D., Hill, R., Johansson, B. (eds.) Proceedings of the 2010 Winter Simulation Conference, pp. 3363–3375. IEEE Press, Piscataway (2010)

Juan, A.A., Rabe, M.: Combining simulation with heuristics to solve stochastic routing and scheduling problems. In: Dangelmaier, W., Laroque, C., Klaas, A. (eds.) Simulation in Produktion und Logistik 2013, pp. 641–649. Heinz-Nixdorf-Institut, Paderborn (2013)

Kengpol, A.: Design of a decision support system to evaluate logistics distribution network in Greater Mekong Subregion Countries. Int. J. Prod. Econ. **115**(2), 388–399 (2008). https://doi.org/10.1016/j.ijpe.2007.10.025

Larson, P.D., Halldorsson, A.: Logistics versus supply chain management: an international survey. Int. J. Logistics Res. Appl. **7**(1), 17–31 (2004). https://doi.org/10.1080/13675560310001619240

Ma, H., Suo, C.: A model for designing multiple products logistics networks. Int. J. Phys. Distrib. Mater. Manag. **36**(2), 127–135 (2006). https://doi.org/10.1108/09600030610656440

Rabe, M., Dross, F.: A reinforcement learning approach for a decision support system for logistics networks. In: Yilmaz, L., Chan, W.K.V., Moon, I., Roeder, T.M.K., Macal, C., Rossetti, M.D. (eds.) Proceedings of the 2015 Winter Simulation Conference, pp. 2020–2032. IEEE, Piscataway (2015)

Rabe, M., Schmitt, D., Dross, F.: Method to model actions for discrete-event simulations of logistics networks. In: Chan, W.K.V., D'Ambrogio, A., Zacharewicz, G., Mustafee, N., Wainer, G., Page, E. (eds.) Proceedings of the 2017 Winter Simulation Conference, pp. 3370–3381. IEEE Press, Piscataway (2017)

Rabe, M., Ammouriova, M., Schmitt, D.: Utilizing domain-specific information for the optimization of logistics networks. In: Rabe, M., Juan, A.A., Mustafee, N., Skoogh, A., Jain, S., Johansson, B. (eds.) Proceedings of the 2018 Winter Simulation Conference, Gothenburg, Sweden (2018a, to be published)

Rabe, M., Schmitt, D., Ammouriova, M.: Improving the performance of a logistics assistance system for materials trading networks by grouping similar actions. In: Rabe, M., Juan, A.A., Mustafee, N., Skoogh, A., Jain, S., Johansson, B. (eds.) Proceedings of the 2018 Winter Simulation Conference, Gothenburg, Sweden (2018b, to be published)

Rabe, M., Schmitt, D., Ammouriova, M.: Utilizing domain-specific information in decision support for logistics networks. In: Freitag, M., Kotzab, H., Pannek, J. (eds.) Proceedings of the International Conference on Dynamics in Logistics (LDIC), pp. 413–417. Springer, Cham (2018c)

Reuter, B., Rohde, J.: Coordination and integration. In: Stadtler, H., Kilger, C., Meyr, H. (eds.) Supply Chain Management and Advanced Planning. Concepts, Models, Software, and Case Studies, 5th edn, pp. 241–256. Springer, Heidelberg, New York, Dordrecht, London (2015)

Rushton, A., Croucher, P., Baker, P.: The Handbook of Logistics & Distribution Management. Kogan Page, London, Philadelphia (2010). http://site.ebrary.com/lib/alltitles/docDetail.action?docID=10404161

Seÿffert, R.: Wirtschaftslehre des Handels. Westdeutscher Verlag, Opladen (1972). https://doi.org/10.1007/978-3-322-83523-9

Wang, N., Mi, Y., Gao, H., Liu, W.: Logistics network model based on matter element node. Procedia Comput. Sci. **91**, 351–356 (2016). https://doi.org/10.1016/j.procs.2016.07.093

Werner, H.: Supply Chain Management: Grundlagen, Strategien, Instrumente und Controlling, vol. 5. Gabler, Wiesbaden (2013)

West, M.: Developing High Quality Data Models. Morgan Kaufmann, Burlington (2011). http://www.eblib.com/patron/FullRecord.aspx?p=630008

Agent-Based Simulation in Logistics and Supply Chain Research: Literature Review and Analysis

Uwe Clausen[1], Matthias Brueggenolte[2], Marc Kirberg[1], Christoph Besenfelder[2], Moritz Poeting[1](✉), and Mustafa Gueller[2]

[1] Institute of Transport Logistics, TU Dortmund University, Leonhard-Euler-Str. 2, 44227 Dortmund, Germany
poeting@itl.tu-dortmund.de

[2] Chair of Enterprise Logistics, TU Dortmund University, Leonhard-Euler-Str. 5, 44227 Dortmund, Germany

Abstract. In complex supply chains decision-makers strive to act quickly and effectively to ensure the efficient operation of their systems. Particularly, at the operational level immediate decision-making is required. In this context, simulation is becoming increasingly important for decision-making in logistics and supply chain. The classic event-discrete simulation paradigm is reaching its limits in the modelling of individual interacting system components of complex socio-technical systems due to a lack of flexibility and adaptability. The agent-based simulation (ABS) paradigm offers the capability to design heterogeneous individuals as agents that interact with each other as well as with the environment. This paper analyses the state-of-the-art of ABS in literature with a focus on operational logistics. We use a multi-level classification framework to provide a literature overview for publications of the operational logistics research field from in the recent years. On the basis of the literature review, categories are identified which may indicate research gaps.

1 Introduction

In today's logistics, a supply chain is a composition of complex physical technical systems, such as factories, hubs and customers, forming a network of interdependent actors. Such systems can only be fully analyzed, if subsystems and their interdependencies are taken into account. Therefore, modern supply chains are typically viewed as a socio-technical system (Behdani 2012). Additionally, the structure of supply chains is shifting away from the chain structure to a network of interacting entities, which further increases the complexity. A supply network is often regarded as a network of (semi-) autonomous organizations that make decisions independently. These organizations usually work in a distributed or decentralized manner and form a complex logistic network (Chan and Chan 2010; Hongler et al. 2010). This makes the modeling of supply chain networks challenging.

U. Clausen et al. (Eds.): ICPLT 2019, LNLO, pp. 45–59, 2019.
https://doi.org/10.1007/978-3-030-13535-5_4

Additionally, in the course of globalization and digitalization, supply chain structures are established to meet the growing diversity of customer demand (Schwemmer 2016). Hence, operational decisions in logistics have to be made at short notice and as profoundly as possible in order to ensure efficient operations (Besenfelder et al. 2017). Major difficulties associated with the modeling of these systems are the presence of non-linear relations between several elements and the level of details to be represented (Sonnessa 2004). Since analytical methods have limitations to handle the degree of complexity inherent to real-world problems, simulation is recognized as an effective tool for detailed analysis and reliable problem solving of complex socio-technical systems (Chen et al. 2013). Advanced modeling and simulation techniques support decision-makers to gain knowledge about the system in highly uncertain environments. Simulation can provide valuable insights about the relationships between important performance indicators of a system and the associated configuration.

Many consider the relatively new agent-based modelling and simulation (ABS) technique as a separate technique that combines characteristics and capabilities of Discrete Event Simulation (DES), System Dynamics (SD), Monte Carlo Simulation, and continuous simulation (Macal 2016). As a matter of fact, other experts regard ABS as a variation or subset of DES that is extended through the consistent application of an object-oriented world view (Law 2015; Pegden 2010). Consequently, there seems to be a lack of consensus in the simulation community regarding ABS.

In 2005, a tutorial as well as a dedicated track of papers on ABS were presented at the Winter Simulation Conference (Law 2015). Since then, indicators for a growing interest in ABS are the number of conferences, workshops, devoted tracks on ABS, and the growing number of peer-reviewed publications in various discipline-specific journals as well as methodological journals on modeling and simulation. In view of the growing interest of the logistics and supply chain research community in ABS, it is reasonable to devote a review paper to recent ABS publications. Reviews of DES and SD, which are widely used techniques, already exist and are not subject of this paper (Tako and Robinson 2012).

The aim of this paper is to identify the state-of-the-art in logistics and supply chain research, utilizing the technique of ABS. Furthermore, this paper provides an overview of certain characteristics of ABS approaches and problems that have been addressed in recent years. Section 2 introduces what is considered as ABS and how it can be distinguished from other modelling and simulation approaches. The research methodology for identifying and selecting relevant publications is presented in Sect. 3. The Subsect. 3.1 presents our framework for classification and evaluation of ABS publications in the logistics and supply chain context. Section 4 focuses on the results of our study with regard to the classification scheme. The frequency distributions of the characteristics in each category are presented. In order to see relations between categories, the frequency of relevant combinations of categories are investigated. Finally, Sect. 5 gives concluding remarks on the key findings and a discussion of possible research gaps.

2 Agent-Based Simulation

Agent-based simulation is a technique for modeling and simulating systems with a focus on individual interacting entities at a micro system-level (Macal 2016; Behdani 2012; Macal and North 2010). In this context, an active entity is represented by an agent that is located within the system environment. Agents represent active system components, as they are aware of their surroundings and are capable of making decisions independently. Over time, the agent can react autonomously to the environment and perceive the resulting changes (Macal 2016). Agents can represent persons, companies, or other entities that actively pursue a specific goal in a system (Abar et al. 2017). Furthermore, Macal and North (2010) identified essential characteristics: An agent is a self-contained, modular, and clearly identifiable entity. It operates autonomously, self-directed, has a state that changes over time, and is capable to make decisions.

The general approach of ABS is to incorporate the internal microstructure of a complex system by describing the behavior and interactions of agents according to the bottom-up approach (Klügl and Bazzan 2012). Thus, the system characteristics emerge from the internal microstructure modeled and correspondingly allows the exploration of the overall system behavior. It is necessary to understand its subsystems and their interconnections. The macro system-level behavior emerges from the sum of the individual agent behavior patterns (Behdani 2012). Thus, the advantages of ABS are based on the ability to explain and evaluate complex adapting systems and their unexpectedly occurring new behavior (Heath and Hill 2010).

In order to enable agents to interact with other agents and influence the environment, they consist of attributes and behaviors (Macal and North 2010). The agent behavior is generally determined by rules ranging from simple reactive if-then rules to more complex behaviors modeled by adaptive techniques such as artificial intelligence (Macal and North 2010). An agent follows these decision rules in a predefined order. However, the modeler needs detailed knowledge about the behavior of each agent and the possible interactions between agents themselves and the environment. Obtaining this knowledge in the context of modelling the microstructure can be seen as a major limitation of ABS.

Usually, ABS is implemented in an object-oriented software, where variables represent attributes and methods represent behaviors (Law 2015). Other structural elements of ABS are the relationships and methods of interaction. How and with whom agents interact is defined by the topology of connectivity (Macal and North 2010). In summary, an agent-based model is determined by the following elements (Macal 2016):

- The individual and autonomous agents themselves, with their attributes and behavior.
- Agent relationships and methods of interaction: It is defined how and which agents interact with each other.
- The environment: In addition to the interaction with other agents, an agent interacts with its environment and therefore has to be able to perceive it.

Accordingly, ABS is a suitable method to model, simulate, and evaluate complex socio-technical systems, if the aforementioned elements are known and can be adequately modeled. A major advantage of this method is that it allows adaptability to environmental changes as the system behavior is based on individual agent decisions. This is of great importance in times of increasingly complex supply chain structures and rapidly changing logistic processes. Currently, ABS is receiving substantial attention from simulation experts in the field of logistics.

3 Methodology

The aim of this study is to explore the use of ABS for decision-making in logistics and supply chain context, specifically looking at modeled issues and covered application fields. The interest of the simulation community is expressed by the increasing number of publications in the last years as well as by the availability of a dedicated track at the Winter Simulation Conference, which is the leading international forum for disseminating the latest advances in system simulation (Law 2015). For this review Elsevier's Scopus, the largest data base for peer-reviewed literature, is used to find publications that meet the restrictive criteria of the search.

Over 1,050 articles were found by searching for the keywords 'ABS' and 'logistics' or 'supply chain' in the database. In order to focus on a distinct field and to reduce the number of search results to a manageable number of publications, this paper specifically addresses the operational level of logistics and supply chain management. Consequently, the ability to deal with uncertainties in an operational short-term decision-making environment is of particular interest. For this reason, we have added the keywords "operational" and "short term" to the search string. The search returned 323 publications that meet our search criteria. Furthermore, we have specified our search for the period from 2012 to 2018 in order to look at the recent activities in the respective area. The limitation in time further reduced the number of publications to 173. In order to ensure the thematic relevance, we have crosschecked title, keywords and abstracts with regard to the thematic focus on production, logistics, and transport. If a paper was available, a full-text reading was carried out. The 68 publications we have reviewed were then classified according to the classification scheme. An overview of the papers included in the investigation is available online (Clausen et al. 2018). The review process is visualized in Fig. 1. In the next chapter, the classification framework is introduced.

3.1 Classification Framework

In order to get an overview of the different problems and applications of ABS in logistics, a classification framework was developed. It is based on a framework by Davidsson et al. (2005) and used to classify and asses the research publications found. For each publication surveyed, a description of the problem, the modeling approach taken to solve it, and the evaluation of the results were carried out.

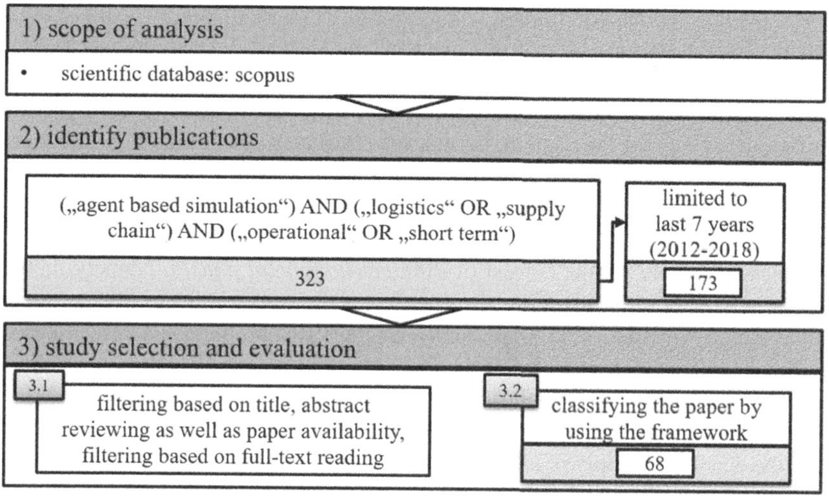

Fig. 1 Procedure of the systematic review process

3.1.1 Problem Description

The problems under study in each of the reviewed publications were classified by the level of decision-making, level of detail, and horizontal phase of the supply chain. With this classification scheme, we systematize the publications on the topic of ABS. The first problem-describing category is the *level of decision-making*, which has three different manifestations. The publications studied are divided into the categories of design tasks, planning tasks, and control tasks, which were similarly introduced by Chopra and Meindl (2016). Problems assigned to the *design tasks* are dealing with tasks such as building new systems or reconfiguring an already existing system design to meet increased performance requirements. Problems related to the class of *planning tasks* were characterized by the decisions that have to be made to efficiently operate an existing system and its resources. The category of *control tasks* includes those publications that focus on short-term decision-making at the execution level. Regarding the restrictions of the search query to publications including the term 'operational', one could assume an inconsistency at this point. However, given the fact that agent systems can organize themselves, it is reasonable to deal with design, planning, and control decisions in an operational system environment. In other words, designing the layout of a network or factory used to be a long-term strategic decision, which had little to do with operational decision-making. But today, one of the reasons why ABS is of great interest, is the vision that autonomous self-organizing agents are going to be capable of changing the layout regularly according to the actual needs of the business environment they are part of. For this reason, strategic, or tactical planning tasks – such as system design – were also considered in the context of publication on operational aspects.

The category *level of detail* pays special attention to the delimitations of the problem. The perspective of looking at logistic systems is not generally of the same granularity. We distinguish a macro, meso, and micro perspective to classify the object of consideration of each paper. The *macro perspective* focuses on multiple logistic nodes and edges, which add up to a whole logistic network or supply chain. It aims to assess the overall performance of this macroscopic system. A single logistic node can represent a distribution center, transshipment terminal, depot, warehouse, or factory, for example. A simulation model of a logistic node, which is modeled with respect to all the relevant aspects for the system, is categorized as *meso perspective*. At the level of *micro perspective*, a simulation model represents processes within logistic facilities (e.g. order picking, assembly, loading) or between facilities (e.g. transport of goods or information).

Typically, supply chains consist of different *horizontal phases*. Namely these are the phases of supply, production, distribution, and reverse logistics (Femerling and Gleißner 2013; Ait-Kadi et al. 2012). In this category, the problem addressed in each of the publications is assigned to the appropriate phase of the supply chain to ascertain in which horizontal phase the application of ABS is most prevalent. The *supply phase* deals with issues concerning procurement logistics. Decisions of this phase aim to ensure the economic supply of all input factors such as materials and commodities. The *production* phase is associated with manufacturing systems. Publications belonging to this category deal with planning and controlling tasks that are related to the material flow, storage, and internal transport (Femerling and Gleißner 2013). Publications concerned with the processes that serve to deliver goods to their destinations belong to the *distribution* phase (Femerling and Gleißner 2013). The *reverse* phase is destined for all planning, implementation, and controlling topics which aim at maximizing the creation of value through clean disposal, a reverse product flow, and an efficient management of raw materials (Aït-Kadi et al. 2012). Since several authors - especially those who take a macro perspective - have a holistic view on the covered problem, multiple assignments to the categories of the horizontal phase are reasonable.

3.1.2 Modeling Approach

The modeling approaches taken to solve the problems described in the publications are classified by the problem solving mechanism, intended usage of the agent application, agent capabilities, and agent attitude.

In order to categorize the *problem solving mechanism*, we focused especially on the utilization of mathematical optimization techniques in combination with ABS. Therefore, a distinction was made between optimization with a *specific heuristic* or a *metaheuristic*. If the problem addressed is solved by a manual process, i.e. without mathematical optimization, we categorized it as solved *through experiments* made by the modeler. Unfortunately, in some publications the approach used is not specified at all.

Another interesting aspect to focus on, when reviewing ABS publications, is that there are many different possibilities of agent system implementations. Each implementation makes use of different *agent capabilities*. The agents' behavior is determined by individually defined rules enabling them to interact within their environment (Macal and North 2010). For some approaches, it might be reasonable to equip agents with very simple behavioral defining structures, which are characterized by *basic if-then rules*. More advanced behavioral structures of an agent differ in relation to the more *advanced algorithms*, which serve as a rule for decision-making. In this case, mathematical optimization techniques can be found within agents. The following examples clarify how the agents and the algorithms work together:

- A routing agent uses an algorithm to identify and select suitable tours (Yang et al. 2017; Sprenger and Mönch 2014).
- A consolidation agent minimizes the number of transportation means needed (Yang et al. 2017),
- A transportation network agent assigns orders to a distribution location (Sprenger and Mönch 2014).
- An agent has to calculate the cost of an order and therefore has to solve the Traveling Salesman Problem with Time Windows using an insertion heuristic (Gath et al. 2012).

Another category named *intelligent algorithms,* subdivides approaches that handle uncertainty by integrating tools like machine learning or knowledge discovery methods to the decision-making processes in the model. Regardless of the particular agent behavior and the complexity of the inherent rules, it was determined whether the agents had the ability to *learn and adjust* their behavior. The last category deals with the *agent's attitude*. The attitude determines the social behavior of the agents and is classified as either *cooperative,* i.e. agents agree to social laws and global goals, or *competitive*, representing self-interested agents pursuing individual goals.

3.1.3 Maturity

The categorization according to the *degree of maturity* of an approach checks whether the agent application is complete and valid. The description of the maturity of an agent application is defined according to Parunak (2000) by the work that remains to carry out the implementation of that specific application. The class with the lowest degree of maturity is a *conceptual proposal* that describes the general characteristics of the agent application. If the application resulting from the conceptual model is demonstrated in a simulated environment, using either real or generated data, it is placed in the class of *simulation experiment.* The most mature level in this categorization is the *field experiment*. In this case, a prototypical demonstration is performed in the environment in which the application is supposed to be used. Theoretically, there may be a higher class of maturity if the application is regularly used commercially as a *deployed system.*

Finally, the evaluation comparison denotes whether the developed approach has been compared with other existing approaches that were previously used to solve similar problems. The comparison could be *qualitative*, by comparing major characteristics of the approaches, or *quantitative*, by different types of experiments. For reasons of clarity and comprehensibility, the entire framework is presented in Table 1.

Table 1. Classification framework

Problem description			Approach				Maturity	
Level of decision-making	Level of detail	Horizontal phase	Problem solving mechanism	Agent capabilities	Agent attitude	Agent coordination	Degree of maturity	Evaluation comparison
Design tasks	*Macro perspective*	*Supply*	*Specific heuristic*	*Basic if-than*	*Competitive*	*Decentralized*	*Conceptual proposal*	*Qualitative*
Planning tasks	*Meso perspective*	*Production*	*Metaheuristic*	*Advanced algorithms*	*Cooperative*	*Centralized*	*Simulation experiment*	*Quantitative*
Control tasks	*Micro perspective*	*Distribution*	*Experiments*	*Intelligent algorithms*		*Not specified*	*Field experiment*	
		Reverse	*Not specified*	*Learn & adjust*			*Deployed system*	

4 Results

The study of ABS publications shows that this modeling and simulation paradigm is actively applied in logistics research by addressing many different problems of several application fields. Problems under consideration ranged from supply chain issues, e.g. inventory systems, supply chain configuration, supplier selection, collaboration, risk management, negotiation mechanisms, or transport issues such as urban distribution, pickup and delivery problems, autonomous vehicles, fleet sizing, vehicle routing problems, or cooperative transportation. The subsequent descriptive analysis of the results is carried out for the categories of the problem description, modeling approach, and maturity. Afterwards the interrelationships between certain main categories are introduced.

4.1 Distribution Among the Categories

The distribution of publications among all categories are shown in Fig. 2. Publications with problems that are addressed from a macro perspective make up the dominant share of publications, which is consistent with the fact that many publications consider several horizontal phases.

Macal (2016) argues that ABS is attractive because it offers the capability to model an entire population of heterogeneous agents in order to gain insights into overall system behavior, based on individual agent behavior. This capability is particularly evident in macro perspective publications. For instance, Amini et al. (2012) studies the impact of alternative supply chain policies on the diffusion of a new product. The approach takes the adoption behavior of 3,000 individual customer agents into account. It is noteworthy that each agent decides individually, based on non-linear interactions and a changing environment, whether to adopt or reject the new product. Examples of similar approaches, in which an agent population was an essential element, deal with the analysis of the disaster relief distribution (Fikar et al. 2017), the decision-making and behaviors of humanitarian logistics actors (Krejci 2015), or the benefits of an open logistic network through physical internet (Sarraj et al. 2013; Yang et al. 2017).

In contrast to the characteristic of considering a large agent population, publications with problems from a micro perspective focus on the interaction of numerically fewer agents in a particular environment. In this case, the functional modes and mechanisms as well as control principles are usually in the foreground. For example, Garro et al. (2015) evaluates a dispatching strategy for the straddle carrier pooling problem. Vojdani et al. (2017) tests negotiation mechanisms to ensure decentralized production control, which is based on the interaction of a smaller number of agents. Lima et al. (2015) tests a procedure which allows an agile planning and control of transport flows in port logistic systems. In accordance with the micro-perspective, the delimitations of the problems are more specific.

The assignment of multiple publications to the three horizontal phases is visualized in the Fig. 3 that also shows the total number of assignments for each phase in brackets. There were 9 publications assigned to the common three phases of supply, production and distribution. The phase representing revers logistics only matched with a single paper. Despite the limited sample size, this allows the assumption that ABS has not been frequently used in the research area of reverse logistics. Nevertheless, also the meso and micro perspectives are adequately represented. This may indicate that ABS is not only applicable for individual decision-making on a process level, but also arbitrary scalable to a supply network level. In view of the level of decision-making, a tendency towards planning and control tasks is discernible.

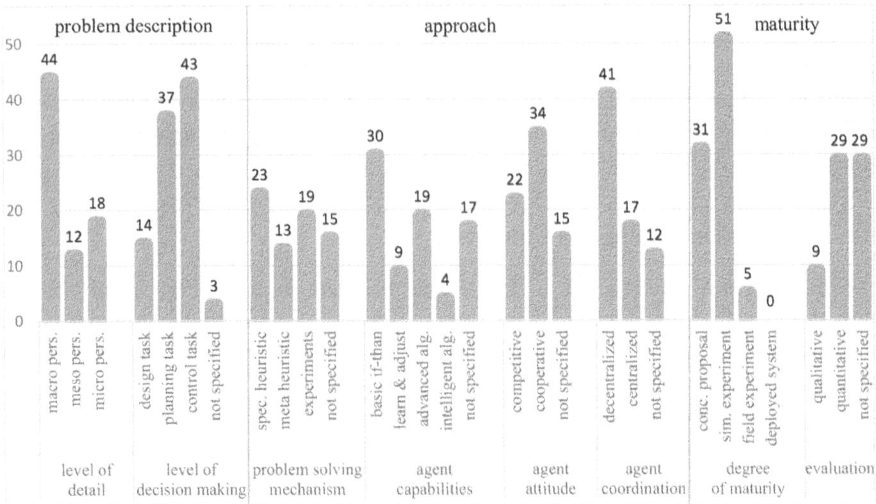

Fig. 2 Distribution of publications among the categories.

A look at the mechanisms that are used to solve the problems and the way the agents are implemented shows, that many publications do not provide detailed information on certain aspects. For this reason, the publications could not always be precisely sorted into the classification framework. This is also reflected in the number of unspecified publications. There is no particularly dominant problem-solving mechanism apparent, but it can be seen clearly that heuristic procedures, whether specific or meta-heuristics, in combination with ABS are the rule rather than the exception. Specific heuristics are used slightly more frequently than metaheuristics, but additionally, knowledge gained by the modeler through simulation results was used to cope with the problems. The number of unspecified publications is not only caused by the lack of a description by the authors, but also due to the fact that some papers were not available as full text and that a handful of papers merely present frameworks for modeling supply chains without focusing on specific problems.

The agent capabilities are determined by the rules of behavior implemented into each agent. It is worth to note that regardless of the complexity of the rules, apparently only nine modelers considered to implement the property of learning and adapting into the agent behavior. Furthermore, the most common behaviors were simple basic if-than rules. Only four publications included machine learning or knowledge discovery components into the agent's behaviors, which can be interpreted as an indication of a research gap. The high number of unspecified publications is mainly caused by unclearly described implementations of the agents. Many authors concentrated more on the frameworks they created and the tasks that the agents have to perform within this framework rather than on describing the agents in detail.

The dominant attitude of the agents is the cooperative attitude, which means that all actions are directed towards a common goal. However, ABS is also used to represent market-based relationships. This is particularly appropriate in view of the fact that agents can represent self-deciding individuals. Systems without any direct coordination between actors result from a high level of uncertainty and absence of direct connections. This was for example the case with publications on the topic of humanitarian relief logistics or competitive distribution agents. The use of ABS is often associated with the idea of decentralized coordination within a system. This is largely confirmed by the vast majority of publications using a decentralized type of agent coordination. Nevertheless, a not insignificant number of publications make use of centralized control, e.g. by defining a single agent as coordination unit for all other agent types.

Regarding the maturity, most of the publications are in the stage of conceptual proposal or simulation experiments. There are only five publications that have conducted field experiments and not a single paper presented a deployed system, which might be taken as evidence for a lack of maturity at the time the publication was published. However, it is also reasonable to assume that the deployed systems are not the subject of academic publications. Most concepts are solely investigated in the simulation with real or generated data, but have not yet been transferred to the system environment. Surprisingly, more publications draw a quantitative comparison of the results rather than a qualitative. Almost 40% did not compare their results at all.

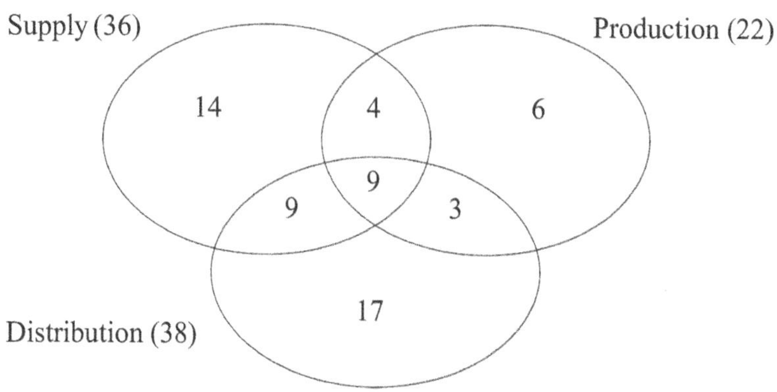

Fig. 3 Distribution of publications by horizontal phase

Figure 4 shows the relationship between different dimensions of the classifications. The main finding is that ABS in combination with specific heuristics has been extensively applied to the distribution phase of supply chain. Furthermore, it can be stated that there is a lack of studies that implement metaheuristics to solve complex production problems. Most of the publications tend to incorporate problem solving mechanisms in their simulation model. At the same time, a general shortcoming of existing studies is that field experiment as an advanced maturity level are neglected. In addition, no application could be found in our analysis that fulfilled the criteria of a deployed system. From the view of horizontal phases, several papers deal with supply, production, and distribution phases, but there is a lack of agent-based approaches addressing reverse logistics. As shown in Fig. 4, no accurate statements can be made regarding the use of certain problem solving mechanisms for the horizontal phases. This finding can also be applied to further comparisons of different dimensions, so that no particular problem type is solved with a typical solving approach. It seems as if there is no standard procedure in the research community yet to be able to deduce proven solutions from similar problem types.

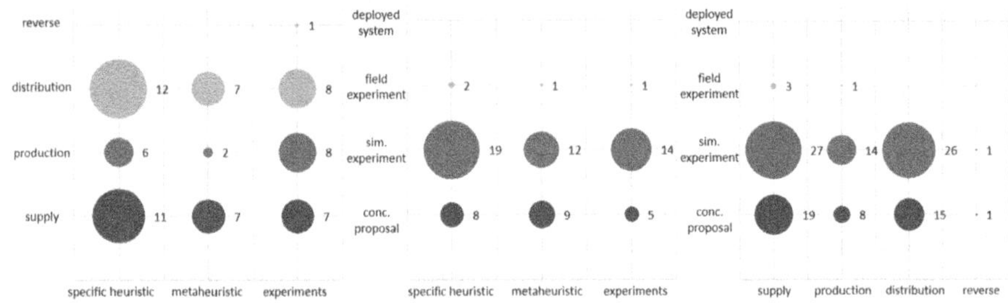

Fig. 4 Combination of the categories

5 Conclusion

This paper presents a conducted study on ABS in operational logistics and supply chain research. The findings of this study provides useful insights about the state-of-the-art of agent-based simulation in that research field. The study comprises a categorization of 68 publications in the time period from 2012 to 2018, with different characteristics for the problem description, modeling approach as well as maturity. These characteristics are captured by means of a developed classification framework. Some essential conclusions can be drawn from the distribution of publication among the categories.

It can be assumed that ABS is more suitable for micro level simulations because of its ability to model individual interacting entities at a low level. However, there are significantly more applications of ABS at a macro level. This implies that ABS is easily scalable to a high level perspective. In addition, ABS is used in current research to exploit the advantages of decentralized coordination on the one hand and the ability to make decisions for planning or control tasks on the other. This matches with the agent system characteristic of self-organization. A comparison of the results was made in 56% of the publications. Surprisingly, there were more quantitative comparisons (42%) than qualitative (14%). This allows the assumptions that research, previously carried out with other methods, was continued or expanded by the ABS method, so that quantitative results were already available. In those cases, ABS was not the only method used to solve the studied problem, but it was considered by the authors to be an appropriate complementary method. However, the number of publications that do not compare their results is quite high (40%).

Moreover, this study identifies certain research gaps for ABS in different manifestations. One gap becomes apparent, due to the lack of paper in the area of reverse logistics. Besides that, the ability to learn and adapt, which is often associated with ABS, should be given much greater consideration in future research. In the modeling of decentralized socio-technical systems, the ability of the actors to learn and adapt should also be mapped accordingly. Only four publications used intelligent algorithms to determine the agent's behavior rules, which could be interpreted as an indication of a research gap. For example, implementing a neural network or other learning algorithms that dynamically determine the agent's behavior can adequately represent individual supply chain actors. In real logistic systems, organizations also change their behavior through experiences, incentives, or other environmental stimuli. Finally, future research should increase the degree of maturity of the concepts by conducting more field studies and testing commercial tools based on ABS.

The findings of this study are defined by the procedure of the systematic review process and the selected sample of publications. The review is based on the search query result of Scopus, which only contains peer reviewed publications. Those publications may not represent the full range and frequency of ABS in a supply chain and logistics context and should therefore be further expanded to other sources of literature (e.g. conference proceedings and non-IEEE journals). In addition, further analysis should revise the keywords to assess papers on strategic and tactical planning tasks. Nevertheless, this study provides the basis for further comparison studies or extension of the classification framework. Future work will aim at a broader classification framework to cover the methods of mathematical optimization in more detail. Additionally, the search areas will be expanded to include publications from conferences such as the Winter Simulation Conference. Furthermore, the field of integration of machine learning, artificial intelligence, and other newly emerging research areas will be examined.

Acknowledgments. This paper is a research result in context of the Leistungszentrum (Center of Excellence) Logistics & IT, an initiative of the Fraunhofer-Gesellschaft and the state of North Rhine-Westphalia, Germany.

References

Abar, S., Theodoropoulos, G.K., Lemarinier, P., O'Hare, G.M.P.: Agent based modelling and simulation tools a review of the state-of-art software. Comput. Sci. Rev. **24**, 13–33 (2017)

Ait-Kadi, D., Chouinard, M., Marcotte, S., Riopel, D.: Sustainable Reverse Logistics Network Engineering and Management. ISTE, Wiley, London, Hoboken (2012)

Amini, M., Wakolbinger, T., Racer, M., Nejad, M.G.: Alternative supply chain production–sales policies for new product diffusion an agent-based modeling and simulation approach. Eur. J. Oper. Res. **216**(2), 301–311 (2012)

Behdani, B.: Evaluation of paradigms for modeling supply chains as complex socio-technical systems. In: Proceedings of the 2012 Winter Simulation Conference (WSC). IEEE, Piscataway, New Jersey (2012)

Besenfelder, C., Brüggenolte, M., Austerjost, M., Kämmerling, N., Pöting, M., Schwede, D.-I.C., Schellert, M.: Paradigmenwechsel der Planung und Steuerung von Wertschöpfungsnetzen. Unter Mitarbeit von Michael ten Hompel, Michael Henke und Uwe Clausen (2017)

Chan, H.K., Chan, F.T.S.: Comparative study of adaptability and flexibility in distributed manufacturing supply chains. Decis. Support Syst. **48**(2), 331–341 (2010)

Chen, X., Ong, Y.-S., Tan, P.-S., Zhang, N., Li, Z.: Agent-based modeling and simulation for supply chain risk management - a survey of the state-of-the-art. In: 2013 IEEE International Conference on Systems, Man, and Cybernetics, pp. 1294–1299. IEEE (2013)

Chopra, S., Meindl, P.: Supply Chain Management Strategy, Planning, and Operation, 6th Global edn. Pearson, Boston (2016)

Clausen, U., Brueggenolte, M., Kirberg, M., Besenfelder, C., Poeting, M., Güller, M.: List of Categorized Publications (2018). https://1drv.ms/f/s!AiEgswzazr3VfLZXuOtkHXHHJ70. Accessed 30 Oct 2018

Davidsson, P., Henesey, L., Ramstedt, L., Törnquist, J., Wernstedt, F.: An analysis of agent-based approaches to transport logistics. Transp. Res. Part C Emerg. Technol. **13**(4), 255–271 (2005)

Femerling, J.C., Gleißner, H.: Logistics Basics - Exercises - Case Studies. Springer, Cham (2013). (Springer Texts in Business and Economics)

Fikar, C., Hirsch, P., Nolz, P.C.: Agent-based simulation optimization for dynamic disaster relief distribution. Cent. Eur. J. Oper. Res. **12**(4), 301 (2017)

Garro, A., Monaco, M.F., Russo, W., Sammarra, M., Sorrentino, G.: Agent-based simulation for the evaluation of a new dispatching model for the straddle carrier pooling problem. Simulation **91**(2), 181–202 (2015)

Gath, M., Wagner, T., Herzog, O.: Autonomous logistics processes of bike courier services using multiagent-based simulation. In: The 11th International Conference on Modeling and Applied Simulation, pp. 134–142. DIME, Univ. di Genova, Genova (2012)

Heath, B.L., Hill, R.R.: Some insights into the emergence of agent-based modelling. J. Simul. **4** (3), 163–169 (2010)

Hongler, M.-O., Gallay, O., Hülsmann, M., Cordes, P., Colmorn, R.: Centralized versus decentralized control—a solvable stylized model in transportation. Phys. A Stat. Mech. Its Appl. **389**(19), 4162–4171 (2010)

Klügl, F., Bazzan, A.L.C.: Agent-based modeling and simulation. AIMag **33**(3), 29 (2012)

Krejci, C.C.: Hybrid simulation modeling for humanitarian relief chain coordination. J. Hum. Logist. Supply Chain. Manag. **5**(3), 325–347 (2015)

Law, A.M.: Simulation Modeling and Analysis, 5th edn. McGraw-Hill, New York (2015)

Lima, A.D.P., Werner de Mascarenhas, F., Frazzon, E.M.: Simulation-based planning and control of transport flows in port logistic systems. Math. Probl. Eng. **2015**(1), 1–12 (2015)

Macal, C.M.: Everything you need to know about agent-based modelling and simulation. J. Simul. **10**(2), 144–156 (2016)

Macal, C.M., North, M.J.: Tutorial on agent-based modelling and simulation. J. Simul. **4**(3), 151–162 (2010)

Van Dyke Parunak, H.: Agents in overalls experiences and issues in the development and deployment of industrial agent-based systems. Int. J. Coop. Inf. Syst. **09**(03), 209–227 (2000)

Pegden, C.D.: Advanced tutorial: overview of simulation world views. In: Proceedings of the 2010 Winter Simulation Conference, pp. 210–215 (2010)

Sarraj, R., Ballot, E., Pan, S., Hakimi, D., Montreuil, B.: Interconnected logistic networks and protocols simulation-based efficiency assessment. Int. J. Prod. Res. **52**(11), 3185–3208 (2013)

Schwemmer, M.: Top 100 der Logistik Marktgrößen, Marktsegmente und Marktführer; eine Studie der Fraunhofer Arbeitsgruppe für Supply Chain Services SCS. 2016/2017. DVV Media Group GmbH, Hamburg (2016)

Sonnessa, M.: Modelling and Simulation of Complex Systems. University of Torino, Italy (2004)
Sprenger, R., Mönch, L.: a decision support system for cooperative transportation planning design, implementation, and performance assessment. Expert Syst. Appl. **41**(11), 5125–5138 (2014)
Tako, A.A., Robinson, S.: The application of discrete event simulation and system dynamics in the logistics and supply chain context. Decis. Support Syst. **52**(4), 802–815 (2012)
Vojdani, N., Erichsen, B., Lück, T.: Nutzung von Produktionsechtzeitdaten - Eine agentenbasierte Feinplanung mittels Simulation. Logist. J. Proc. **2017** (2017)
Yang, Y., Pan, S., Ballot, E.: Freight transportation resilience enabled by physical internet. IFAC-PapersOnLine **50**(1), 2278–2283 (2017)

Acceptance of Technology Implementation in Industrial Intralogistics

Martin Eisenmann(✉) and Alfredo Virgillito

Sozialforschungsstelle (sfs), TU Dortmund University, Dortmund, Germany
{martin.eisenmann,alfredo.virgillito}@tu-dortmund.de

1 Introduction

According to the assessment of rather pessimistic publications (Frey/Osborne 2013; Bowles 2014; Bonin et al. 2015; F.A.Z.-online 2018; PWC 2018) regarding the effects of digitization and industry 4.0 - in the Anglo-Saxon and Anglo-American area also known as "the second machine age" (Brynjolffsson/McAfee 2014) or "third industrial revolution" (Rifikin 2011) - which have a decisive influence on the international and national discourses on the "fourth industrial revolution" (Forschungsunion/acatech 2013; Bauernhansl 2014, p. 5–36), a disruptive change is taking place in conjunction with numerous job losses - i.e. substitution of work. The change to the digital world of work is characterized, among other things, by polarization tendencies in the sense of "lovely and lousy jobs" (Goos/Manning 2007; also Hirsch-Kreinsen 2015, pp. 11–13) and increased control potentials through technology. These often technology-driven perspectives are contrasted by publications that tend to assume a quantitative increase in the number of jobs (BCG 2015; Vogler-Ludwig et al. 2016, p. 80) and qualitative changes in some activity elements (Ittermann et al. 2016, p. 13–18; BMAS 2016, p. 100ff.) as well as the emergence of new occupations (IGBCE 2017, p. 22–32) as future developments in industrial and industry-related sectors. It is undisputed, however, that the digital transformation will bring opportunities with it but also risks. This results in changes in work and employment structures for different groups of employees.

This paper is based on a socio-technical understanding of innovations in business contexts. This assumes that technical changes always have interactive consequences for the overall system and do not only affect the respective subsystems of people, technology and organization. A necessary core element of interdependent digitization processes is the acceptance of those who are directly or indirectly affected by the implementation of technology in their work processes. For the involved persons, the "dismantling of worries or reservations" (DGB 2016, p. 1)[1] will be a central moment in order to succeed in reducing corresponding barriers and tapping into potentials in a sustainable, economic and socially acceptable way (BMAS 2016, p. 89). Furthermore, (in)direct participation plays a central role in change processes, especially when

[1] All citation from German sources as well as the empirical data and the figures presented have been translated by the authors.

U. Clausen et al. (Eds.): ICPLT 2019, LNLO, pp. 60–74, 2019.
https://doi.org/10.1007/978-3-030-13535-5_5

questions of control potential of digital technologies are brought into play (ibid., p. 147). The central concern of this article is the identification of possible acceptance barriers and drivers in case of implementation of digital technologies.

The article is structured as follows: The second chapter presents various models, dimensions and acceptance factors described in the literature as well as the underlying definition of acceptance. In addition, the concept of acceptance and what it means in the context of digitization are given. In the third part of the paper, a brief description of empirical findings and methods is given, followed by (Sect. 4) the empirical findings and the three main factors that promote or acceptance: Inform, Include, and Intermingle. The conclusion (Sect. 5) contains a short summation of the results and a presentation of ideas for applying the socio-technical approach to findings on technology acceptance.

2 Acceptance

The following chapter presents the development and outlining of basic concepts of acceptance and the visualization of results from acceptance research that are relevant for this article. In the following, the elements of acceptance are described and a definition of the concept of acceptance appropriate to the article is given. The different acceptance dimensions, acceptance models and acceptance factors are described with the focus on the empirical data presented later.

2.1 On the Concept of Acceptance

In principle, it can be stated that there is no uniform, generally accepted definition for the concept of acceptance in the scientific literature (Schäfer/Kepppler 2013, p. 11; also Lucke 1995, p. 45–50). One example would be that acceptance can be defined as "the willingness of the population to agree to a political measure, e.g. the construction of nuclear power plants […]" (Fuchs-Heinritz et al. 2007, p. 27). However, the moment of action or intention of action is missing in the definition (see below). Acceptance is closely linked to acceptability, which is to be understood more in connection with social values and therefore to be distinguished from acceptance. Acceptability focuses more on the fundamental possibility, i.e. the "expected willingness of those affected by decisions […] to accept their consequences voluntarily". (ibid. p. 27) Overall, the concept of acceptance can be applied "to almost any (material or immaterial) object […]" (Schäfer/Keppler 2013, p. 5), although this paper focuses on the acceptance of technology in work contexts. In their study, Sauer et al. (2005) deal with the acceptance of nature conservation areas; in their report they also provide a definition that integrates attitude and action dimensions (see below). The introduction and use of new technologies in work processes affects both of these dimensions.

Accordingly, acceptance is to be understood in this paper: "as a positive attitude of an actor towards an object, whereby this attitude is connected with consequences of action (also by omission)" (Sauer et al. 2005).

2.2 Core Elements of Acceptance

The three main elements of acceptance are acceptance subject (1) and acceptance object (2), which are embedded in an acceptance context (3). Figure 1 illustrates the interaction of the acceptance elements:

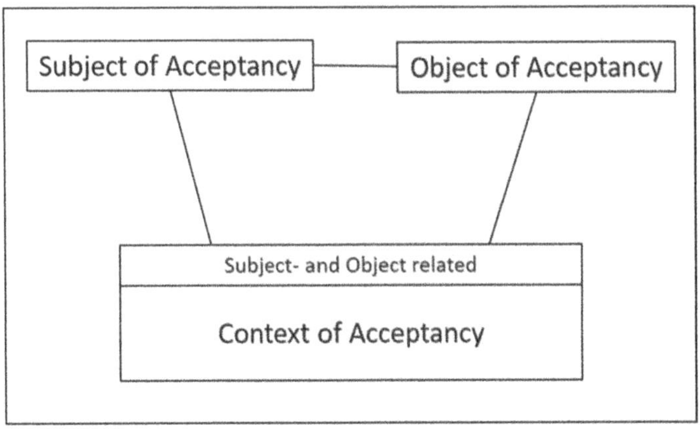

Fig. 1. "Object, subject and context-relatedness of acceptance", Schäfer/Keppler (2013), p. 17; extracted data: Lucke (1995), p. 89

So what is the significance of this trio of acceptance in the context outlined here? It is the interactions between the elements of acceptance, because these categories are identified in operational practice with every introduction of technology. The "acceptance subject develops an attitude related to the object of acceptance [...], including a behavioral disposition that can be activated on a case-by-case basis, which possibly leads to observable behavior" (Schäfer/Keppler 2013, p. 17; see Lucke 1995, p. 90–91). The subject can consist of both a person and a group of people. In the context of the paper, it is the groups of employees who are confronted with digital technologies, the object of acceptance. According to Kollmann (1998), organizational acceptance is thus taken into account by company decision-makers, those affected and/or their representatives.

The second central category is the object of acceptance. "Acceptance always refers to an object. It refers to the acceptance of something offered, existing or proposed. The object reference is less directed to its immanent properties than to its socially defined and ascribed meanings." (Schäfer/Keppler 2013, p. 19) Acceptance objects do not necessarily have to be of a material nature. Thus they can also have an immaterial character, e.g. in planning, strategies, concepts, political measures or acceptance is directed at persons or their actions. In the sense of the objects of acceptance presented here, these are technological innovations in the course of digitization processes in companies, a material type of whatever design, which was described by Renn (2005, p. 32) as "work technology".

The last core category in outlining the concept of acceptance is the acceptance context. "Acceptance also varies depending on the social and cultural context in which the subject of acceptance perceives and evaluates an object of acceptance and in which it may act in relation to it. In general, the context of acceptance can contain all factors or conditions that are neither subject nor object of acceptance, but affect the process of acceptance genesis, in other words, are relevant for it. (Schäfer/Keppler 2013, p. 22) The acceptance genesis in a particular context is thus strongly influenced by the interactions between the first two categories. With regard to the investigations of the companies in which new technologies are used, the context of acceptance in the framework of this article has to be limited to the individual and organizational level.

2.3 Dimensions of Acceptance

Following on from the description of the core elements of acceptance, the dimensions of acceptance (Schäfer/Keppler 2013, pp. 11–16; Kollmann 1998; Lucke 1995, p. 82–83) must also be examined. The attitude dimension focuses on the positive or negative attitude of the perceived object of acceptance by the subject of acceptance, e.g. the attitude towards electricity generation by nuclear energy. A concrete example, which already implies an action component, is the basic willingness (i.e. positive attitude) of a driver to be able to use the navigation device in certain situations - namely when he does not know the way. Acceptance contains another dimension: the action dimension. Some publications (see among others: Kollmann 1998, Schweizer-Ries et al. 2010) approach the concept of acceptance of action in different ways, so that it can express itself in different activities; e.g. through the purchase of an artefact, its use, the promotion of a technology or the support of (political or entrepreneurial) decision-making processes with the aim of introducing technologies (see Schäfer/Keppler 2013, p. 13).

Figure 2 illustrates the interaction of the action dimension and the attitude dimension, since elements of both dimensions are integrated. At the same time, it opens

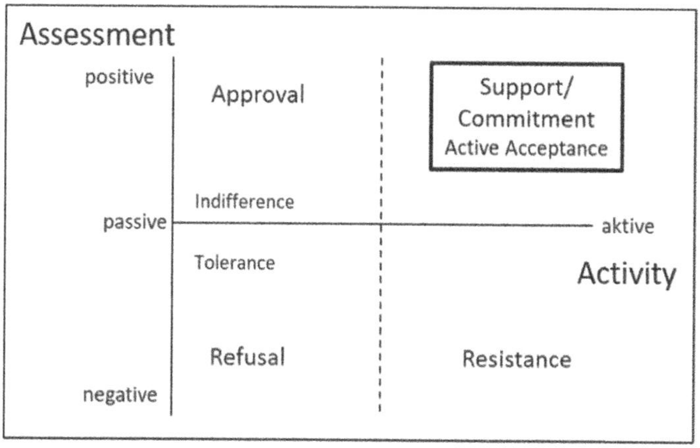

Fig. 2. "Dimensions of the Concept of Acceptance", Schäfer/Keppler (2013), p. 13; data according to: Schweizer-Ries et al. (2010), p. 11.

up the possibility of locating the acceptance or rejection level for one or more acceptance subjects with regard to an acceptance object.

A further acceptance dimension is the value dimension, which usually comprises a positive evaluation and is often assigned to the attitude dimension, since these two dimensions also have an interdependent relationship to each other (Schäfer/Keppler 2013, p. 14f.). Pressure (e.g. from a superior) exerted on subjects of acceptance can lead to an *acceptance of adaptation* in which the subjective values for a technical innovation do not match with those of the target or overall system. On the other hand, there is the *acceptance of adoption* in which the (positive) values are internalized and in harmony with the target system. Thus the value dimension is to be mentioned rather in the context of the acceptability.

2.4 Models and Systematics for Acceptance

The following part of the chapter gives a rough overview of the colourful landscape of different acceptance models, which mostly analyze the process of acceptance genesis. Which system was chosen as a point of orientation for the further explanations, in order to systematically record different forms of expression and in particular factors of acceptance, is also explained.

Lucke (1995, p. 218), for example, contrasts different types of acceptance that follow a dichotomous logic, such as latent and manifest acceptance, true and false acceptance, or charismatic and objective acceptance, etc. These typologies originate from action-theoretical approaches. Overall, a distinction can be made between input, output, feedback and process models (Schäfer/Keppler 2013, pp. 28–41). *Input models* for acceptance merely refer to influences that have an effect, for example, on the introduction of technology. These model types are similar to a black box to which a certain amount of input is fed in and in the end (in-)acceptance - in whatever form - can be observed as a result. Models of this kind can be found in Schönecker (1995), Kollmann (1998), Schnell (2009) or Huijts et al. (2012).

On the other hand, the *input-output models* focus not only on the influences, but also on the result (output) - and this is where the difference lies - also on the systematization of these. Hiblig's model (1984), which also records the output as a consequence of acceptance, is to be listed here as an example.

Feedback models (Reichenwald 1978; Schönecker 1995) record mainly the effect of the acceptance genesis on the influencing factors previously considered. Reichwald (1978, p. 32) differentiates in his 'Basic Model of the Acceptance of New Technology' between situation-independent categories: Technical system, user and organisational environment, along the above-mentioned elements of acceptance subject, object and context, each of which is divided into further sub-categories. At the same time, the relevance of situation-related factors and the situation-independent factors determine the level of acceptance of a technology in the basic model. The reactions result from the primary effects that technology takes on the individual and determines the respective degree of acceptance; these effects in turn have consequences (secondary effects) at the organisational and individual level (see also Schäfer/Keppler 2013, pp. 36–37).

The last acceptance model variant are the *phase models* (see Fig. 3), which focus on the acceptance process. Rogers (1995, p. 170), for example, differentiates between five phases (knowledge, conviction, decision, implementation, confirmation), which connect the single phases with each other via communication channels and thus represent the process of an acceptance genesis.

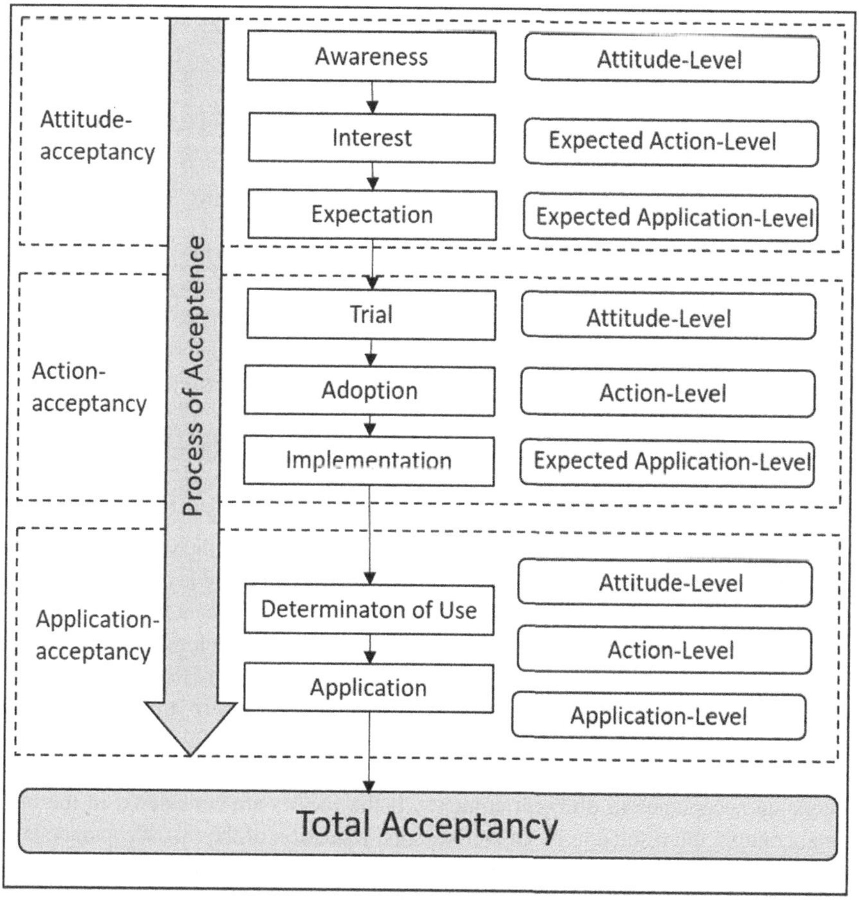

Fig. 3. "The dynamic acceptance model of Kollmann"; Kollmann (1998) p. 108; data taken: Schäfer/Keppler (2013) p. 40

Kollmann (1998) provides a further example of a phase model (see Fig. 3), which summarises the various phases of the acceptance process, which gradually get closer to total acceptance, in the following groups: acceptance of attitude, action and application. In line with the definition used here by Sauer et al. (2005), in which a distinction is made between acceptance subject, acceptance object and acceptance context as central categories, the "Nonacceptance - Acceptance Scale" (see Fig. 4) by Sauer et al. (2005) was also identified as the system that most clearly illustrates the various forms of acceptance.

It offers the possibility to filter out different levels of acceptance and at the same time allows the integration of the attitude and action components with regard to acceptance. Furthermore, the scale makes it possible to classify and compare the respective empirical findings and to make statements about the intensity of acceptance or nonacceptance.

Step 1	Step 2	Step 3	Step 4	Step 5	Step 6	Step 7	Step 8
Active Opposition	Refusal	Ambivalence	Indifference	Tolerance	Conditional Acceptance	Approval/ Benevolence	Engagement
Nonacceptance			Acceptance				

Fig. 4. "Inacceptance Acceptance Scale", Schäfer/Keppler (2013), pp. 23–24; Data according to: Sauer et al. (2005).

2.5 Factors Influencing the Formation of (in-)Acceptance

In the case of technical change processes in company contexts - or rather in a company acceptance context - attention is often paid to the individual acceptance factors, because acceptance of the innovations appears as "the […] central challenge in the context of industry 4.0." (Kagermann 2014, p. 610) In order to reduce prejudices and fears against digital technologies, it is important to understand that "acceptance […] is not generated from outside […], but must be gradually adjusted […]." (acatech 2016, p. 10) According to the German Academy of Engineering Sciences, this requires "a positive or even inspiring user experience […]" (ibid. p. 10) and a "participative design of technology" (ibid. p. 57) seems to promote acceptance. There are a number of publications (see e.g. acatech 2016; Thim 2017, p. 71–74; Schnell 2009, p. 7; Lucke 1995, p. 361ff; Hüsing et al. 2002, p. 24–26; Sauer et al. 2005) which approach the factors influencing acceptance in different contexts. If the factors are considered in the operational context, the result can be an increase or a reduction of the (in-)acceptance level, whereby acceptance remains an unstable construct (Schäfer/Keppler 2013, p. 25). Schäfer and Keppler (2013, pp. 25–27) provide a list of the influencing factors in terms of technology acceptance based on the three core elements of acceptance (see above) without claiming completeness:

Influencing factors based on psychological approaches rather focus on the attitude of the subject of acceptance (such as attitude, norms and values as well as emotions); in addition to these, the present article will also focus on influencing factors oriented towards the action dimension:

- Perceived influence on technology implementation and result
- Assessment of experienced and perceived possibilities for action
- Personal attitude to behaviours regarding the new technology

"Influencing factors related to the object of acceptance (here: technology) are based on characteristics of the object of acceptance" (Schäfer/Keppler 2013, p. 26). The influencing factors of the acceptance object are also part of and closely linked to the valuation and perception of the acceptance subject and therefore the object-related influencing factors vary from case to case. If the majority of object-related influencing factors are subsumed, the following can be filtered out:

- Costs and utilization of the implementation of a technology
- Risks of the use of technology, performance and reliability
- Usability in the use of technology
- Effectiveness of the technology
- Aesthetics of technical design

When examining the acceptance context, a number of influencing factors are also taken into consideration, when adequately implementing technology. Also the context-related factors are not to be considered separately from the object or subject. The context-related influencing factors vary depending on the acceptance system:

- Work task (simplified/complicated)
- Actor- and group-related processes in the organization as well as expected behavior patterns and routines in the social environment
- Social, cultural, societal context (legal/political/economic conditions; models, participation culture)
- Method and technique of technology implementation: participative, communicative, accompanying

3 Concept and Method of the Research

Methodically, the contribution is based on own empirical surveys from logistics, in particular intralogistics, and other industry-related economic sectors. On the basis of the analysis of 12 guided interviews in 3 companies, in which the experiences of different operational disciplines and hierarchical levels (operational level, corporate administration, department manager, business unit manager, works council, HR manager) could be captured, well-founded statements can be made on possible inhibiting and promoting influencing factors regarding the acceptance in the case of implementation of new technologies. The range of the sample extends into the following areas: Maintenance of intralogistic systems as well as lifting and conveying technology, production logistics and trade logistics.

The conceptual basis of the analysis is the socio-technical approach (Trist/Bamforth 1951; Sydow 1985; Hirsch-Kreinsen 2016), which addresses the interdependencies between technology, organization and the personnel subsystems of a company. It shows how opportunities of modelling and shaping at its interfaces (e.g. through technical innovations), generate ever new negotiation processes. In this sense, the introduction of technology in the course of digital transformation is understood as "complementary innovation" (Brynjolfsson/McAfee 2014, p. 102; see also Ittermann et al. 2016, p. 10), which means that not only the technical subsystem, but also work processes, the organisational structure and social relations are focused as an overall system of industrial 4.0 innovations.

Our study identifies some salient features that could lead to an improvement in the acceptance of industry 4.0 solutions in the enterprise. Three hierarchical moments that promote the acceptance of the introduction of digitization solutions in enterprises are found which also generate tangible added value. In our opinion, it is above all the dissemination of information, including the workforce, and intermingling of employees and thus creating diversity in the project teams that drive the moments of acceptance within the company. We present these moments in the following with small empirical examples and also name the respective success factors that require these moments in operation.

4 Empirical Findings

The results of the study are, on the one hand, various barriers to the introduction of digitization solutions and, on the other hand, possible solutions. At the center are participation of the affected workers and transparency in carrying out the Industry 4.0 introduction. Furthermore, it is important to find and promote potential multipliers. In particular, the lower management positions (group or shift leaders) promise great leverage in this respect, since they can be a solution carrier as they are very close to the workers on the shop floor (see also Wienzek/Virgillito 2018). But on the other hand, group leaders are often drafted from the shop floor workers based on their working expertise and not necessarily because of their leadership skills. We also found that diversity was decisive for the positive reception of Industry 4.0 solutions. The 'mixing' of the employees with regard to age, qualification, departments and hierarchy not only brought new perspectives and fruitful discussions regarding the planned change, but also through the exchange the understanding of the respective action logics and constraints of individual areas as well as new suggestions for improvements.

4.1 Inform

In a first step it is necessary to inform the staff or the acceptance subjects. Posters and company newspapers are often used here. However, a personal approach is usually better and initial difficulties in understanding can be clarified immediately. It is therefore important to recruit suitable multipliers such as team leaders, shift leaders, division heads, works councils or management for a joint and, in particular, transparent change process. The multipliers play a key role in the introduction of digital technologies and therefore multipliers who are negative towards the change process are to be mentioned as an acceptance inhibiting factor in the sense of the above categories "conflict" or even "active opposition":

> *"The problem is that we also have team leaders of different strength/weakness. One is very extensive, the other very emotionless, who says: "Here, you have to do it".*

Chairman of the Works Council, Trade Logistics

To raise the level of acceptance and at least achieve tolerance or even higher levels of acceptance, appropriately conducted meetings of the multipliers in which the acceptance context is concretized are helpful. Through the strategically planned use of the multipliers as well as the information and involvement of the operative employees in team meetings or similar meetings, acceptance can increase:

> *"You don't have to worry about that." So also the signal, works council and management together did that and also explained the background. [...] Transparency is written in capital letters."*

Chairman of the Works Council, Trade Logistics

It is also essential to create a formal framework for the presentation of information on the implementation of technology to the operational level, so that each employee, or the subject of acceptance, feels "picked up" and thus reservations and non-acceptance are prevented:

> *"Without getting great instructions, because for some it's intuitive, but for others it's necessary. And also to explain backgrounds, why we actually do that, that's also important. And the one doesn't want that at all, but the other maybe wants it much more and then maybe mistakes happen. Is it perhaps only the missing information or are it indeed reservations?*

Chair of the Works Council, Trade Logistics

The scope and timing of the information often represent a balancing act. At an advanced stage of planning, speculations, half-truths or reservations may already be circulating. However, if, on the other hand, if information starts too early, the change project is still not concrete and tangible enough for the employees. In such early phases of technology implementation, the acceptance level moves in grey areas (ambivalence or indifference). A change process should not give the impression of "top secret" planning, as reservations and skepticism may have arisen at this point:

> *"We are already trying t o inform very extensively. That is, I always find that very important, also with background that the people can understand that. And not to let any rumors come up."*

Chair of the Works Council, Trade Logistics

4.2 Include

In general, the information of the workforce is necessary, but not sufficient enough for a successful introduction of industry 4.0 solutions in the company. Furthermore, for a goal-orientated implementation of technologies and the establishment of acceptance (i.e. the achievement of acceptance levels: approval or engagement) the correct involvement of all those participating in the process - i.e. the acceptance subjects - is essential. The involvement of the workforce is to be understood as a central part in the overall concept. Thus, the most promising step seems to be the participation of employees in the introduction of technology. One important aspect should be emphasized: Employee participation does not mean introducing basic democracy in the company or questioning the owner's right to direct. Most employees are more concerned with information, communication and, where appropriate, participation. It is a widespread misbelief that employees want to have a voice in every business decision;

but they want to participate where their workplace is concerned. In principle, this participation should also be in the interests of management because much can be gained (better acceptance, better organizational culture etc.). However, the participation must be meaningful. When entrepreneurially decided technology implementations are realized without obtaining and implementing the relevant experiences and knowledge of the operative employees it can lower acceptance levels:

> *"But for us, it's not yet clear that the other scanners (conventional scanners) will go away. So we do indeed want to leave a choice, because in this very test we found out that there are colleagues who say: "I can't cope with it (scan glove). I do not want that. I find that it is cumbersome. I get stuck with it." But (the company) has already decided for itself that it will now be introduced."*

Chair of the Works Council, Trade Logistics

In the same case study, the alternative to the scan glove is also considered to be ergonomically very much in need of improvement. With regard to the concrete involvement of employees at the operational level in the implementation of technology, there is clear potential for optimization from an ergonomic point of view:

> *"What we noticed with the cuff, which we have optional for people who don't want a glove, these are not so good ergonomically. It has the knobbel down here, which always disturbs when wearing, because it's just a plastic part, which is convex and it's hard to trigger it [...]. They find this alternative very bad. Especially when carrying parcels, this is a handicap."*

Deputy Chairman of the Works Council, Trade Logistics

In many larger companies in Germany, co-determination is institutionalized by law through works council, whereas in SMEs the spread of employee representation, however regulated, is rather rare. Especially in these smaller and often more agile companies, a large part of the organization of participation opportunities remains on the shoulders of executives. Elsewhere, for example, a contract logistics company relies on the use of shop stewards (trade union members) to sensitize employment groups that show a fundamental rejection or at least a conflict with any kind of technology:

> *"Augh, Computer? For God's sake! In my life never I sat at a computer before. For God's sake, that's not possible! No, no, no!' [...], "We always do rounds there, especially in hall six, also with shop stewards."*

Chair of the Works Council Contract Logistics

The selection of team leaders is often only decided due to their professional competence. The extent to which the person also brings along actual leadership skills, organizational and social competences is often left to chance. Overlooked or underestimated are then these social and personal competencies, which play an important role especially with change processes, since executives act as moderators and multipliers for changes. Thus, even with an employee-friendly actor - the works council - with regard to the role of rather unsuitable multipliers, the impression can arise that "[…] something went wrong with the social behavior […]." (Chair of the Works Council, trade logistics). It is important to focus more on these aspects when it comes to determining the suitability of managers - especially at the lower management level - and, if necessary, help them through training and further education.

The issue of participation thus comes to with important antecedents. For those who (can) decide to introduce a digitization solution in the company it is important to listen to the information and knowledge they can get from shop floor workers. And for superiors who are to function as multipliers (lower management) need to be able to do so, i.e. they need the appropriate leadership skills.

4.3 Intermingle

The right mixture of persons in project teams is particularly important. Diversity - in relation to the acceptance subjects - is an important success factor, both in the conception of integration and information formats as well as in their implementation. During concept development, mixed teams contribute different points of view to the process; in addition, target group-specific topics and concerns can be anticipated and considered. Thus not only the operative level is to be focused, but the attention is to be put to the same extent and at a relatively early time of the process on the strategic level, i.e. the management. Experience at the operational level in the work process is also of particular importance, as some of this knowledge must be taken into account and integrated in the introduction of digital technologies:

> *"Let me give you an example, the customer, where I was last week, where we held a workshop on the subject. There, the control station level and the working level are integrated in order to be sure that the processes are displayed properly."*

Management Customer Service, Intralogistic Systems

In the case of a trade logistics company, feedback is constantly used so that employees at the operational level are continuously involved in improvement processes and their further development:

> *"We focus on long, monotonous tracks (executed by AGV's), because this has always been feedback from the drivers. That's why they developed it together with the manufacturer. [...] The employees always gave us feedback and then we did something.*

Deputy Customer Distribution Center, Trade Logistics

Looking back, it was confirmed that diversity was a decisive factor for the positive acceptance (see above: adoption acceptance) of the planned changes. The 'mixing' of employees with regard to age, qualification structure, functional areas and hierarchy not only brought new perspectives and fruitful discussions with regard to the planned change but also, through the exchange, an understanding of the logic and pressures of action in individual areas as well as new suggestions for improvements and an easier transfer of knowledge within the company. The condition for this is a functioning participation culture in the company. If one does not dare to express one's opinion or to oppose one's superior, diversity is also of no use. The participation of employees in digitization solutions in a company fundamentally requires a participatory corporate culture. Especially if the introductions are developed in mixed groups.

5 Outlook

The empirical findings can be sorted into three hierarchical moments on how to achieve acceptance of a new implemented Industry 4.0 solution: Inform, Include, and Intermingle. When trying to bind these findings back to the literature which we reviewed in Sect. 4, we can make a complementary contribution to the existing acceptance elements that are effective in technology implementations. It becomes apparent that with regard to acceptance issues in technology implementation (acceptance object), the existing systematics and models for reaching acceptance levels provide clues, but must nevertheless be adapted to the operational conditions (acceptance subject) and contexts (acceptance context). In the perspective considered here, acceptance is then the result of the "communication and action process between acceptance subjects and acceptance objects" (Schäfer/Keppler 2013, p. 5), which arises from dynamic negotiation processes. Our data suggests that it is not enough to subsume the most important negotiations, i.e. between the acceptance subject and his or her supervisor, in to the relation between acceptance subject and context. The findings show that it is more important to change or influence the decider (company decision-makers and/or the lower management) then the acceptance subject: Nearly all of the advice for successful implementation (i.e. reaching acceptance) is actually dependent on the decider: Again and again it has been shown that information and participation promise the most success in reaching acceptance. That in turn actually means the deciding factors which are to be influenced and worked upon to make an implementation process more probable are the deciders not the acceptance subjects. This is to say the old adage "the fish rots from the head down" still applies: The deciders are the people who must be open to participative methods, these are the people who need to know how to communicate effectively and these are the people who need to be open to suggestions.

References

acatech (Hrsg.) Innovationspotenziale der Mensch-Maschine-Interaktion. In: acatech IMPULS April 2016, Herbert Utz Verlag GmbH, München (2016)

Bauernhansl, T.: Die Vierte Industrielle Revolution – Der Weg in ein wert-schaffendes Produktionsparadigma. In: Bauernhansl, T., ten Hompel, M., Vogel-Heuser, B. (Hg.): Industrie 4.0 in Produktion, Automatisierung und Logistik, pp. 5–36. Springer Vieweg, Wiesbaden (2014)

Beeger, B., Heeg, T.: Sind wirklich 3,4 Millionen Arbeitsplätze in Gefahr?, in FAZ-Online vom 02.02.2018 (2018). http://www.faz.net/aktuell/wirtschaft/diginomics/achim-berg-bitkom-bekommt-widerspruch-zu-warnung-vor-arbeitsplatz-verlust-15429637.html. Accessed 19 Oct 2018

BMAS – Bundesministerium für Arbeit und Soziales (2016): Weissbuch Arbeiten 4.0. Berlin

Bonin, H., Gregory, T., Zierhan, U.: Übertragung der Studie von Frey/Osborne (2013) auf Deutschland. Endbericht im Auftrag des Bundesministeriums für Arbeit und Soziales, Mannheim (2015)

Boston Consulting Group (BCG): Man and Machine in Industry 4.0: How Will Technology Transform the Industrial Workforce Through 2025? (2015). https://www.bcgperspectives.com/content/articles/technology-businesstransfor-mationengineered-products-infrastructure-man-machine-industry4/?chapter=7#chapter7. Accessed 16 Oct 2018

Bowles, J.: The computerisation of European jobs – who will win and who will lose from the impact of new technology onto old areas of employment? (2014). Internet: www.bruegel.org/nc/blog/detail/article/1394-the-computerisation-of-european-jobs/. Accessed 19 Oct 2018

Brynjolfsson, E., McAfee, A.: The Second Machine Age. Work, Progress, and Prosperity in a Time of Brilliant Technologies. Norton & Company, New York (2014)

DGB-Vorstand: DGB-Bundesvorstand Januar 2016 Grundlegende Anforderungen an ein Weißbuch 'Arbeiten 4.0'; In: DGB-Arbeitspapiere (2016)

Doppler, K., Lauterburg, C.: Change Management. Campus, Frankfurt am Main/New York (2005)

Forschungsunion/acatech: Deutschlands Zukunft als Produktionsstandort sichern. Umsetzungsempfehlungen für das Zukunftsprojekt Industrie 4.0. Ab-schlussbericht des Arbeitskreises Industrie 4.0. Frankfurt/M (2013)

Frey, C.B., Osborne, M.A.: The future of employment: how susceptible are jobs to computerisation? Technol. Forecast. Soc. Chang. **114**(1), 254–280 (2017)

Fuchs-Heinritz, W., Klimke, D., Lautmann, R., Rammstedt, O., Staheli, U., Weischer, C., Wienold, H. (Eds.): Lexikon zur Soziologie, VS Verlag für Sozialwissenschaften, Wiesbaden, p. 27 (2007)

Goos, M., Manning, A.: Lousy and lovely jobs: the rising polarization of work in Britain. Rev. Econ. Stat., Jg. 89 (2007), H. 1, 118–133 (2007)

Hirsch-Kreinsen, H.: Digitalisierung von Arbeit: Folgen, Grenzen und Perspektiven; Soziologisches Arbeitspapier Nr. 43/2015, Hirsch-Kreinsen, H., Weyer, J., Wilkesmann, M., Dortmund (2015)

Hirsch-Kreinsen, H., ten Hompel, M., Ittermann, P., Niehaus, J., Dregger, J.: Social Manufacturing and Logistics Konturen eines Leitbildes digitaler Industriearbeit, In: AUTONOMIK für Industrie 4.0 VDI / VDE Innovation+Technik GmbH, Alfons Botthof (Eds.), Berlin (2016)

Hirsch-Kreinsen, H.: Industrie und Arbeit 4.0. Entwicklungstrends und Gestaltungsperspektiven. In Vassiliadis, M. (Ed.): Digitalisierung und Industrie 4.0. Technik allein reicht nicht, IGBCE, Hannover (2017)

Hirsch-Kreinsen, H. (2017a): Arbeiten 4.0 – Qualifikationsentwicklung und Gestaltungsoptionen. In: Wirtschaftsdienst. Zeitschrift für Wirtschaftspolitik. Jg. 97, Heft 7, July 2017

Hilbig, W.: Akzeptanzforschung neuer Bürotechnologien. Office Manag. **3**, 320–323 (1984)

Hüsing, B., et al.: Technikakzeptanz und Nachfragemuster als Standortvorteil. Faunhofer-Institut ISI, Karlsruhe (2002)

Huijts, N.M.A., Molin, E.J.E., Steg, L.: Psychological factors influencing sustainable energy technology acceptance: a review-based comprehensive framework. Renew. Sustain. Energy Rev. **16**(2012), 525–531 (2012)

IGBCE: Digitalisierung und Industrie 4.0 – Technik allein reicht nicht; Michael Vassiliadis (Eds.) (2017)

Ittermann, P., Niehaus, J., Hirsch-Kreinsen, H., et al.: Social Manufacturing and Logistics Gestaltung von Arbeit in der digitalen Produktion und Logistik; Arbeitspapier Nr. 47, p. 13ff (2016)

Kagermann, H.: Chancen von Industrie 4.0 nutzen. In: Bauernhansl et al., pp. 603–614 (2014)

Kollmann, T.: Akzeptanz innovativer Nutzungsgüter und-systeme: Konsequenzen für die Einführung von Telekommunikations - und Multimediasystemen. Reihe Neue betriebswirtschaftliche Forschung Band 239. Gabler Verlag (1998)

Lucke, D.: Akzeptanz: Legitimität in der "Abstimmungsgesellschaft". Leske+Budrich, Opladen (1995)

PWC: Will robots really steal our jobs? An international analysis of the potential long term impact of automation (2018). https://www.pwc.com/hu/hu/kiadvanyok/assets/pdf/impact_of_automation_on_jobs.pdf. Accessed 19 Oct 2018

Reichwald, R.: Zur Notwendigkeit der Akzeptanzforschung bei der Entwicklung neuer Systeme der Bürotechnologie. Hochschule der Bundeswehr, München (1978)

Renn, O.: Technikakzeptanz: Lehren und Rückschlüsse der Akzeptanzforschung für die Bewältigung des technischen Wandels. In: Technikfolgenabschätzung – Theorie und Praxis 3/14, pp. 29–37 (2005)

Rifkin, J.: The Third Industrial Revolution. How Lateral Power is Trans-forming Energy, the Economy, and the World, Basingstoke (2011)

Rogers, E.M.: Diffusion of Innovations, 3rd edn. University of Michigan, Free Press, New York (1995)

Schäfer, M., Keppler, D.: Modelle der technikorientierten Akzeptanzforschung. Zentrum Technik und Gesellschaft, TU Berlin, Discussion paper 34/2013 (2013)

Sauer, A., Luz, F., Suda, M., Weiland, U.: Steigerung der Akzeptanz von FFH Gebieten. BfN-Skripten 144 (2005). http://www.bfn.de/fileadmin/MDB/documents/skript144.pdf. Accessed 18 Oct 2018

Schönecker, H.G.: Kommunikationstechnik und Bedienerakzeptanz, In: Picot, A., Reichwald, R. (Hrsg.), Forschungsprojekt Bürokommunikation (Band 6), CW-Publikationen (1995)

Schnell, M.: Einführung in die Akzeptanzforschung am Beispiel von Web- TV. Wissen Heute **1** (62), 4–12 (2009)

Schweizer-Ries, P., Rau, I., Zoellner, J., Nolting, K., Rupp, J., Keppler, D.: Aktivität und Teilhabe – Akzeptanz Erneuerbarer Energien durch Beteiligung steigern. Magdeburg & Berlin, Projektabschlussbericht (2010)

Sydow, J.: Der soziotechnische Ansatz der Arbeits - und Organisationsgestaltung, Frank-furt, New York (1985)

Trist, E., Bamforth, K.: Some social and psychological consequences of the long wall method of coal-getting. Hum. Relations **4**(1), 3–38 (1951)

Thim, C.: Technologieakzeptanz in Organisationen Ein Simulationsansatz, (Dissertationsschrift), Universität Potsdam (2017)

Vogler-Ludwig, K., Düll, N., Kriechel, B.: Arbeitsmarkt 2030 Wirtschaft und Arbeitsmarkt im digitalen Zeitalter Prognose 2016. Studie im Auftrag des Bundesministeriums für Arbeit und Soziales, unter Mitarbeit von Vetter, T., München (2016)

Wienzek, T., Virgillito, A.: Leise Innovation, nicht radikaler Umbruch. Die Einführung einer Industrie 4.0-Lösung bei einem Möbelhersteller - ein Fallbeispiel. In: WSI Mitteilungen 3/2018, pp. 204–210 (2018)

Premises for Truck Sharing in General Cargo Cooperatives – An Exploratory Case Study Research

Wolfgang Stoelzle and Victor B. Wildhaber(✉)

Institute for Supply Chain Management at the University of St.Gallen, Dufourstrasse 40a, 9000 St. Gallen, Switzerland
{wolfgang.stoelzle,victor.wildhaber}@unisg.ch

Abstract

Overview

The road freight transportation-industry in the German speaking countries (Germany, Austria and Switzerland) faces high market pressure. Besides a high competition between the logistic service providers, shippers are exerting pressure too (Cordes 2018).

The market situation urges especially the small and medium enterprises to take action. This is because of their relatively small size and thereby minor possibilities than large enterprises to foster efficiency, e.g. truck-utilization (DLK 2016). Through an innovative form to increase truck-utilization such as a sharing economy-application, an important facilitation for small and medium enterprises during the downtimes could be achieved. As the sharing economy offers various advantages such as fostering utilization of idle capacities or the reduction of the ecological footprint, these should be levered.

Methods

To enable sound exploratory research results about the premises (e.g. enablers and barriers) of truck sharing in general cargo cooperative, two different research fields should be addressed. First, a short literature review provides insight into relevant literature about truck sharing. Second, case-interviews and -analyses with general cargo cooperatives share the practitioners view on truck sharing. Therefore, a multiple case-design, enabling semi-structured interviews to set the basis for data-collection and evaluation.

Results

The (cross-)case analysis shows that, among others, the mindset of sharing (own) resources (i.e. truck branding), data security, legal liability, financial contribution system, reputation of the sharing platform, trust and the incentive program are relevant premises for truck sharing in general cargo cooperatives. These, however, are re-categorized after their relevance into the architecture, market and property-requirements.

Conclusion

To lever the potentials such as advancements in truck-utilization and environmental pollution of the rather new concept of truck sharing, premises have to be met. The concept is enabled, respectively challenged by the trust, reputation, participant's mindset, security, liability, contribution system and incentives.

U. Clausen et al. (Eds.): ICPLT 2019, LNLO, pp. 75–89, 2019.
https://doi.org/10.1007/978-3-030-13535-5_6

Keywords: Sharing economy · General cargo cooperative · Premises · Resource sharing · Case study research

1 Motivation

1.1 Road Freight Transportation – An Industry Under Pressure

The road freight transportation-industry in the German speaking countries (Germany, Austria and Switzerland [GAS]) faces high market pressure. On the one hand, there is a high level of competition among the market participants. On the other hand, shippers exerting downward pressure on prices with their market power in the market segment of general cargo (Cordes 2018). Moreover, skilled labor, especially drivers are scarce and cause shortages in hauling capacities. However, besides the market situation the GAS-region was chosen due to its high logistics performance (Semmann 2018) and thus strong emphasizes on general cargo cooperatives.

The market segment is distinguished between the following market players (logistics service providers): large enterprises, small and medium enterprises and general cargo cooperatives. The market situation urges especially the Small and medium enterprises to take action. This is because of their relatively small size and thereby lesser possibilities than large enterprises to foster efficiency, e.g. less downtime through truck-utilization (DLK 2016).

1.2 Road Freight Transportation – An Industry that Could Benefit of Sharing Economy Showcases

Applications of the sharing economy can be found in various areas such as accommodation- and car sharing offer the asset owner a higher utilization-rate (Kölling 2018). Besides companies as Airbnb (accommodation sharing) or car2go and Stadtmobil (car sharing) also FLEXE (warehouse sharing) provide a proven track of fostering utilization rates through sharing economy-applications (car2go 2018; FLEXE 2018; Stadtmobil 2018).

The car- and warehouse sharing, however, are solutions that are related to the road (freight) transportation as they rely on sharing mobile assets with a similar function as trucks (bridging places) respectively sharing immobile assets to handle or store goods (bridging time). Therefore, their properties (e.g. digital platform for bookings, settlements and ratings) and form of interaction (e.g. a breach of the terms and conditions can be punished via the user rating) have a reference function and can be transferred with some limitations.

1.3 Truck Sharing – Providing the Industry a Versatile Range of Advantages

First, the term "truck sharing" has to be introduced. It can be defined via the following constitutive attributes:

Truck sharing…

- provides the possibility to share…
- underutilized trucks (as a physical asset),…
- that are owned by logistic service providers…
- via a community-based (digital) platform…
- with other logistic service providers…

(Botsman 2015; Zervas et al. 2016). The community could e.g. be a general cargo cooperative or network (both are closed platforms). If the platform would allow third party members (e.g. subcontractors) to engage in the truck sharing, it would become an open platform. However, the truck sharing understands the sharing as providing a temporary access to the asset for a fee (Belk 2014). We understand truck sharing as the sharing of the physical asset, not sharing transport capacity (service) such as a freight exchange does. Sharing transport capacity, however, in the case of a general cargo cooperative can lead to improved resource (truck as physical asset) utilization when e.g. two logistic service providers are hauling on the same relation.

Using the innovative form "truck sharing" to increase the truck utilization rate, an important facilitation for small and medium enterprises could be achieved. We, however, focus on general cargo cooperatives (dispatching large fleets with many opportunities to improve utilization – e.g. two trucks hauling on the same relation) that carry out their orders via their system partners (large enterprises and small and medium enterprises) expecting higher facilitation by truck sharing in the logistic service provider-community than in independent logistic service providers (dispatching a smaller fleet and thus having lessor opportunities to improve utilization). In general cargo cooperatives, sharing economy-applications to foster a higher truck utilization have yet not come to a use.

As the sharing economy offers various advantages such as fostering utilization of idle capacities or the reduction of the ecological footprint (higher utilization of trucks leads to a reduction of total distance driven and trucks needed to carry out the same volume of shipments), these should be levered. However, truck sharing brings relief to the hauling-capacity-shortage, as the truck can be shared in combination with the driver.

The road freight transport-industry is an eligible field for truck sharing as there still are possibilities to optimize a trucks utilization (as long as the truck can be used alternatively during its downtime).

The above given information shows that there still is a non-irrelevant need for action regarding truck sharing in general cargo cooperatives. However, general cargo cooperatives provide the sharing-ability and readiness, but have neither developed nor implemented truck sharing-solutions. Because of the inexistent knowledge about truck sharing-solutions, it is recommendable to first research the premises for truck sharing in general cargo cooperatives. We next provide a short literature review about truck sharing to highlight the scholarly research.

2 A Short Review of Relevant Literature Leading to the Research Gap

This section identifies and discusses relevant literature regarding truck sharing. It sets the basis for truck sharing in a general cargo cooperative, but also identifies the research gap.

The general cargo cooperative[1] practices appeared in the 1990s and quickly became catalyzers for the optimization of road and rail transports (Serrano-Hernández et al. 2017). In the meantime, general cargo cooperatives were thoroughly studied by scholars and widely discussed by practitioners (Caputo and Mininno 1996; Schmoltzi and Wallenburg 2012).

general cargo cooperative-Initiatives have grown steadily and are critical for the system partners'[2] competitiveness (Singh and Power 2009; Naesens et al. 2009).

New models considering multiple goals and an increasing computational power continuously optimize general cargo cooperatives. As this form of optimization in the recent year only adds marginally to the general cargo cooperatives performance, other approaches such as re-engineering logistics (processes) and interactions gain momentum – also truck sharing-solutions (Christopher 2005; Serrano-Hernández et al. 2017).

However, truck sharing addresses economic (e.g. higher truck utilization), ecologic (e.g. reduction of emissions) and social (e.g. reduction of traffic jams) advantages. Two of them (economic and social) cover important goals (cost reduction, social relevance, improved productivity, better resource management) of the general cargo cooperative (Pomponi et al. 2015).

As system partners already share shipment-orders in general cargo cooperatives and thus demonstrate cooperation ability and readiness, truck sharing would be a logical further development to increase the general cargo cooperatives efficiency.

Long before Dervojeda et al. (2013) used the term "sharing economy", the Charalambides (1984) identified the potential of "shared capacity resource" in decentralized service systems. However, in the meantime the literature on the (road) freight industry just provided fragments of applied sharing economy-concepts.

The publications cover rather generic contributions such as potentials and risks of a company's material flows mitigated by conceptual simulation studies (Freitag et al. 2016). Some highlight potentials, contributions and constraints of hinterland trucking (Islam and Olsen 2014) or freight matching-approaches for empty truck trips (Islam 2017). Other focus on facilitators of resource sharing such as visibility. The latter increases the willingness to share resources and achieves a higher performance (Maghsoudi and Pazirandeh 2016).

[1] The general cargo cooperative is a form of the horizontal logistics cooperation related to road freight transportation. Cruijssen (2006) defines the horizontal logistics cooperation as an "active collaboration between two or more firms that operate on the same level of the supply chain and perform a comparable logistics function on the landside".

[2] The system partner is a general cargo cooperative-affiliated logistic service provider that carries out the transport-orders within a general cargo cooperative.

However, the literature demonstrates that sharing economy concepts in the road freight industry still constitute a non-addressed gap in research.

Since this paper highlights a non-addressed research gap through an explorative-inductive approach (for details compare Sect. 3) and literature only gives little insight into premises of truck sharing, the authors identified the following areas that could constitute relevant premises for truck sharing:

- The mindset of the road freight transport-industry, as it still has potential, but the Logistic service providers rather show a reluctant attitude concerning e.g. new technologies or solutions such as truck sharing and a paradigm shift remains open.
- The data security has become more and more important during the recent years. The general data protection rights have brought more attention and thus gained importance.
- The legal liability becomes more important, as an logistic service provider in some cases cannot resolve a liability case (e.g. damaged truck) with itself and its drivers, but has to resolve it with the sharer or user.
- The financial contribution system is important, as a sharing concept within a general cargo cooperative only would work as long as benefits (and losses) are allocated fairly (Cruijssen et al. 2007).
- The reputation is important to attract and bond the (potential) users to a (sharing) platform (Huang et al. 2014).
- However, trust is essential to provide a mutual sentiment among the transaction partners (Huurne et al. 2017).
- Moreover, should the incentive program be designed to attract as many platform users as possible.

These partially literature-based, possibly constitutive premises are used in later sections to provide a framework for the evaluation.

This section shows that there is only little research done on sharing economy applications in the road freight transportation. However, some companies implement(ed) sharing economy-solutions in markets such as car sharing (e.g. Stadtmobil, car2go, Zipcar), freight warehousing (e.g. FLEXE) or road freight transportation (e.g. Uber-Freight). The latter, however, only matches freight with a single trucker with an empty truck. The authors yet could neither identify sharing economy applications in the road freight transportation (i.e. truck sharing as a fleet-scalable, business to business solution for logistic service providers) in practice nor find discussions in scholarship.

3 Methodology

As it is one of the first research papers, we provide insights into the existing, and for truck sharing relevant literature. In this step, we also identify the research gap and areas that could constitute relevant premises for truck sharing. To conduct this research, we use the inductive research logic applying our explorative and quantitative approach.

Therefore, we apply Eisenhardt's (1989) structure of case study research design provides the basis for internal validity via e.g. pattern matching and explanation building. By the latter and the application of Yin's (2018) embedded single case-design to analyze multiple units (a series of interviews), we first identified and selected appropriate units: six interviews with senior management personnel of four innovative logistic service providers in the GAS-area. Second, we crafted a data collection method that offers a certain flexibility regarding the answers: interview guideline embedding definitions, the constitutive premises (compare Sect. 2) and an outlook to conduct semi-structured, open-ended interviews. The transcripts, however, were checked and released by the interviewees. Third, we overlapped collected data and analyzed field notes to fourth, analyze data in terms of pattern searching.

For the latter two steps, we made use of the Strauss's and Corbin's method of open coding (1998) to strengthen the methodology at analysis stage (compare Sect. 4). To reach a mature interview-evaluation we considered the following Strauss's and Corbin's steps:

- data splitting (splitting interview results into statements),
- phenomena identification (identifying important statements for truck sharing),
- labelling (labelling of phenomena's as a first step towards concept and group-building),
- concept identification (building concepts through linking of phenomena's),
- concept grouping (building logical groups of the concepts found) and
- formation of categories (providing categories to summarize groups of concepts).

Fifth, we shape propositions (compare Sect. 5). Due to the yet made little research and only reduced validity of our results, we cannot provide hypotheses but provide propositions. Sixth and seventh, we enfold literature (if available) to reach closure (compare Sect. 6).

The following Sect. 4 shows an extract of the interview-evaluation (Table 1). It gives insight into the premises for truck sharing.

4 Analysis

This section highlights the interviews-findings regarding premises for truck sharing. The evaluation after Strauss's and Corbin's (1998) method led to seven categories. The interview evaluation shown hereafter is broken down into groups, labels and phenomena's (extracts of the interviews are presented). The evaluation is sorted labels and their frequency of occurrence.

The content of the above presented (Table 1) categories, coded groups and labels are analyzed embedded in the road freight industry-context in the subsequent paragraphs.

Table 1. Extract of the interview-evaluation

Categories	Groups	Labels	Phenomena's
Mindset of sharing resources	Third parties	Third party carriers	*The branding of the trucks is not a challenge, unless we do not operate with an open platform that would enable our main competitors to carry out our shipments*
	Sharing readiness & experiences	Sharing of services	*Sharing should be possible among the CargoLine-logistic service providers without major difficulties. The Logistic service providers already share shipment-orders/-revenues (and thus also shipment- and transportation-data)*
	Sharing readiness & experiences	Sharing of tangibles & intagibles	*As well we are used to share trailers and swap bodies*
	Digitalization & platforms	Digitalization & platforms	*The digitalization is necessary*
Data security	Data sharing	Sharing of shipment resp. transport data	*Only limited data are given to another logistic service provider in the cooperation that carry out the shipments. If we would open up the platform, we would have to define with which Logistic service providers we would not work together*
	Third parties	Third party carriers	*When the data would be shared with third parties (open platform), security measures should be taken that trust can be ensured*
	Law	Data security laws	*When an app would be used, data could be analyzed, but with respect to data security (drivers are made transparent) the driving-behavior or speed can be identified*
Legal liability	Third parties & protection	Third party carriers	*When using a closed platform trucks are always operated by the dispatching logistic service provider (with exception of the truck pool), so there would not occur a legal liability-issue. In such a case each logistic service provider is responsible for its own resources and respective damages*
	Policy experience & transfer	Existing policies	*Possible cases of damages should be managed via the terms and conditions*

(*continued*)

Table 1. (*continued*)

Categories	Groups	Labels	Phenomena's
	Third parties & protection	Inspection & insurance	*In an open platform or central truck pool insurance or other securities have to be applied to enable lessor difficulties*
	Policy experience & transfer	Transferability of policies	*If needed a (driver pool or use of shared trucks) car sharing-proven legal frameworks could be transferred to the logistic resource sharing-solution*
Financial contribution system	Contribution elements	Contribution elements	*A price-model could work, when relating to the truck/trailer (chartered for a day or hours). If the resource only is loaded partially, distance, weight and loading could be used for calculation*
	Contribution models	Contribution models	*The financial contribution system should be developed that it can calculate prices dynamically*
	Transfer of existing systems	Transferability of contribution systems	*However, one could transfer the already existing pricing-structures of delivery zones and shipment weights*
Reputation of the sharing platform	Reliability	Critical volume & reliability	*The platform needs a critical mass of approx. 10'000 shipments per day. Thereby a dispatcher can guarantee to get a shipment to be delivered on time*
	Additional services	Digitalization & additional services	*... it could build a reputation via additional service such as a chatroom for drivers or information about resting areas*
Trust in the transaction partner	Known carriers	Trusted carriers	*The trust is very important to enable truck sharing initiatives*
	Unknown carriers	Non-trusted carriers	*If the LRS-platform would be used as an open platform, the third party truckers would have to undergo a check to be trusted*
Incentive program	Benefits	Expected benefits	*The advantages of LRS (higher utilization) are the incentives itself*
	Promotion	Promotion, subsidies & penalties through the cooperation	*However, the LRS-solution would be promotes by CargoLine*
	Additional services	Additional services	*The app providing various additional services (as part of an ecosystem), that are as well an incentive (especially for drivers)*

4.1 Mindset of Sharing Resources

Sharing readiness and experience leads to system partners being open for a further development concerning truck sharing in a closed platform. The system partners, however, are used to share shipment-orders (and shipment/transportation-data as well) to carry them out in the general cargo cooperative-area. However, some system partners even have experience in sharing of trailers and swap bodies.

System partners principally do not trust **third parties** and are not open to share trucks in an open platform. System partners do not business with third parties[3] as the sharing of shipment-orders could lead to a revenue-loss or data-leak due to missing policies and trust. A certain form of quality check of third parties could lower this barrier. Furthermore, the truck branding could be a challenge in the mindset of system partners and third parties.

Digitalization and platforms have a minor[4] but still important[5] impact on truck sharing. Digital tools can constitute a challenge for late adopting logistic service providers[6]. Due to perceived transparency and customer intimacy logistic service providers are not welcoming platforms as those could erode their margins and wiping out the competitive advantage.

4.2 Data Security

Besides being a hurdle in the mindset of system partners **data**[7] **sharing** is also a data security-issue. **Third parties** could use customer[8] data (e.g. shippers' contact information and freight information) unauthorized for their advantage (e.g. taking over of a system partners shipments) and damage the system partners. System partners thus would like define which third parties could receive the data shared. Additionally, one could install data security measures to prevent an unauthorized use. However, contractual and financial data would not be exchanged.

Further premises are:

- Ensuring **data quality** when system partners receiving data from third parties.
- Respecting **laws** such as general data protection rights regarding data analyzing (driver made transparent and thus vulnerable).

[3] In this paper, we understand the "third party" as third party-Logistic service providers that are Logistic service providers with are not participating in a general cargo cooperation.

[4] Many small and medium logistic service providers still work paper-based or digital, but are only little digitally interconnected with other logistic service providers and shippers.

[5] Digitalization offers not little room for improvement at logistic service providers and is understood as an important factor to stay competitive in the future.

[6] Logistic service providers can be either a system partner of a general cargo cooperation or an independent third party.

[7] Truck data such as telematics information or freight data.

[8] The shipper as a customer of the system partner.

4.3 Legal Liability

Usually the **carrier**[9] is responsible for the insurance and possible damages (**policies** govern). In case of organizing truck sharing centrally (e.g. the general cargo cooperative is overseeing and managing the truck sharing), terms and conditions would have to govern the liability of the carrier (not the truck owner), notwithstanding who the (temporary) carrier is. Therefore, existing **sharing-policies** for simple tangible assets in general cargo cooperatives could partially be **transferred**. Moreover, the truck provider should install **protection measures** such as truck inspection, oblige or include an insurance.

4.4 Financial Contribution System

The contribution should be based on various **components** such as fixed and variable. Moreover, it should account for the distance and time run with full truck loads (FTL) or a mixture of distance, weight and load utilization when carrying less than truck load (LTL). The contribution system could be designed by either a dynamic-, revenue- or cost-oriented **compensation model**. The contribution system could, with some adaptions be **transferred** from an existing one.

4.5 Reputation of the Sharing Platform

Reliability is critical to operate a sharing platform successfully. Therefore, the platform needs to carry out the critical volume of approximately 10'000 transactions per day (average execution period of about a half an hour).

Furthermore, attractive **additional services** could foster platform growth and bond users. By offering services such as a driver-chatroom, traffic and resting area-notifications or advanced digitalization jobs would be eased and the platform enriched.

4.6 Trust in the Transaction Partner

The **system partners** know themselves well and have experience in sharing shipment-orders and thus the respective data. Trust however, is highly important for these sensitive topics. **Third parties** are generally non-trusted and must undergo a quality-check before being able to build up trust and share trucks and respective data among the system partners.

4.7 Incentive Program

The expected **benefits** of the truck sharing are highly attractive and thus are an incentive itself to attract the system partners to join. However, to understand the benefits a fundamental understanding of truck sharing is required. Moreover, the general cargo cooperative-management would **promote** the truck sharing-program

[9] The shipment carrying logistic service provider.

among the system partners. Optionally, a **subsidy and penalty-system** could be installed to motivate the system partners to involve in the truck sharing-program.

As mentioned before (compare reputation) **additional services** such as an app building an eco-system could attract e.g. drivers by easy to use value adding-information.

Following to the analysis of the premises, the results match them with important requirements regarding the truck sharing (platform).

5 Results

To provide insightful results, the most important (according to interviewees) requirements of truck sharing are structured and highlighted in the context of the premises analyzed earlier (compare Sect. 4). To structure important premises, phenomena's again were being labelled and grouped through the method of Strauss and Corbin (1998). Table 2 presents the premises-examination (sorted by the frequency of label occurrence) with exemplary phenomena's (extracts of the interviews are presented).

Table 2. Requirements of truck sharing

Category	Group	Label	Phenomena's
Requirements	Property	Usability	A simple handling must be ensured (internal platform via the transport management system)
	Property	Reliability	Matches (shipment and trucks) must be provided quickly (approx. 30 min)
	Architecture	Data security	Data security
	Market	Feasibility	Set up and implementation at a reasonable price
	Market	Acceptance	Providing a high acceptance among the logistic resource sharing-platform-participants. Therefore, one could build a success story to attract participants
	Architecture	Neutral organization	The platform needs to be run by a neutral organization
	Architecture	Forecasting	Forecasting data is key to know how high the capacity shortage or surplus is
	Architecture	Reputation & trust	Platform reputation
	Architecture	Mode & design	The application must be driver and logistic service provider-centric to enable a max. value added and thus must be socialized among users

5.1 Property-Requirements

Usability in terms of eased platform/app handling and less administrative efforts through more digitalization installments is highly important. It can lead to a return of the digitally less affine logistic service provider to use the platform again or bond him in the first place. However, a platform establishes its reputation when it is considered as **reliable**. Therefore, it must top a critical transaction volume quickly that timely matches are enabled.

5.2 Market-Requirements

Qualified measures that support the **feasibility** of the truck sharing should be realized. After addressing key aspects such as governance and business models of truck sharing, a feasibility study could give first insights. Thereafter, a minimal viable product should follow to prove the concept to then adapt the truck sharing-concept and implement it at a reasonable price.

To foster the truck sharing-**acceptance** among the system partners, early measures such as trust and a legal framework especially when dealing with third parties. Policies could regard the use of shared data or truck protecting measures (e.g. inspection or insurance). When setting up and running truck sharing, a success case should be spread in order to spur the system partners interest to participation.

5.3 Architecture-Requirements

However, to ensure the above-mentioned reliability a **neutral organization** (thus without any conflict of interest) such as university, general cargo cooperative or even a superordinate organization to various general cargo cooperatives. However, logistic service providers could not take the role of providing and operating the platform because of their conflict of interest. The neutral organization would set the configuration of the truck sharing. Under the neutral organization, it is easier to build **trust and reputation**, as fewer Logistic service providers take in a critical position. Trust and reputation are an advantage when collecting shipment volume and load capacity-data. The forecasting (-tool) requires the data as early as possible and in a high data-quality and consistency to ensure an accurate **forecast**. The latter is necessary to enable truck sharing (day or hours) ahead of the dispatching slot. However, **data security** issues must be taken into account, as sharing order, shipment and transportation-data is highly sensitive. The concerned parties should be protected respectively prevented from an unauthorized use of data.

Furthermore, the **mode** defines the platform's characteristics: either necessary (core-business) functions or additional services (value adding core and non-core businesses). Both should be logistic service provider but primarily driver-centric. Besides stakeholder centricity, especially the **design** of additional services is important, as it can be a fertile soil for an ecosystem and create a form of incentive.

As the research only provides first insights into the truck sharing and only little literature is available, evidence can derive propositions. There, however, cannot be derived closure in terms of hypotheses that are well iterated with evidence and compared with existing literature. Thus, we derive the following propositions regarding truck sharing:

- As general cargo cooperatives is the main catalyzer for optimizations among the system partners, it has a particular position and is responsible for enabling fertile grounds for truck sharing (Serrano-Hérnandez et al. 2017). This means the general cargo cooperatives must foster trust, enabling readiness and willingness to share resources, setting up the framework for legal liability, data security, financial contribution system and incentives. It moreover is responsible for ensuring market feasibility and acceptance.

- The platform operator has to ensure the reliability and a proper implementation of the framework that is been set by the general cargo cooperatives. The platform itself needs to ensure data security and high usability, which is facilitated by the chosen mode and design of the necessary functions and additional services. However, the platform has to ensure a proper forecast of shipments and capacities.
- The additional services could either be set up and run by the platform operator or an independent IT-provider. As the additional services aims to establish an eco-system it is of high relevance. It thereby should attract system partners and third parties.

6 Conclusion, Limitations and Outlook

This paper discusses with "truck sharing" a rather new area of research and the addresses the premises of truck sharing. Among other the premises "mindset of sharing", "financial contribution system" and "trust in the transaction partner" were analyzed (Sect. 4) and requirements regarding the truck sharing "properties", "market" and "architecture" were highlighted with the associated premises (Sect. 5).

Due to the limited validity of the research paper, we could not provide hypothesis. Although, we provide propositions that are highlighted below in brief:

- General cargo cooperatives, as catalyzers of optimizations of system partners are enablers of an innovative and utilization improving solution such as truck sharing.
- The neutral platform operator has to ensure usability and reliability of the sharing platform.
- Additional services most possibly an important incentive, but also can shape a truck sharing solution by providing value adding information.

The papers structure and method led to limitations, of which the most important are listed as follows:

- The research in terms of collected data covers six qualitative interviews. The interviewed companies were general cargo cooperatives and system partners from two countries. We however, also could have interviewed suppliers. The latter would add another perspective, but as would be only a shipment-related perspective, it probably would not add much (insights into) premises on truck sharing.
- Also in the dimensions of regions, we focused on a particular region (GAS). We could have compared the premises for truck sharing between different regions (countries and continents). The latter advancement would be useful in terms of getting insight into the similarities and differences of further regions. As we do aim a better understanding of possible applications of truck sharing, it is not an aim to generate sophisticated insights into multi-regional truck sharing solutions.

Thereby we could have reached a broader picture and probably could have highlighted further aspects of the premises for truck sharing.

The sharing economy-concepts prompted up in various applications (compare Sect. 1) in recent year. The concepts gained momentum as they lead to higher resource utilization and thus cut costs and even cover aspects of sustainability. Within the road freight industry, these concepts are of interest as the industry faces a competitive market and margin-pressure. Thus, sharing economy-concepts such as truck sharing should be fostered. In a next step, hypotheses should be set up based on the propositions of this paper and thereafter validated. As further steps in researching truck sharing, e.g. business models or functionalities of sharing-platforms should be highlighted.

References

33. Deutscher Logistik Kongress [DLK]: Den Wandel gestalten (2016). http://dlk.dvz.de/data/BVL_Tagungsband_33_Deutscher_Logistikkongress.pdf. Accessed 30 June 2018

Belk, R.: You are what you can access: sharing and collaborative consumption online. J. Bus. Res. **67**(8), 1595–1600 (2014). https://doi.org/10.1016/j.jbusres.2013.10.001

Botsman, R.: Defining the Sharing Economy (2015). https://www.fastcompany.com/3046119/defining-the-sharing-economy-what-is-collaborative-consumption-and-what-isnt. Accessed 20 Nov 2018

Caputo, M., Mininno, V.: Internal, vertical and horizontal logistics integration in Italian grocery distribution. Int. J. Phys. Distrib. Logistics Manag. **9**, 64 (1996). https://doi.org/10.1108/09600039610149101

car2go (2018). https://www.car2go.com/. Accessed 14 Nov 2018

Charalambides, L.C.: Shared capacity resource reallocation in a decentralized service system. J. Oper. Manag. **5**(1), 57–74 (1984). https://doi.org/10.1016/0272-6963(84)90007-X

Christopher, M.: Logistics and Supply Chain Management. Financial Times/Patience Hall, London (2005)

Cordes, M.: Frachtraten steigen nur leicht, 3 April 2018. https://www.verkehrsrundschau.de/nachrichten/vr-index-frachtraten-steigen-nur-leicht-2147567.html. Accessed 30 June 2018

Cruijssen, F.: Horizontal cooperation in transport and logistics. Tilburg, Univ., Diss., 2006, CentER Dissertation Series, 176. – VII (2006)

Cruijssen, F., Dullaert, W., Fleuren, H.: Horizontal cooperation in transport and logistics: a literature review. Transp. J. **46**(3), 22–39 (2007)

Dervojeda, K., Verzijl, D., Nagtegaal, F., Lengton, M.: The Sharing Economy. Accessibility Based Business Models for Peer-to-Peer Markets. European Commission, Directorate-General for Enterprise and Industry, Brussels (2013)

Eisenhardt, K.M.: Building theories from case study research. Acad. Manag. Rev. **14**(4), 532–550 (1989). https://doi.org/10.5465/AMR.1989.4308385

FLEXE (2018). https://www.flexe.com/. Accessed 14 Nov 2018

Freitag, M., Kück, M., Becker, T.: Potentials and risks of resource sharing in production and logistics. In: Delfmann, W., Wimmer, T. (eds.) Proceedings of the 8th International Scientific Symposium on Logistics, BVL, Karlsruhe, Germany, pp. 199–209 (2016)

Huang, Q., Davison, R.M., Liu, H.: An exploratory study of buyers' participation intentions in reputation systems: the relationship quality perspective. Inf. Manag. **51**(8), 952–963 (2014). https://doi.org/10.1016/j.im.2014.09.003

Huurne, M., Ronteltap, A., Corten, R., Buskens, V.: Antecedents of trust in the sharing economy: a systematic review. J. Consum. Behav. **16**(6), 485–498 (2017). https://doi.org/10.1002/cb.1667

Islam, S., Olsen, T.: Truck-sharing challenges for hinterland trucking companies: a case of the empty container truck trips problem. Bus. Process Manag. J. (2), 290 (2014). https://doi.org/10.1108/BPMJ-03-2013-0042

Islam, S.: Empty truck trips problem at container terminals: a review of causes, benefits, constraints and solution approaches. Bus. Process Manag. J. **23**(2), 248–274 (2017). https://doi.org/10.1108/BPMJ-06-2015-0086

Kölling, M.: Wie Japan AirBnB und Uber ausbremst, 13 June 2018. https://www.handelsblatt.com/politik/international/weltgeschichten/koelling/weltgeschichte-wie-japan-airbnb-und-uber-ausbremst/. Accessed 30 June 2018

Maghsoudi, A., Pazirandeh, A.: Visibility, resource sharing and performance in supply chain relationships: insights from humanitarian practitioners. Supply Chain Manag. Int. J. (1), 125 (2016). https://doi.org/10.1108/SCM-03-2015-0102

Naesens, K., Gelders, L., Pintelon, L.: A swift response framework for measuring the strategic fit for a horizontal collaborative initiative. Int. J. Prod. Econ. **121**(2), 550–561 (2009). https://doi.org/10.1016/j.ijpe.2007.04.004

Pomponi, F., Fratocchi, L., Tafuri, S.R.: Trust development and horizontal collaboration in logistics: a theory based evolutionary framework. Supply Chain Manag. **20**(1), 83–97 (2015). https://doi.org/10.1108/SCM-02-2014-0078

Schmoltzi, C., Wallenburg, C.M.: Operational governance in horizontal cooperations of logistics service providers: performance effects and the moderating role of cooperation complexity. J. Supply Chain Manag. **48**(2), 53–74 (2012)

Semmann, C.: Deutschland ist abermals Logistik-Weltmeister, 16 July 2018. https://www.dvz.de/rubriken/logistik/detail/news/deutschland-ist-abermals-logistik-weltmeister.html. Accessed 25 Nov 2018

Serrano-Hernández, A., Faulin, J., Juan, A.A., Perez-Bernabeu, E.: Horizontal collaboration in freight transport: concepts, benefits, and environmental challenges. SORT **41**(2), 1–22 (2017). https://doi.org/10.2436/20.8080.02.49

Singh, P.J., Power, D.: The nature and effectiveness of collaboration between firms, their customers and suppliers: a supply chain perspective. Supply Chain Manag. Int. J. (3), 189 (2009). https://doi.org/10.1108/13598540910954539

Stadtmobil (2018). https://www.stadtmobil.de/. Accessed 14 Nov 2018

Strauss, A.L., Corbin, J.M.: Basics of Qualitative Research. Techniques and Procedures for Developing Grounded Theory. SAGE Publications, Thousand Oaks (1998)

Yin, R.K.: Case study Research and Applications. Design and Methods. SAGE, Thousand Oaks (2018)

Zervas, G., Proserpio, D., Byers, J.W.: The rise of the sharing economy: estimating the impact of Airbnb on the hotel industry. J. Mark. Res. **54**(5), 687–705 (2016). https://doi.org/10.1509/jmr.15.0204

Influence of a Reasonable Allocation of Pallets in the Pallet Exchange System

Ralf Elbert and Roland Lehner(✉)

Department of Law and Economics, Technische Universität Darmstadt, Darmstadt, Germany
lehner@log.tu-darmstadt.de

Abstract. In the pallet exchange system, consignees, shippers and forwarders all bear large financial burden regarding the exchange of pallets. Thus, efficiency and costs are strongly influenced by the individual behaviour of the participants. The aim of this study is to focus on the influences of these factors. Of special interest is the reasonable allocation of pallets. The pallets shall be kept in good condition as long as possible to avoid costs for repairing and replacing defective pallets. Thereby, the predictable depreciation in value can be reduced by a reasonable selection of pallets depending on the wear trait of the transported goods.

Keywords: Human factors · Reasonable pallet allocation · Pallet exchange system

1 Introduction

Pallets come in different shapes and sizes. The pallets are made of wood (soft/hard), cardboard or plastic [1], which have to be managed specifically. The most common one is the wooden Euro-pallet, 550 million of these pallets are used in Europe [2]. When the transported goods are handed over, the actors in the supply chain exchange the pallets. After the transfer of the palletized goods, they are returned as empty pallets. The exchange can be structured in different ways. It can be performed directly or deferred [3]. In practice, however, the general pallet exchange procedure leads to many disputes and sometimes high costs for the individual actors [4, 5]. According to [6], the costs are around 4.96 € per circulation, with more than 60% of the costs incurred in Germany being induced by forwarders, more than a quarter by shippers and only 10–15% by consignees. In a study on pallet exchange in Switzerland, a share of 40% each for shipper and forwarder was found, while consignees accounted for 20% [7]. The level of incurred costs depends largely on the process design of the actors and their interaction [7, 8]. The behaviour of the individual actors thus influences the efficiency of the pallet exchange system. In the scientific literature, however, the focus has been on the overarching choice of the pallet management system so far [1, 9, 10]. The economic perspective generally results in a comparison of pallet management systems with the aim of selecting the most cost-effective system [11, 12]. Operational considerations are less in the focus of scientific research than strategic decision-making aids for the selection of a pallet management system.

U. Clausen et al. (Eds.): ICPLT 2019, LNLO, pp. 90–101, 2019.
https://doi.org/10.1007/978-3-030-13535-5_7

An operative behaviour is the reasonable allocation of pallets to products. This means that an appropriate pallet is used depending on the wear and tear of the product. In this way, foreseeable depreciation, e.g. in the form of staining caused by construction materials transported on the pallet, can be prevented. Accordingly, high quality pallets should be used for products with low wear, e.g. food. Two factors in particular have a strong influence on the life expectancy of a pallet: the load caused by the loaded mass and the handling of the pallet. These factors are confirmed by the results of a corresponding series of tests in which the effects of handling and loaded mass were systematically investigated. In the case of relatively light loading and careful handling, in extreme cases the difference was a six times longer pallet life expectancy [10]. The influence of a needs-based selection of pallets depending on the products with different wear factors in pallet exchange have hardly been discussed so far. This paper addresses precisely this research gap. Since the costs of replacing damaged pallets are the largest cost component of pallet exchange [6], it makes sense to take a closer look at this topic. The aim of this paper is to clarify why reasonable allocation is not carried out in practice and to what extent reasonable allocation has a positive effect on the pallet exchange system. Different behaviours of the individual actors are to be taken into account and thus a contribution to the investigation of pallet exchange systems from a behavioural perspective is to be made. As a result, the contribution ties in with both identified research gaps, the insufficient consideration of the individual actors behaviour and the reasonable pallet selection as influencing factors on the advantageousness or efficiency of an exchange system. The following research questions arise from this:

1. Which factors and behaviours in current practice prevent a functioning pallet exchange system in general and a demand-oriented allocation in particular?
2. To what extent can the life expectation of the pallets be increased with a reasonable allocation?

In the remainder of this article, first an overview of the current state of research on the subject of pallet exchange is given. The focus here is on the identification of the general influencing factors and behaviours that lead to problems in the practice of pallet exchange. In the next step, the behaviours are identified from this, which prevent a reasonable allocation. Subsequently, the influence of the reasonable allocation of pallets is analysed with the help of a simulation model. The paper then ends with a conclusion.

2 State of Research

Euro-pallets are a form of reusable packaging and count as returnable assets [13]. In a classic flow of goods within a supply chain, goods pass through stations from the supplier to the customer. This is referred to as open-loop supply chain. Pallets as returnable assets initially flow through the supply chain together with the goods to the customer, but are then taken back to an upstream location in the supply chain. This closes the pallet cycle and the pallets can be used for transporting other goods again. This is referred to as a closed-loop supply chain [14].

The exchange of pallets can be designed in different ways. In case of a step-by-step exchange, pallets of the same quality and quantity are exchanged. When the goods are delivered on pallets, the same number of empty pallets is returned to the shipper via the forwarder [15]. In the case of a pallet account, the number and quality of the delivered pallets is noted when the goods are delivered. At a later date, pallets of corresponding quality and quantity can be returned. The pallet account is similar to the step-by-step exchange, but the pallets will be handed over to the owner at a later date [8]. Pallet receipts will be issued on site to document the pallet debts. Another possibility is pallet pooling. The management of the pallet including procurement, repair, efficient distribution and disposal will be outsourced to an external service provider, the pallet pool operator [16].

In scientific literature, the concrete topic of pallet exchange can be mainly separated into two research streams: economic efficiency and environmental considerations. In recent years, the use of radio frequency identification (RFID) in pallet management has also been discussed. The aim is to analyse the applicability of the technology and to show potentials for cost reduction [17–19]. Assessments from an ecological point of view are also becoming increasingly important. Only a few of the sources already mentioned in this paper point out that pallets are subject to heavy wear due to their use as load carriers for a wide variety of products and under changing environmental conditions [10]; [16]. There are various criteria for assessing the wear of a pallet. According to European Pallet Association, pallets are divided into two classes: exchangeable and non-exchangeable pallets [20]. A method widely used in practice is the criteria of GS1 [21]. A distinction is made between new pallets, A-pallets, B-pallets, C-pallets and defective pallets. A model to determine the loss in value of a pallet is also based on this categorisation. With the so-called Cologne depreciation key, the pallet loses value per circulation [22]. Wang et al. [23] even developed eight criteria to determine the value of pallets. However, these criteria are of a rather academic nature, and the use of these criteria in practice could not be proven.

Even if the behaviour of individual actors in pallet exchange and the emerging problem of quality loss do not receive much attention in the scientific literature, they are at least discussed in the specialist trade press [2, 24, 25] and in practical studies [6, 15, 21]. These sources are used to identify the factors and behaviours that cause problems in the pallet exchange system and prevent reasonable allocation. The results of this literature analysis will be validated in a focus group interview. This group consists of people who deal with the topic of pallet exchange on a daily basis. Among them were two representatives from the forwarding company, one from manufacturing industry and two representatives from logistics service providers. The interview took place on 25 October 2018 at TU Darmstadt.

The general influencing factors resulting from a literature analysis are divided into different categories. First a superordinate, general category is formed, which describes general weaknesses of the current system in the pallet exchange. Classified in this category is the time pressure, the existence of individual evaluation criteria and a lack of a formal, contractual basis. There is usually time pressure in the delivery process [2, 7]. Unloading often has to be carried out quickly, as other trucks are waiting to be dispatched. The loading of empty pallets is often not desired, so that pallet receipts are often used. Often the forwarder can only retrieve these at another location, so that additional trips are sometimes necessary to collect pallets. Another overriding problem is the contractual constellation of the various actors. In general, there is no contractual relationship between the forwarder and the consignee; instead, the shipper orders the forwarder to deliver goods to the consignee [15, 26]. Thus, there is no prior agreement between the forwarder and the consignee, so that in practice disputes often arise when exchanging pallets, e.g. with regards to quality. In particular, if the forwarder awards the contract to a subcontractor, the situation becomes even worse [21], as there is no contractual relationship with the supplier. Furthermore, the quality determination of pallets is often subjective. Although organisations such as EPAL and GS1 have set standards for the quality determination of pallets, in practice the pallets are evaluated according to subjective criteria, which can be very different depending on the actor [2, 7].

For the individual actors, there are further factors influencing the exchange of pallets at two levels: at the company level and at the employee level. It should be classified in the company level category that, in practice, there is hardly any relevance for shippers to allocate pallets according to product requirements [27]. Currently, such an assignment is only required if the consignees use automation technology, such as an automatic warehouse. Pallets with a high quality standard are an essential requirement for automation. Another factor discussed in literature is the forwarder's concession regarding his disadvantage when exchanging pallets. This is due to the characteristic high competition on the transport market [8]. With the aim of not losing any orders, it is often accepted that pallet exchange is a loss-making business for the forwarder [28]. This is particularly the case when the consignee deliberately only hands over worse pallets to the forwarder than those received. This strategy is also being demonstrated by forwarders and shippers [21]. However, since the consignee has no contractual relationship with the forwarder and is not interested in service quality for the customer, e.g. like the shipper, the exchange is assigned to the consignee at this point for his own benefit. This is also reflected in the fact that the forwarding agent usually bears the highest costs when exchanging pallets. In the focus group interview it was noted that the consignee is the main influencing factor of the pallet exchange system. Depending on how he exchanges, he decides on the efficiency of the trade. In the case of the consignee, especially if it is a retailer, it can also happen that not enough pallets are available for the procedure. Due to seasonal fluctuations in particular, there are times all pallets of the company are in use and only pallet receipts are issued to the forwarders [21]. Further factors that can be assigned to all actors at the company level is the frequently lacking awareness of the return or exchange processes [29]. There is often no standard process [7] and the costs for the exchange are unknown to the actors [28, 29]. Another aspect is the lack of communication between the individual participants. Agreements between them are an essential factor for a working return logistics system [30]. However, the lack of contractual relationships already mentioned and a lack of awareness of return logistics make such an exchange difficult. Another aspect is the sorting of pallets according to quality. In reality, sorting often does not take place. If a sorting does take place, then often according to the two criteria good and bad [21].

Table 1. Influencing factors and behaviours to the pallets exchange system

	Shipper	Forwarder	Consignee
General factors	Time pressure [2, 7, 15]		
	Individual evaluation criteria [2, 7]		
	Lack of contractual constellation [15, 26]		
Company level	No relevance for reasonable allocation [21, 27]	Goodwill [28]	Exchange to own benefit [21]
			Not enough pallets [5, 21]
	Lack of awareness of return logistics [28, 29]		
	No training for employees [21]		
	Lack of communication [15, 21]		
	No pallets sorting [21]		
Employee level	Illegal sales of pallets [21]		
	Not sorting out defective pallets [21]		
	Defects are deliberately hidden during the exchange [21]		
	Preferences of the individual employee [21]		

Further factors are added at the level of individual employees. These behaviours are often not defined in the company guidelines. Literature also speaks of deviant behaviour [31]. In the context of pallet exchange, illegal sales of pallets [21] also occur. The good quality pallets are exchanged for bad ones and the difference is retained by the employees involved. Furthermore, some defective pallets are not sorted out, but continue to be exchanged. The costs for repair and replacement are thus transferred to another actor. The same applies if certain defects are deliberately hidden during the exchange [21]. Furthermore, the preferences of the individual employee also influence the exchange. For example, some drivers receive good pallets from one employee, while others always receive bad pallets [21].

The results of the literature analysis are summarised in Table 1. It should be noted that general influencing factors can be traced back to the general conditions of pallet exchange. However, there are multi-layered behaviours of the individual actors that characterise the pallet exchange system. The separation carried out between company level and employee level can be seen as fluid. For example, the sorting of pallets can be arranged on an administrative level or the individual employees can conduct it without specific instructions. Since the influencing factors and behaviours are initially related to the overall pallet exchange system, the parameters that also specifically counteract the reasonable allocations need to be mentioned. A lack of awareness of the processes in the pallet cycle as well as untrained employees prevent reasonable allocation. Without appropriate sorting of the pallets according to quality, reasonable allocation is also made more difficult. Furthermore, there is currently no relevance for the shipper for a reasonable allocation. Usually, customers do not require this and there is no incentive to select a range of certain quality products.

It should be noted that the individual factors were only collected on the basis of statements in the literature. A qualitative and quantitative evaluation of the influence of individual behaviours and influencing parameters did not take place. After the analysis of literature and the interview of experts in the context of a focus group interview identified influencing factors and behaviours that affect the efficiency of the pallet exchange system, the next step is to determine the influence of a reasonable allocation of pallets.

3 Modelling and Development of the Simulation Model

Simulation is used as a methodology to investigate the impact of reasonable allocation, especially in production and business processes [32], as well as in the scientific field of logistics and transport [33, 34]. An agent-based simulation is used, which is ideally suited for use in pallet logistics: Different actors with individual behaviour interact with each other and with the environment [35]. With the help of the simulation, stochastic influences such as the fluctuating demand in a specific case and the random selection of pallets of the actors can be taken into account. The simulation model is created using Anylogic, a software from AnyLogic Company. The systematic procedure for the development of the simulation model is based on the development model according to [36].

The conceptual model takes into account the actors forwarder, shipper and consignee. A step-by-step exchange is carried out at the transfer points, as many companies prefer direct exchange on the spot because of its simplicity [3]. The central objective of the model is the life expectancy of the pallets. This is recorded as the average number of cycles a pallet can last from a new condition to a defective condition. Another target parameter is the quality of the pallets in the system. The input parameters are the quality and quantity the individual actors are having at the beginning. Furthermore, the wear factor of the products transported on the pallets must also be assigned to the previously defined parameter. Different behaviours of the actors are also defined as input parameters.

In the simulation model, the consignee orders two different products daily from the shipper. Between 0 and 30 pallets can be ordered for the two products, with an average value of 26. This was chosen so that the order can be carried out with a truck that can transport between 20 and 34 pallets depending on the type of truck and trailer [37]. The shipper prepares the products for pickup and informs the forwarder of the order. The forwarder loads the quantity of empty pallets corresponding to the order into his truck and drives to the forwarder. Once there, he unloads the empty pallets and hands them over to the shipper. The palletised goods get picked up and transported to the consignee. The goods are delivered and the forwarder receives empty pallets back. With the empty pallets he returns to his base, where he unloads the pallets and waits for the next transport order. This process repeats daily. The system is closed, which means that no new pallets enter the cycle. Defective pallets are also not repaired. The simulation stops as soon as there are not enough pallets left and further pallets have to be added into the system. This is a terminating simulation [38].

The loss in value is based on the Cologne depreciation key [22]. For the first product, the following losses are assumed per circulation: A new pallet has a value of 7.20 € and loses 0.80 € per circulation. An A pallet has a value of 7.00 € and loses 0.27 € per circulation, a B pallet (value 6.20 €) loses 0.30 € per circulation and the C pallet (value 5.30 €) loses 0.71 € per circulation. A wear factor between 0.9 and 0.1 is assumed for the second product. The factor describes the difference in wear between the products. A wear factor of 0.5 means that the second product has 50% less loss in value per circulation.

The shipper can allocate the pallets according to requirements. Depending on the product and its wear factor, different pallet qualities are selected. Alternatively, the shipper randomly selects a pallet for the two products from his stock. In the focus group interview, it was emphasized that the efficiency of pallet exchange depends on whether the players exchange fairly. Therefore, this behaviour is also implemented in the model. The forwarder and recipient also select the pallets randomly or fairly for the exchange. In a fair exchange, the forwarder and the consignee try to hand over pallets with the quality levels to the exchange partner.

The verification and validation of the simulation model took place at the same time as its development. Animations and monitoring, extreme value tests and validation in dialogue with experts from practice were used. For more detailed information on the verification and validation types mentioned, see [39].

4 Results of the Simulation Study

For the evaluation of the simulation experiments, a confidence level of 95% was chosen, which is often used in science in order to achieve valid results [40]. Therefore, a corresponding number of replications of the individual simulation experiments were performed. In the first test series, the wear factor is varied between 0.9 and 0.1. The shipper selects the pallets reasonably. The forwarder and the consignee randomly select their pallets. The life expectancy of the pallets with the parameter variations is shown in the table. It can be seen that a reasonable allocation increases the life value of a pallet. This result was to be expected, the value ranges are interesting here. Even with wear differences of 30%, the number of possible circulations in the model can be increased by round about 4%. If the wear factors of the products differ greatly, the life expectancy can even be increased by about 20% (Table 2).

Table 2. Pallet life expectations depending on allocation and wear factors

Wear factor	Pallet circulations		Percentage deviation
	Random allocation	Reasonable allocation	
0.9	12.7	12.8	0.5%
0.8	13.3	13.4	1.1%
0.7	14.2	14.8	4.3%
0.6	15.3	16.8	8.9%
0.5	16.1	18.8	14.2%
0.4	17.2	20.1	14.4%
0.3	18.4	21.5	14.5%
0.2	20.3	25.1	19.2%
0.1	22.3	28.6	22.1%

Table 3 shows the number of average pallet circulations up to the defect pallet with an average wear factor of 0.5 as dependent on a fair exchange. In the model, a fair exchange leads to a longer service life of the pallets. Which is why a fair exchange has an influence on the life expectancy of the pallet can be explained by the following evaluation.

Table 3. Pallet life expectations depending on exchange behaviour

Wear factor	Shipper	Forwarder/Consignee	Pallet circulations
0.5	Reasonable allocation	Fair	20.3
		Random	18.8

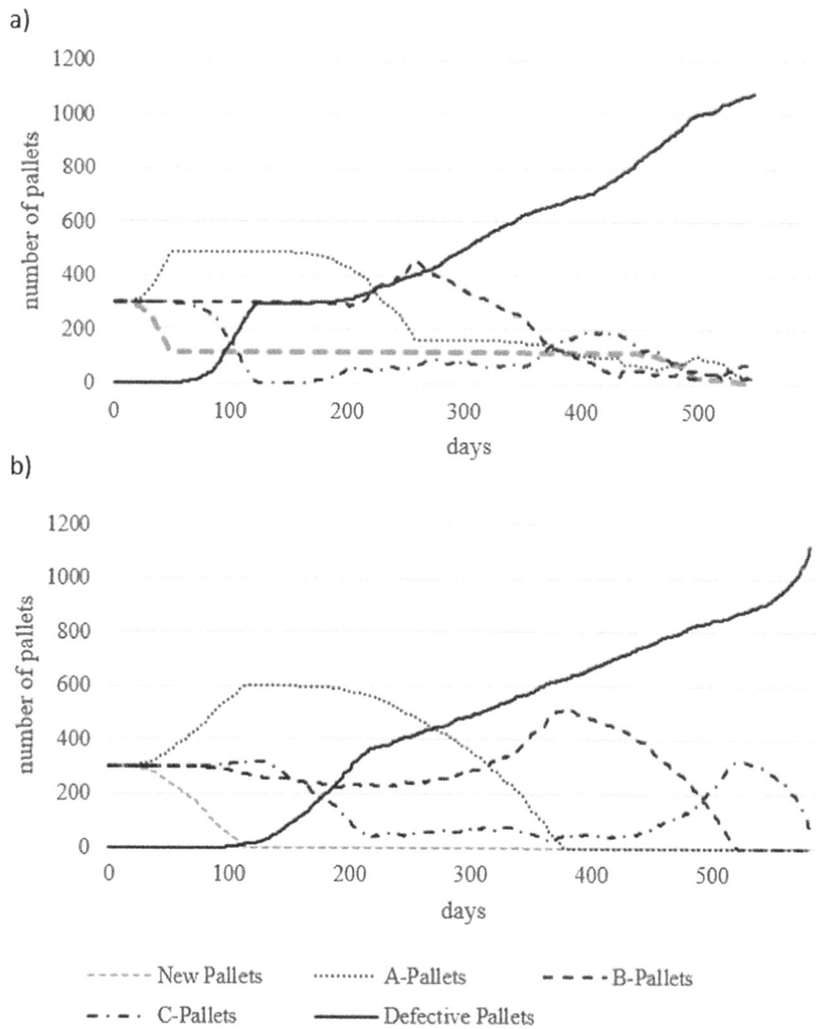

Fig. 1. Number of pallets with different qualities in the pool depending on specific pallet exchange behaviours (a) shipper and forwarder choose pallets for exchange fairly (b) shipper and forwarder choose pallets randomly

Figure 1 shows the number of different pallet qualities in the system in relation to the simulation time. In the variant in which the shipper allocates the pallets according to requirements and the forwarder/consignee exchanges them fairly, it can be seen that pallets of all quality classes are still in the system until the end of the simulation run. The parameter variation, in which a reasonable allocation is made and the shipper/consignee randomly exchanges pallets, shows a different picture. As of day 400, only B- and C-pallets are left in the system. As a result, the shipper can no longer optimally allocate the pallets in line with requirements, and as a result the number of pallet runs to sorting out is reduced.

Considering the results, it should be noted that not only a reasonable allocation is responsible for the wear of the pallet, but also the handling of the pallet. This topic was also discussed in the focus group interview. During the interview, industry concerns were also expressed. The reasonable allocation of pallets can lead to significantly higher costs in in-house logistics, since in some cases the pallet passes through production from the raw material to the finished good. The reasonable allocation is ultimately an additional expense for the shipper, of which the shipper has no direct advantage. Here the shipper must also be given incentives to make such an allocation. Although reasonable allocation can contribute to improving the pallet exchange system, there are a number of other factors influencing the system. Ultimately, the major problem is how the actors interact with each other and the trend towards optimising themselves and increasing efficiency for their own benefit by exchanging pallets.

5 Conclusion

In summary, it can be said that the topic of pallet exchange leads to challenges in practice due to a variety of influencing factors and behaviours. These can be divided into general influencing factors and behaviour patterns specific to actors, which can be assigned to a company level or an employee level. Some of the behaviours found in practice prevent the reasonable allocation of pallets to products with different wear factors. In detail, these are the lack of awareness of the costs and the further processes in the pallet cycle, no corresponding training of the employees, no sorting of the pallets according to quality and no concrete incentives for a proper allocation. The experiments in the simulation model show, however, that the service life of the pallets can be increased by a reasonable allocation of pallets to goods with different wear properties.

The costs for repair and replacement could thus be reduced, which would benefit the entire pallet exchange system. In practice, therefore, a reasonable allocation is advisable. However, caution must be taken here to ensure that this does not result in any greatly increased additional internal costs. The behaviours identified in this article also offer starting points that must be taken into account for implementing the concept.

It should be noted that only the costs of the quality shift of the pallets as well as only the direct step-by-step exchange are considered. Other costs, such as those arising from the use of pallet receipts or accounts, from the use of a pallet service provider or costs for sorting, were not taken into account in this article. Furthermore, only three individual actors with a limited number of behaviours in a closed pallet exchange system were considered. However, the analysis of the influencing factors has shown many different behaviours of the actors. In the future, the different behaviours of the individual actors and the effects of these behaviours should be investigated, as they influence the efficiency of the pallet exchange system to a large extent. In particular, mechanisms for fair trade (incentives and sanctions) should be developed to ensure cooperative behaviour by all actors and, in particular, incentives for shippers to actually carry out reasonable allocation.

References

1. Mazeika Bilbao, A., Carrano, A.L., Hewitt, M., Thorn, B.K.: On the environmental impacts of pallet management operations. Manag. Res. Rev. **34**(11), 1222–1236 (2011)
2. n.n.: Wie der Palettentausch in Europa funktioniert. In: Disponaut by DVZ (2016). https://www.disponaut.de/verkehrstraeger/strasse/single-view/nachricht/wie-der-palettentausch-in-europa-funktioniert.html
3. Bottani, E., Montanari, R., Rinaldi, M., Vignali, G.: Modeling and multi-objective optimization of closed loop supply chains: a case study. Comput. Ind. Eng. **87**, 328–342 (2015)
4. von Lindern, J.: Die Paletten-Kette. In: DVZ (2018). https://www.dvz.de/rubriken/logistik/detail/news/die-paletten-kette.html
5. Wild, C.: Paletten mitverkaufen, trans aktuell 01/2016, 18 December 2015
6. Lange, V., Hoffmann, J.: Empirische Kostenanalyse des Tauschverfahrens im Europalettenpool. In: Fraunhofer-Institut für Materialfluss und Logistik (2009). https://www.iml.fraunhofer.de/content/dam/iml/de/documents/OE%20150/Flyer1/Produktblatt_Palettenstudie.pdf
7. Grimm, J., Hofstetter, J.S.: Optimierungspotenzial: Palettentausch. Handelszeitung Special, S. 48, 21 April 2011
8. Gnoni, M.G., Lettera, G., Rollo, A.: A simulation comparison analysis of effective pallet management scenarios. Paper Presented on the 18th IEEE International Conference on Industrial Engineering and Engineering Management, Singapore, pp. 1228–1232, 6–9 December 2011
9. Bengtsson, J., Logie, J.: Life cycle assessment of one-way and pooled pallet alternatives. Procedia CIRP **29**, 414–419 (2015)
10. Carrano, A.L., Pazour, J.A., Roy, D., Thorn, B.K.: Selection of pallet management strategies based on carbon emissions impact. Int. J. Prod. Econ. **164**, 258–270 (2015)
11. Elia, V., Gnoni, M.G.: Designing an effective closed loop system for pallet management. Int. J. Prod. Econ. **170**, 730–740 (2015)
12. Roy, D., Carrano, A.L., Pazour, J.A., Gupta, A.: Cost-effective pallet management strategies. Transp. Res. Part E Logistics Transp. Rev. **93**, 358–371 (2016)
13. Glock, C.H.: Decision support models for managing returnable transport items in supply chains: a systematic literature review. Int. J. Prod. Econ. **183**, 561–569 (2016)
14. Govindan, K., Soleimani, H., Kannan, D.: Reverse logistics and closed-loop supply chain: a comprehensive review to explore the future. Eur. J. Oper. Res. **240**(3), 603–626 (2015)
15. Hagenlocher, S., Wilting, F., Wittenbrink, P.: Schnittstelle Rampe - Lösungen zur Vermeidung von Wartezeiten. In: Schlussbericht für das Bundesministerium für Verkehr, Bau und Stadtentwicklung (2013). https://www.bmvi.de/SharedDocs/DE/Anlage/VerkehrUndMobilitaet/laderampe-schlussbericht-schnittstelle-rampe.pdf?__blob=publicationFile
16. Twede, D., Mollenkopf, D., Guzman-Siller, C.: A market-driven investigation of pallet trends in grocery chains. J. Food Distrib. Res. **38**(1), 161–169 (2007)

17. Duraccio, V., Elia, V., Forcina, A.: An activity based costing model for evaluating effectiveness of RFID technology in pallet reverse logistics system. In: AIP Conference Proceedings, vol. 1648 issue 1, p. 570005 (2015)
18. Gnoni, M.G., Rollo, A.: A scenario analysis for evaluating RFID investments in pallet management. Int. J. RF Technol. **2**(1), 1–21 (2010)
19. Qinghua, Z., Jun, W., Guoquan, C., Zhuan, W., Dawei, Y., Shanshan, Z.: Pallet rental information system based on RFID. Paper Presented on the 4th IEEE Conference on Industrial Electronics and Applications, University of Science and Technology Beijing, Beijing, pp. 886–891, 8–11 December 2009
20. Grimm, J., Stölzle, W., Hofstetter, J.: EUR/EPAL-Palette-Marktstudie Schweiz. Cuvillier Verlag, Göttinen (2010)
21. Löw, M.: Analyse zum Zug-um-Zug Palettentausch, GS1 Germany GmbH, Köln (2008)
22. Hector, B.: Palettenwert richtig kalkulieren. In: DVZ (2013). https://www.dvz.de/rubriken/management-recht/detail/news/plettenwert-richtig-kalkulieren.html
23. Wang, Y., Li, Y., Liang, Q., Sheng, J., Shang, W.: A hybrid evaluation method for pallet quality evaluation. In: Proceedings of IMCIC – ICSIT, vol. 1, pp. 96–100 (2016)
24. Hector, B.: Paletten verkaufen statt tauschen - Erfahrungen mit Tauschgebühren. In: DVZ (2016a). http://www.dvz.de/rubriken/management-recht/single-view/nach-richt/paletten-verkaufen-statt-tauschen-erfahrungen-mit-tauschgebuehren.html
25. Wild, C.: Palettentausch: Den Warenträger im Blick behalten. In: Eurotransport.de (2015). http://www.eurotransport.de/news/palettentausch-den-warentraeger-im-blick-behal-ten-6761049.html
26. Iassinovskaia, G., Limbourg, S., Riane, F.: The inventory-routing problem of returnable transport items with time windows and simultaneous pickup and delivery in closed-loop supply chains. Int. J. Prod. Econ. **183**, 570–582 (2017)
27. Jüngst, I.: Pläne für die Palette: Effizientes Palettenmanagement bei Dachser – neue Plattform für alle Beteiligten in der Logistik. In: Trans aktuell, Nr. 16/17, S. 11, 17 August 2018
28. Breen, L.: Give me back my empties or else! A preliminary analysis of customer compliance in reverse logistics practices (UK). Manag. Res. News **29**(9), 532–551 (2006)
29. Lau, K.H., Wang, Y.: Reverse logistics in the electronic industry of China: a case study. Supply Chain Manag. Int. J. **14**(6), 447–465 (2009)
30. Guzman-Siller, C.F., Twede, D., Mollenkopf, D.A.: Differences in the perception of pallet systems between US and Canadian grocery retailers. J. Food Distrib. Res. **41**(3), 84–97 (2010)
31. Franzke, F.: Der Mensch als Faktor in der manuellen Kommissionierung: Eine simulationsbasierte Analyse der Effizienz in Person-zur-Ware-Kommissioniersystemen. Gabler, Dortmund (2018)
32. Jahangirian, M., Eldabi, T., Naseer, A., Stergioulas, L.K., Young, T.: Simulation in manufacturing and business: a review. Eur. J. Oper. Res. **203**(1), 13 (2010)
33. Davidsson, P., Henesey, L., Ramstedt, L., Törnquist, J., Wernstedt, F.: An analysis of agent-based approaches to transport logistics. Transp. Res. Part C **13**(4), 255–271 (2005)
34. Tako, A.A., Robinson, S.: The application of discrete event simulation and system dynamics in the logistics and supply chain context. Decis. Support Syst. **52**(4), 802–815 (2011)
35. Law, A.: Simulation Modeling and Analysis, 5th edn. McGraw-Hill Education, New York (2014)
36. Manuj, I., Mentzer, J.T., Bowers, M.R.: Improving the rigor of discrete event simulation in logistics and supply chain research. Int. J. Phys. Distrib. Logistics Manag. **39**(3), 172–201 (2009)

37. Martin, H.: Transport- und Lagerlogistik: Planung, Struktur, Steuerung und Kosten von Systemen der Intralogistik. Springer Vieweg, Wiesbaden (2009)
38. Bandyopadhyay, S., Bhattacharya, R.: Discrete and Continuous Simulation: Theory and Practice. CRC Press, Boca Raton (2014)
39. Rabe, M., Spieckermann, S., Wenzel, S.: Verifikation und Validierung für die Simulation in Produktion und Logistik - Vorgehensmodelle und Techniken. Springer, Heidelberg (2018)
40. Martin, R.: Scientific Thinking. Broadview Press, Peterborough (1997)

Automated Trucks in Road Freight Logistics: The User Perspective

Stephan Müller(✉) and Felix Voigtländer

Institute of Transport Research, German Aerospace Center (DLR), Rutherfordstr. 2, 12489 Berlin, Germany
stephan.mueller@dlr.de

Abstract. A current topic of high relevance in the transport sector is the automation of driving. Automation affects not only passenger transport, but also freight transport. While industry players are promoting automation to achieve the next level of driver assistance systems as a selling point, policymakers are supporting this innovation in the expectation of reducing congestion, fuel consumption and accidents. Hence, publicly funded projects deal with activities on infrastructure issues, legal aspects, technology development, connectivity aspects, information technology security and data protection. However, the user side of automated driving in freight traffic, the logistics and forwarding companies so far had little involvement in designing the technology and the framework conditions according to their needs. The exclusion of the user perspective can however lead to non-market-oriented solutions. Against this background, the article aims to convey the perspective of potential users of automation technology and to demonstrate the conditions of technology acceptance.

The main result of the analysis is clear: The great benefit potential for users will be unleashed if the driver is replaced by the technology, due to the increasing scarcity of drivers, the increasing cost pressure and low margins and, the increasing need for efficient logistics processes, which drives the demand for automated trucks. In the opinion of the logistics industry, it is very unlikely that the driver will take on other tasks while driving for labour force qualification and economic reasons. The novelty of this finding contradicts the current use cases considered by the manufacturing industry: an automated truck with a driver who can follow other activities on the tour. There are signs, that the development of a user-oriented product therefore means the development of a driverless truck.

1 Introduction

Automatic and connected driving (ACD) is currently a topic of high relevance in transport. A vision of the future of transport without automated vehicles hardly seems to be consistent any more - not only for passenger traffic but also for freight transport. This technology is being strongly fostered by vehicle manufacturers and suppliers because, starting with the use of driver assistance systems, automation is the logical continuation of the development of an intelligent vehicle. Politicians support the technology against the background that traffic jams, accidents and fuel consumption

U. Clausen et al. (Eds.): ICPLT 2019, LNLO, pp. 102–115, 2019.
https://doi.org/10.1007/978-3-030-13535-5_8

can be reduced. Another political goal by the promotion of ACD is to maintain and secure the German automotive industry market leading worldwide. The promotion of ACD by public efforts comprises activity points on infrastructure issues, legal aspects, technology development, connectivity aspects, information technology security and data protection (BMVI 2015). Hence, the introduction to market of automated and connected driving can be expected to the near future.

However, the user side of ACD, the logistics and haulage companies, have so far been little involved in shaping the technology and the framework conditions according to their needs. Manufacturers and politicians must therefore more or less assume that the products, however automated, will be purchased and that the design of suitable framework conditions required from a technological point of view will be sufficient. But experience from the promotion and development of electro mobility, for example, has shown that the involvement of the user side and user-oriented product development are decisive for technology success.

Against this background, the article aims to provide the perspective of potential users of ACD technology and highlight the conditions of technology acceptance. Interviews with logistics companies have been conducted and evaluated for this purpose. The article is structured as follows: After providing insights in the methodology chapter three addresses the need for ACD technology by its users, their hurdles to adopt the technology, use cases of automated trucks and, the framework conditions of an ACD use case. Finally the article provides implication of the findings for technology development and transport policy.

2 Methodology of Interviews

As buyers of trucks, potential users of truck automation technology are logistics and forwarding companies. The aim of the investigation was to determine their perspective on the new technology in terms of drivers and barriers of technology, use cases and framework conditions of use cases. Because this has not yet been published in the literature, a qualitative research design by in depth interviews was chosen. This choice was driven by two aspects: firstly to discover the scope of opinions and secondly to collect the opinion of high level managers instead of anonymous respondents.

The in depth interviews were guided by questions. This means that 23 questions were prepared (see appendix) and moderated in a one-hour session. The often not direct questioning in the course of the conversation has the advantage that even extensive and previously unnoticed aspects on the subject of automated and connected driving around the core question are discovered, since there is more freedom in the course of the conversation. However, it may have the disadvantage that not all questions could be asked during the interview, as the interviewee focused on a particular aspect during the interview, which was then discussed in more detail.

Supported by the LogistikNetz Berlin-Brandenburg e.V. (LNBB), a logistics competence centre, our research team got access to members of the LNBB. By this support we could gain five high level managers of logistics companies in the region Berlin-Brandenburg for in depth interviews. One additional interview with an interest group was made to consider whether company oriented opinions differ to supposed collective opinions. The interview partners are listed in Table 1.

Table 1. Companies included in the interviews

Company	Internet presence for detailed information
Kroll Internationale Spedition GmbH	www.spedition-kroll.com/
Hermes Schnellverkehr Gebrüder Rieck GmbH & Co. KG	www.rieck-logistik.de
Lekkerland Deutschland GmbH & Co. KG	www.lekkerland.de
Integral logistics GmbH & Co. KG	www.integral.de
eCom Logistik GmbH & Co. KG	www.ecomlogistik.de
Fuhrgewerbe-Innung Berlin-Brandenburg e.V.	www.fuhrgewerbe-innung.de

All companies were guaranteed complete anonymity with regard to the contents of the survey. This ensured open communication in the interviews and independence from corporate policy goals or similar in the discussion. For this reason, the statements contained in the next chapter of the article are presented in such a way that it is not possible to assign them to companies (including the frequency of same answers). Against this background, please note that not all statements are equally supported or rejected by all companies surveyed. The evaluations should produce a spectrum of opinion in the logistics sector. Thus, the investigation in this article can serve as a preliminary study for a future quantitative investigation, for example, to clarify if there is enough potential for specific ACD use case in the logistics industry as a whole.

Logistics is a very inhomogeneous market as far as service integration, position in the value chain, customer profiles etc. are concerned. The business profiles of logistics companies vary accordingly. However, in order to make clear what kind of potential users of automatic vehicles are presented below, the following analysis is offered.

The survey took place in medium-sized companies in the Berlin-Brandenburg area. This is due to the members of the LNBB which could be gained for interviews. These companies are active in the area of general cargo forwarding with their own vehicles and warehouses, without their own vehicle fleet and warehouses, as consulting companies in the area of process planning, as logistics service providers in the function of intermediaries and as trade associations. To characterise the provision of services, the companies surveyed can essentially be described as companies in the following logistics submarkets (Schwemmer 2017):

General Land Transport: Transportation of dry and stacked goods from "ramp to ramp" or "door to door" with standard trucks and rail in the weight range from 3 tons to approx. 25 tons consignment weight. Transports take place in local or long-distance traffic as well as cross-border. Partial loads are organized in efficient multi-stop tours (milk run principle). The most strongly represented business model in this logistics submarket is depot-bound cargo traffic with the respective fleet at the location. At the location itself are typically the customers and the driver residences.

General Cargo Transports: Transports of individually labelled dry and stacked goods without special handling requirements in the weight range from approx. 31 kg (consignment sizes above the CEP market) up to approx. 2,500 kg, which are bundled and invoiced in regional depots (branches in Germany) and transported by land in non-specialised standard trucks and loading containers in national general cargo networks. The transit times within the network within Germany are typically 24 h–48 h and organized in depot-to-depot trips as well as the previous collection or delivery in the regional 'milk run' principle.

Consumer Goods Distribution and Contract Logistics Including Temperature-Controlled Transports: In this submarket, specialised transport, warehousing and value-added services are provided for the distribution systems of the consumer goods industry and the retail trade, in particular for food and other consumer goods as well as durable goods for private use. 62% of food is transported in the submarket. Special requirements are often the complete compliance and control of cold chains as well as a high demand on reliability of delivery.

With 66.1 billion euros, these three submarkets cover about a quarter of the total German logistics market volume (253 billion euros in 2017). In addition, only the core markets of the companies were named - in many cases logistics companies are active in more than one submarket. For example, 55% of the companies active in consumer goods distribution also serve the submarket of industrial contract logistics (Schwemmer 2017).

After providing the methodology and the description of the potential users of ACD covered in this survey, in the next session the outcomes of the interviews are shown.

3 The Results of the Interviews

The answers to the interviews were collected in a first step to identify several answers. Particularly frequent answers can therefore be classified as trends or an opinion in the industry. The questions and answers were then visualized in the form of mind maps. The following analysis of the business survey highlights four key issues: (1) the need for automation, (2) possible barriers to technology adoption, (3) use cases of automated vehicles and (4) conditions for technology adoption in logistics and transport services.

3.1 The Need for Automation Technology

The first topic of the evaluation is the question whether there is a need for automation of driving among the users. According to the surveyed companies, there are three main drivers for the automation of the driving process in logistics.

The most frequently mentioned driver is the prevailing shortage of truck drivers. Existing driving costs are generally close to retirement age and will therefore disappear from the labour market in the coming years. The number of young drivers is insufficient, as low wages and poor working conditions make the profession unattractive for career starters. In addition, the abolition of compulsory military service has also abolished a large driving school. Driving licences must now be obtained on the private market, which is another obstacle to entering the labour market. The high and rising demand for transport and the lack of drivers increase transport costs.

The second key driver is the targeted cost reduction that companies expect from automation. The potential for cost reduction is due to the savings in driver salaries. Increasing profit margins in the industry is a desirable effect of automation. Today the margins for standard logistics processes are very low and the possibilities for product differentiation are limited to additional services and prices. The reduction of process costs through automation and thus of production costs for logistics providers opens up a new window of opportunity for higher margins.

The third aspect is increasing efficiency. By switching from man to machine, companies expect to be able to plan their production more reliably and efficiently. Intralogistics in particular is currently being highly automated, as technology prices have fallen sharply over the past ten years and investment costs are therefore low. Business processes are also currently being digitized. To a certain extent, automation of the driving process is also a consistent penetration of automation technologies into logistical processes in order to achieve high process efficiency. Further improvements in terms of availability and flexibility result from the independence of the drivers. Better planning options reduce storage capacities, fewer vehicles are needed (especially to cover order fluctuations) and throughput times are shortened.

Figure 1 shows the answers to the question "What is the driver for integrating automated trucks into your business processes?". The frequently given answers can be seen in the extension of the connecting lines in Fig. 1: The bolder the line is displayed, the more often it was mentioned. For a better understanding of the answers and an insight into the opinion picture, some answers were accompanied by an original statement in the form of a quotation from the interviews.

This is why ACD technology is in great demand in logistics, as market developments exist that promote the need for a technological solution. The advantages of automated road transport in the logistics sector serve to counteract the lack of drivers, reduce costs and thus increase the profit margin and achieve greater process reliability.

3.2 Potential Obstacles for Technology Adoption

ACD is aimed directly at the job description of the truck driver. Some of the activities of truck drivers, however, go beyond the mere driving task. It can be stated that the drivers are responsible for the safety of the load and the vehicle. Customs clearance and container exchange are also part of the driver's tasks. On the first and last mile, the driver is responsible for the service to the customer, such as unloading the goods, transporting the goods to the warehouse or salesroom or accepting complaints. Although we saw a strong need for the ACD technology, it was necessary to discuss possible obstacles to the integration of the technology. Against this background, the companies were asked what they see as the greatest risks of automation and in which of their business areas automation is excluded.

It can be deduced from the answers that companies consider automation to be out of the question, especially in urban areas or for customers. This is due to the lack of infrastructure, which would also have to be created by the customer (the "dockable ramp" on the sender or recipient side of the freight), as well as the service task of the

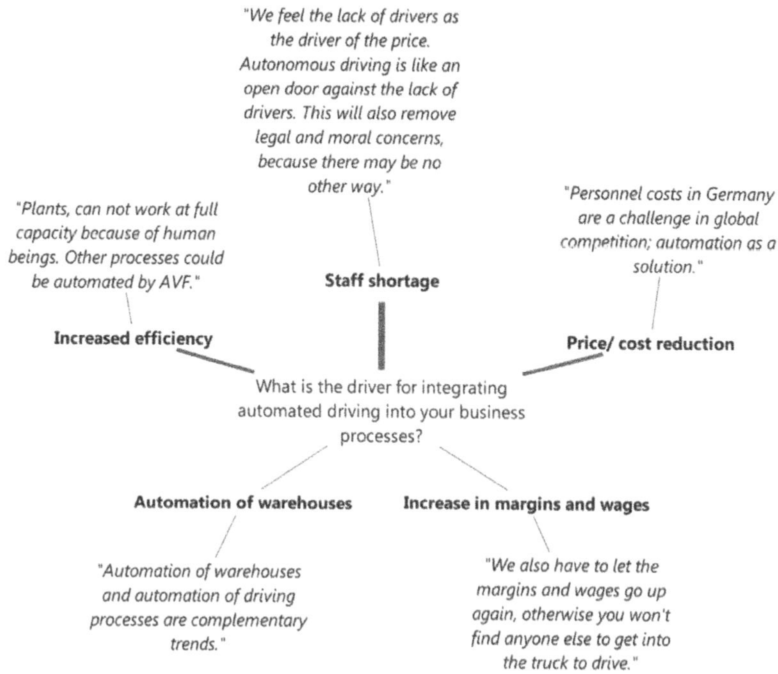

Fig. 1. Factors driving the need for automated trucks

driver on site at the customer. In the case of automated vehicles, the integration of interfaces must be taken into account. Special transports are considered problematic for automation because they require special monitoring and load securing, for which a driver would be indispensable.

One of the trends to be considered is that depots are increasingly establishing themselves in urban areas. The increasing trade via "eCommerce" platforms is also leading to a change in the industry orientation from industrial customers to private customers. This means that transport companies are increasingly operating within the "last mile". As a result, people's activities are becoming increasingly important, especially for the end customer. Automation is unlikely if a person performs a service activity on the tour or if special requirements are placed on the transport task.

The implementation of platooning (economies of scale) may result in customer restrictions in handling and storage capacity. Customers are prepared to handle a container or truck instead of two or four, as can be the case with a truck platoon.

The question of coping with so-called mixed traffic, i.e. when automated and conventional vehicles share the infrastructure, is also of great importance for companies. There is uncertainty as to how accident potential and congestion can be reduced to a minimum, e.g. without a lane for automated vehicles.

In cases where transport services are purchased from logistics service providers, automation is less important than with self-employment. The purchased service must remain the same, if necessary at a lower price. Not every freight forwarder will be able to invest in automated vehicles - but the market will set new prices if automated vehicles are used more frequently. Accordingly, customers will develop a new willingness to pay at the price level of automated vehicles.

Data security is also an important issue for respondents. On the one hand, this includes the expected protection against cybercrime and clear legal regulations on who may read which data and how (e.g. police, Federal Office for Goods Transport, etc.). On the other hand, automated vehicles will also be data suppliers. This is accompanied by the expectation that new companies will enter the logistics market and try to provide logistics services with available transaction data. Existing logistics companies must adapt to this new business model. However, it may result in small service providers in particular not being able to maintain a market segment.

The mind map, to the answers to the question "Where is the deployment of automated trucks not possible?" is shown in Fig. 2. In summary, while there are clear benefits of ACD technology, there are also general barriers for customers to take advantage of it. The main risks for technology are seen in the fact that interfaces are not integrated and that autonomous and non-autonomous vehicles will not have separate infrastructures.

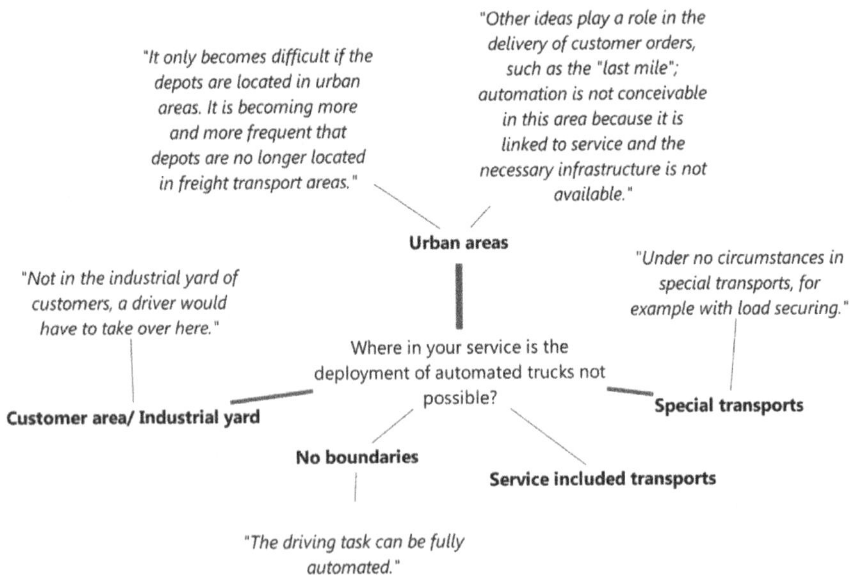

Fig. 2. The limitation of deployment of ACD trucks

3.3 Use Cases of Automated Vehicles

From the user's perspective on general drivers and barriers of ACD technology, it is important to know how the user is likely to integrate the technology into logistic processes and what features the technology is expected to provide. The companies were asked how they would envisage a logistics world with anyway automated trucks, what kind of automation would be advantageous and what impact this would have on the industry.

As a probable field of application for ACD, companies consider standardised transports, such as transports between depots and warehouses, to be realistic. The trips often take place at night and most depots are located outside the city limits. Many warehouses are already fully automated, which creates some efficiency pressure to automate the entire transport chain. In addition, the handling process is another conceivable and advantageous process for automation.

Particularly in the case of physical transport, the distance travelled is limited by the legally prescribed driving and rest periods. The use of ACD promises an increase in efficiency through higher utilization of the vehicles for service provision (not payload). Drivers today make decisions only on the basis of system information and existing contracts. For some of the respondents, the handling of the driving task by automated vehicles does not represent a restriction in this respect.

According to the companies interviewed, platooning in particular is regarded as a conceivable first application of the ACD. With the help of this technology, several vehicles can be coupled together. However, it should be stressed that in this case only the leading vehicle would be manned by a driver and the following vehicles would be electrically coupled trailers. This allows economies of scale such as order peaks or ‘an additional pallet’ that no longer fits on the truck to be handled. Thus the concept of platooning differs significantly from the previously often associated composition of vehicles on the track for shorter distances and relief of drivers. An extended aspect of this was to make the driver’s profession more attractive again by expanding the technical “know-how” - so to speak as a convoy companion. That there would be a driver in every vehicle has been ruled out as it negates the economics of an automated vehicle. The fact that the driver could undertake other activities during the journey is also excluded, as he is not qualified for this, the possibilities for further qualification would only exist in exceptional cases and existing specialist personnel in logistics companies would not have to be mobilised. Platooning has the already mentioned advantage of a higher security by the human instance on site. However, the problem with platooning is that many companies on the receiving side do not usually have the capacity to handle a convoy at the ramp. The availability of personnel at the ramp was also mentioned: “It doesn’t help much if the train is parked outside the industrial yard at four o’clock in the morning, but the unloading personnel don’t start work until six o’clock in the morning”. (one interviewee, literally translated)

Driverless vehicles are most advantageous in the industry as they are the most effective against driver shortages and rising transport costs. The ideas of a completely driverless vehicle and the transport of automatic transport machines as “trailers” (platooning) play a role here. Both systems pursue the core idea of counteracting personnel bottlenecks as far as possible and at the same time achieving cost savings. The completely driverless system offers advantages in terms of route planning and cost optimization. The technical maturity and the necessary infrastructure are decisive. An automated system does not have to be limited by driving times and rest periods. A cross-interface, automated transport chain could thus lead to considerable increases in efficiency.

Figure 3 shows the answers to the question "What is a beneficial scenario for the use of automated trucks?". Most important, therefore, is the realization that driverless trucks are needed by potential users and not automated trucks as mobile offices.

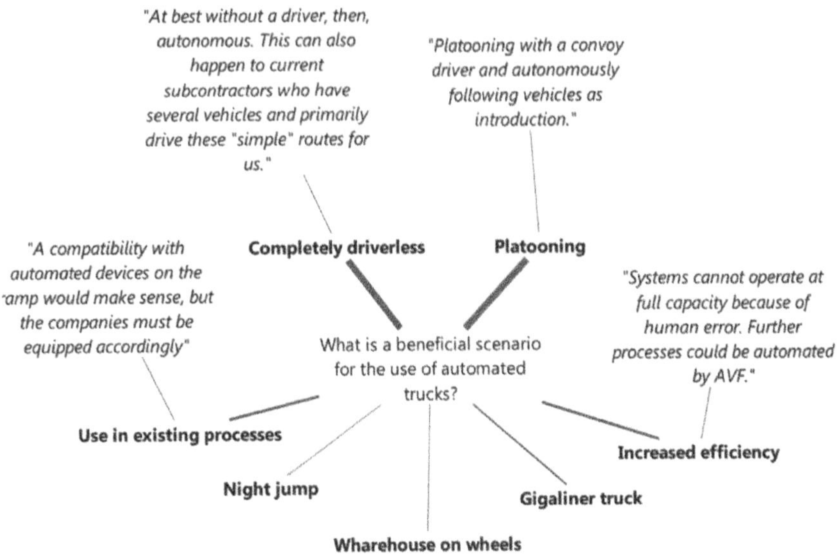

Fig. 3. Beneficial use cases of automated trucks

3.4 Conditions of the Technology Adoption in Logistic and Transport Services

For the establishment of autonomous trucks, the large e-commerce companies and global players in the logistics sector can be considered as pioneers. Market power, advanced digitisation of processes and the necessary capital enable them to integrate these vehicles into the processes most easily and to bear the learning costs for the new technology. The sector, which is characterised by medium-sized and small enterprises, lacks the financial resources to invest in innovation or to absorb bad investments. The guarantee of existing processes and the economic manageability of innovations is thus the starting point for every change in medium sized companies. The surveyed companies see a need for action in legal questions concerning the use of ACD. Politicians are therefore expected to adopt clear rules on liability and the adjustment of driving and working times. Other legal aspects are control sovereignty and data sovereignty.

The question of a possible adaptation of the infrastructure also remains to be clarified, in particular for mixed traffic (automated and conventional vehicles). In addition, the general conditions of use need to be clarified: Any temporal or spatial restriction would be an additional hurdle for the economic efficiency of ACD. The companies therefore want consistent support for the technology - the political process for the approval of the Gigaliner truck was cited as a negative example in some interviews.

The companies surveyed assume or expect that the technology will be available on the market in the near future, about 10 years. It was assumed that it would not be the technology, but the other framework conditions that could delay the broad application of ACD in logistics.

Figure 4 provides answers to the question “What legal, regulatory and financial framework conditions would you like to have for beneficial deployment?”. It becomes obviously that clear rules and framework conditions have to be designed which do not reduce the economic efficiency of the ACD technology and ensure the reliability of the logistics processes.

Fig. 4. Conditions of technology adoption in logistic processes

4 Implications for Technology Manufacturers and Transport Policy

The subjects of the strategy of automated and connected driving at present are infrastructure issues, legal aspects, technology development, connectivity aspects, IT security and data protection (BMVI 2015). Through interviews with managers of logistics and forwarding companies, we have discovered that these topics are important aspects for potential users. But what we also noted was a discrepancy between the vision of automated and connected trucks promoted by the manufacturing industry and currently being demonstrated on test fields, and the user-oriented automated truck expected by potential users. The mainstream vision of an automated and connected truck in the manufacturing industry is an automated truck with a driver, where the driver can use the technology for more comfortable working hours or business activities. A platoon is configured by several automated trucks, each with a driver on board. However, an analysis of the requirements on the user side shows that this scenario brings little benefit. The great potential for benefit for users will be unleashed when the driver is replaced by the technology, due to the increasing scarcity of drivers, increasing cost pressure and the growing need for efficient logistics processes, which are driving the demand for automated trucks.

The consideration of the demand side, the needs of the logistics and forwarding industry for the new technology is an important step to develop the technology user-oriented and to set the framework conditions accordingly. From the survey the following key points for a realistic application scenario of ACD in road freight transport can be summarised:

- There is a high demand and a strong interest in ACD in the logistics industry. ACD has the potential to solve one of the industry's most critical challenges: the increasingly dramatic shortage of drivers.
- The pioneering achievement can be expected from the global players in the industry and the e-commerce retailer. Although small and medium-sized enterprises show a high interest in the technology, the currently realisable profit margins, the entrepreneurial risk calculation and the learning costs generally do not allow these enterprises to play a pioneering role.
- ACD technology is expected to initially be utilized to (cost) optimize existing processes. An expansion of services or a significant reorganisation of logistics processes is therefore not to be expected in the short to medium term.
- The more standardized the logistics processes are, the more likely it is to use ACD. This opens the native application potential for the technology in the logistics submarkets served by hub and spoke network. On the first and last mile, where customer service has a high priority and is becoming increasingly important for the business development of logistics companies, the use of ACD technology is to be ruled out as a driver or customer personnel would still be required. The decisive reason for this is that in this case the costs of the technology and the benefits are at best the same.
- The foreseeable field of application for ACD technology is the tour between depots resp. logistic centres. This concerns also the distance between the motorway and the depot resp. logistic centre.
- It is unrealistic for the driver to take over other business activities while driving. Tasks that were previously performed by specialist staff in the offices of logistics companies could only be transferred to the driver with a considerable amount of retraining. However, the qualification potential of today's drivers, the lack of drivers and the knowledge and experience of existing specialist personnel contradict this notion.
- This means that driverless vehicles must be available for use on federal motorways and beyond in order to offer advantages in logistics. Vehicles in which the driver would quasi have a mobile office would not be user-oriented. Against this

background, autonomous trucks are expected by users. This requires not only technical development, but also the political and legal framework conditions to make corresponding products possible.

- The use of autonomous long-distance trucks and the facilitation of platooning would increase the benefits. However, platooning in the sense of the interviews differ from the general definition of platoons. Beneficial would be if vehicles do not drive with drivers at reduced distances, but that a platoon leader can move several electronically coupled trailers (driverless). This would allow absorbing peak order volumes and achieving extended economies of scale.
- The companies surveyed did not show a long-term scenario for ACD deployment. This can be explained on the one hand by the pressure of short-term day-to-day business and the lack of strategy and R&D units in small and medium-sized enterprises. On the other hand, the required technology is not available, i.e. it is not testable and relatively abstract for the players in the operative business. In addition, innovation scenarios such as long trucks and electric mobility have also left traces of discrepancies between political decisions, technical implementation and practical application. It therefore cannot be excluded that new services and applications for ACD will be developed in the long run. However, the actors initially concentrate on the first step of innovation, which can be reflected from a business perspective.
- The advantageous deployment scenario - fully automatic (long) trucks on the motorway without a driver - requires not only clear political support with regard to the framework conditions (adaptation of the Vienna Agreement, data protection, customs issues, etc.), but also an extended catalogue of requirements for vehicle manufacturers so that user-oriented products can be launched on the market in the next decade.

The concluding point of this paper is that the perspective of potential users opened new and different issues to what is in the scientific and public discourse so far. Hence it is very important to involve the users for "technology-success" in future by broadening the public discourse on automated and connected driving. It moreover needs further research to specify and quantify the qualitative research done in this paper to obtain representative status.

Acknowledgement. The project team would like to take this opportunity to thank the interview partners and the companies for their valuable time, the openness in the conversation and the pleasant atmosphere. We have collected a lot of valuable information and for the first time we have been able to create an opinion poll that show the potential user side of automated vehicles. With your support, these findings can provide valuable impetus for user-oriented vehicle development and policy setting. Ultimately, a contribution to the social discourse on the automation of road traffic is also possible.

The project team wants to thank the German Federal Ministry of Transport and Digital Infrastructure (BMVI) which promoted the project ATLaS in their research programme on "Automation and Connectivity in Road Transport".

Finally, the authors want to thank the full project team (LogistikNetz Berlin-Brandenburg e.V. (LNBB), Technische Universität Hamburg-Harburg, Dronemasters Future Mobility Initiative) for discussion, support and fruitful collaboration.

Appendix

Prepared questions resp. aspects for the interviews (translated from German)

Company-specific starting point

1. What is the specific logistic service of your company?
2. Where in the value chain is your company located (predecessor, successor)
3. What transport logistics structure does your logistics have?
4. In which processes are the driver involved and what activities does he perform?
5. Is the business segment (new services, services, products) currently being changed?

Company-specific expectations

6. Where in this logistics chain/your service is an automation of the driving task conceivable?
7. What would this automation look like in concrete terms (driverless, vehicle attendant, centrally controlled)?
8. What role would be given to the driver (Other/extended tasks)
9. Where in this logistics chain/your service is automation excluded? What's the matter with you?
10. Can new services be developed by automating the driving task?

Enterprise-specific scenario

11. What is a useful scenario for the use of automation technology?
12. What improvements do you expect in this scenario? (tours, costs, quality…)
13. What risks do you expect for this scenario? (compatibility, market position…)
14. What legal, regulatory and financial framework conditions would you like to see for this scenario (and its introduction)?
15. When is the realisation of this scenario realistic?
    a. From your point of view, is the required technology available for the outlined scenario?
    b. Which diffusion pathway do you expect?
16. What is the driver for integrating automation technology into your business processes? (Competition?, optimization?, new service possibilities?)

Industry-specific aspects

17. Who will take the early adopter role?
18. To what extent will the automated truck (as defined above) become the standard in the logistics industry?
19. How would this change the logistics industry? (logistics chains, actors, tasks)
20. Which vehicle technologies, apart from automation, are other future-oriented for your business? (electrification of vehicles, software, hardware, aerodynamics)
21. What would you expect from vehicle manufacturers as the next innovation?
22. What other technological innovations do you expect to change your logistics processes and structures?
23. Which aspects that are important to you have not been discussed?

References

BMVI: Strategie automatisiertes und vernetztes Fahren - Leitanbieter bleiben, Leitmarkt werden, Regelbetrieb einleiten. Bundesministerium für Verkehr und digitale Infrastruktur, Berlin (2015)

Schwemmer, M.: Top 100 der Logistik: Marktgrößen, Marktsegmente und Marktführer. Jahrgang 2016/2017. Studie der Fraunhofer Arbeitsgruppe Supply Chain Services. DVZ (2017)

The Impact of Driverless Vehicles on Transportation Costs in Road Haulage: A System Dynamical Approach

Sandra Lunkeit(✉), Heike Flämig, and Kerstin Rosenberger

Institute for Transport Planning and Logistics,
Hamburg University of Technology, Hamburg, Germany
{sandra.lunkeit,flaemig,kerstin.rosenberger}@tuhh.de

Abstract. Costs are a decisive competitive advantage in the logistics industry. For this reason, companies are constantly striving to reduce these costs. Driverless driving offers an opportunity to support this endeavor and also to counteract the increasing shortage of drivers. However, the interactions between logistics costs and driverless vehicles in road haulage that exist have not been considered much at the moment. The aim of this study is to identify the interactions in the system of automated driving with logistics costs by using a system dynamical approach. The use of a system dynamics approach serves not only to carry out constant calculations, but also to record dynamic changes in the system. For this purpose, a theoretical and a simulation model are developed to identify qualitative as well as quantitative impacts in the system of driverless road haulage. The theoretical model shows not only the elements that are relevant in the system, but also possible rebound effects that can stabilize or even destabilize the system. In addition, the simulation model serves to take a closer look at the individual costs that can be relevant for transportation. Both models showed possible advantages and disadvantages that can occur due to automated driving. It is therefore important to put the elements of a system in an overall context to understand the dependencies between them.

Keywords: Automated driving · Logistics costs · Effects · Freight transport

1 Introduction

The logistics industry is characterized by strong competitive pressure, which is becoming a challenge especially for small and medium-sized companies. Among others, costs are therefore an important success criterion in logistics, more specified, the provision of a desired service at the lowest possible price. By implementing autonomous driving in road haulage, the biggest cost driver, personnel costs, could be reduced or even eliminated. In addition, the technology is considered potentially safer than the human driver (Federal Statistical Office 2015). While the driver is strongly influenced by emotions, monotony and fatigue, which causes accidents, the technology works according to the same technical principle. By implementing autonomous driving, savings up to 47% of current transport costs are being forecasted in 2030 (Nowak et al. 2018).

U. Clausen et al. (Eds.): ICPLT 2019, LNLO, pp. 116–127, 2019.
https://doi.org/10.1007/978-3-030-13535-5_9

Engblom et al. (2012) examined self-reported logistics costs of manufacturing and trading companies operating in Finland and the differences and interdependencies of these costs. The study showed that there is a need for caution in interpreting changes in logistics costs. Furthermore, they identified the need for controlling the effects of background variables. Additionally, Stępień et al. (2016) focused on the problem of identification and measurement of logistics cost parameters in the accounting information system in production companies. They tried to learn and assess in interdependencies between the identification of logistics costs in the analyzed areas and the efficiency of management. Both papers focused on the interdependencies between logistics costs and other parameters. However, the interdependencies and effects of the individual logistics costs and how they affect the system of driverless vehicles in road haulage have so far been not enough considered. As the mentioned papers showed, interdependencies between background variables have to be considered in the analysis as well as just the obvious logistics costs. Of course, personnel costs have a huge impact when talking about driverless driving in freight transport but there might be other (cost) variables which furthermore have an important impact on the system.

The aim of the study is therefore to depict a system of influencing factors of automated driving and in particular to work out the interdependencies with logistics costs. The value of this study is in both the presentation of the structure of effects and in a deep understanding of the effects of costs within road haulage. For logistics companies, the results can be a way to positively influence costs and reduce not only personnel costs but also other cost factors. The focus here is particularly on transportation costs, as these are of great importance for logistics companies and have a fundamental influence on the adaptation of the technology. If autonomous driving does not provide any cost or benefit advantages, the transport companies are very likely to reject it. Only if the introduction of driverless driving can bring benefits to companies it is possible that the technology will be enforced.

For doing so, a theoretical system dynamics model is developed. The advantage of mapping the variables using a dynamic model is that the interaction of different parameters can be mapped in a simple and understandable way that is normally in a complex context. In addition, it provides the ability to incorporate changes over time into the model. The simple and intuitive models do not require precise and numerical values to map the variables, but are intended to represent possible dynamical trends (Meadows 1980).

In this study, a dynamic model was selected because autonomous driving in road freight transport has not yet been implemented. The model should therefore represent potential variables that could influence transport costs in a simple and understandable way. The study shows one possible set of variables, but other scenarios are conceivable. The extension by a mathematical simulation model serves to map the potential changes caused by driverless driving on the transport costs also in quantitative terms.

To create the model, system variables have to be identified. By using the sensitivity analysis according to Vester (1991) a theoretical system can be created. Hence using a dynamic model of autonomous driving in freight transport helps to identify and to interpret possible rebound effects. Including data, a mathematical model can be developed that shows quantified tendencies of the system.

Based on these methods, changes in costs due to autonomous vehicles as well as their development over time can be named.

2 System and Subject of the Study

To analyze the impact of autonomous driving in road haulage it is necessary to define a research system. The definition serves to give a better understanding of the research subject, and also to differentiate the research subject from other systems. According to Flämig et al. (2002) and Bossel (2004), a system consists of elements and relations which lead to a system purpose. The definition can be divided into a factual, spatial or temporal view of the system (Kühn 2016).

The factual view on the system is the external logistics chain itself. The chain forms the typical tasks of logistics - transport, transshipment, storage - which take place from a source via a logistics node to a destination. In this study the transportation process is focused. The system consists of individual elements that are connected by relations based on Flämig et al. (2002). The freight is transported by driverless trucks from one facility to another one via a logistics node on the road. The system is impacted by the volume of transport and the market penetration of driverless trucks. The overall spatial view is on the Federal Republic of Germany.

The temporal view is future-oriented. Driverless driving is currently not legal in Germany and other European countries. By changing the Vienna Convention on Road Traffic in 2016 the use of automated systems to control the vehicle has been legalized, but a driver is still required who is able to take control of the vehicle at any time. The temporal view on the system is therefore set from day one of legalization of autonomous driving up to 50 years. In this case it is assumed that driverless driving is legal in Germany and all other countries.

Due to the fact that the system is based on the future and little information is given, data from other countries than Germany is used to create a most accurate picture of the future.

The subject of the study is on transportation costs. According to Wittebrink (2011) transportation costs are all personnel, material and imputed costs which incurred for the transportation of goods. These include, for example, costs for driver, fuel, service, repair, tires, insurance and depreciation. Furthermore, he differentiates between fixed truck costs (depreciation, taxes, fees and miscellaneous) and variable truck costs (fuel, tires, repair, maintenance) as well as personnel costs with ancillary costs.

Among others, technology costs and safety costs are important in the case of automated driving. Technology costs are currently not enough included in transportation costs but might be of importance for investment costs as they rise with the level of automation. The higher the level of automation the more technology is installed in the vehicle. Safety costs are also special in the case of automation. Normally they are not included into cost consideration, although the loss of productivity (in this case: loss of time and resulting delays at the recipient or delivery failures) and damage costs might affect the logistics companies. Due to automation the flow of traffic might improve and accidents decrease as technology is potentially safer than the human driver.

In this study the transportation costs as well as technology and safety costs are included in the consideration.

3 The Theoretical Approach

A system dynamical approach is used to analyze transportation costs in the system of driverless driving in road haulage. The system consists of several elements which are related to each other and displayed in a causal loop diagram (see Fig. 1). This diagram is created based on the sensitivity analysis according to Vester (1991). The identified elements of the system are shown in Table 1.

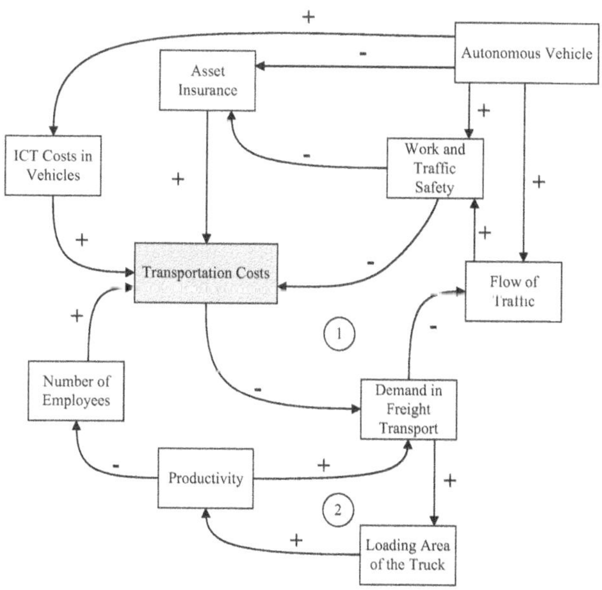

Fig. 1. Causal loop diagram of transportation costs in the system of driverless road haulage

The elements are defined within an iterative process of experts and based on a literature review. The analysis showed the following findings: The transportation costs (marked in grey in Fig. 1) are impacted by information- and communication technology (ICT) costs in the vehicle, asset insurance, work and traffic safety and numbers of employees. These variables have an effect on the amount of transportation costs. At the same time, transportation costs have an impact on the demand in freight transport.

The Fig. 1 shows the direction of impact and the expression of the effect. In example, an increase in work and traffic safety leads to a decrease in transportation costs, as there are less accidents on the road and in the working environment. It is assumed, that all variables have an increasing effect, e.g. the increase in transportation costs results in a decrease in the demand in freight transport. As this study focuses autonomous driving in freight transport, a variable of autonomous vehicles is also

Table 1. Elements of transportation costs in road haulage

Element	Definition
Asset insurance	Sum insured for vehicle and cargo damage
Information- and communication costs in the vehicle	Costs for technology for internal and external communication and provision of information
Autonomous vehicles	Number of driverless vehicles on public infrastructure
Work and traffic safety	Number of accidents in industrial and traffic environment
Flow of traffic	Road traffic flow in terms of congestion and average speed
Demand in freight transport	Quantity of goods in demand in road traffic
Loading area of the truck	Volume of loading area
Productivity	Utilization of trucks (in volume)
Employees	Number of employed truck drivers
Transportation costs	Sum of shipping costs and total cost of ownership

included in the diagram. This variable has an indirect effect on transportation costs through ICT costs in vehicles, asset insurance, work and traffic safety and flow of traffic. Furthermore, the causal loop diagram shows feedback loops and therefore, possible rebound effects. One feedback loop has a stabilizing effect on the system (see no. 1 in Fig. 1). With an increase in transportation costs, the demand in freight transport is decreasing which leads to an improvement in the flow of traffic. This results again in an increase in work and traffic safety with a positive effect on transportation costs. Because of the mixture of strengthened and weakened effects, the system is balanced. However, a destabilizing feedback loop is identified in the system as well (see no. 2 in Fig. 1). An increase in the demand for freight transport leads to an increase in the loading area of the truck, because the vehicle has to transport more goods. This results in an increase in productivity and again in an increase in the demand in freight transport. This feedback loop swings up to a certain limit value and can lead to negative effects on the whole system.

To sum it up, the causal loop diagram shows the interdependencies in the system of transportation costs in driverless road haulage. Based on these findings possible guidelines for the behavior in the system can be given, e.g. the change of one variable might lead to a change of other impacted variables.

4 The Simulation Model

The causal loop diagram shows the interdependencies of the system and how the variables interact with each other on a qualitative level. To get a measureable impact of the variables a simulation model is created by using the software VENSIM. As mentioned before, the model is developed as a dynamic model. The aim of system dynamics is to gain an understanding of the dependency between structure and behavior. It starts with the structural origins of a given temporal development and follows with the behavioral consequences of changes in the system (Randers 1980).

A mathematical simulation of a dynamic system offers an opportunity to build up specific interactions and quantify them. Since no concrete changes due to autonomous driving could be identified at present, this mathematical model should help to show a possible tendency.

The following simulation is developed from a baseline scenario. The baseline scenario has the aim to show the system at its' status in year 2015. To display the quantified effect of autonomous driving in road freight transport a variable with market penetration of automated driving is added and a second model is developed.

In comparison to the causal loop diagram, other variables or more specified, costs, had to be created and calculated to show the quantity effects of autonomous vehicles in freight transport on the behavior of transportation costs. The Fig. 2 shows the structure of calculation to gain the development of transportation costs over time. The origin variables "ICT costs in vehicles", "number of employees", "work and traffic safety" and "flow of traffic" had to be transferred to indicators to measure them. The variable "ICT costs in vehicles" is calculated by technology costs. They describe the costs for implementing automation technology in the vehicle. The number of employees is transformed to drivers' costs, to see how a change in the number of employees influences the overall transportation costs. The variable asset insurance is calculated by insurance costs. Work and traffic safety is split into safety costs and repair and maintenance costs which focuses on the one hand potential costs through accidents and on the

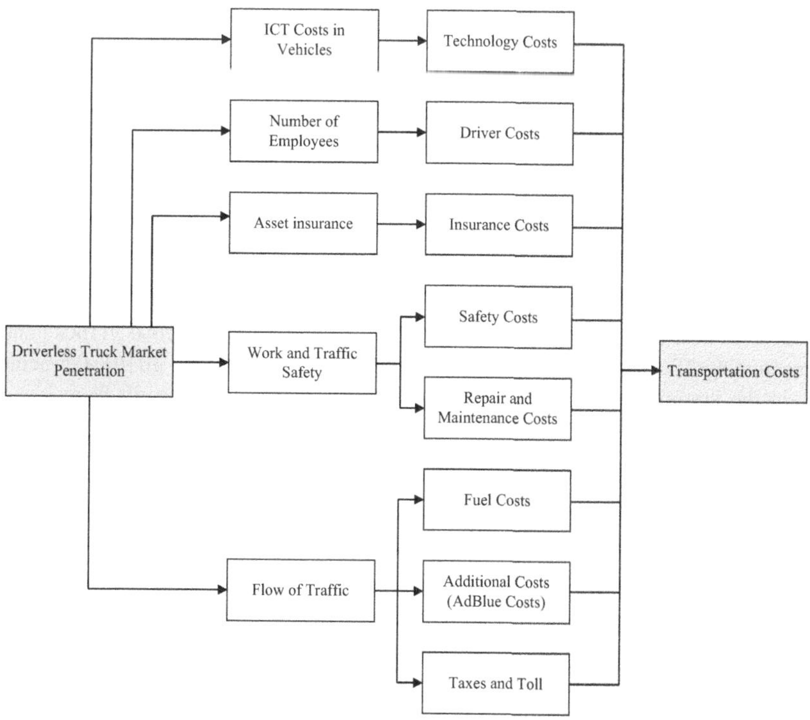

Fig. 2. Structure of the simulation model

other hand recurring costs as repair and maintenance. The variable flow of traffic is built up by fuel costs, additional costs for driving as costs for AdBlue as well as taxes and toll.

The combination of all costs parameters result in transportation costs. By adding the variable "driverless truck market penetration" to the baseline model, the influence of a percentage increase in market penetration of driverless commercial vehicles over time can be identified. Therefore, the model is run first without any automation influence and second by including the market penetration to compare the baseline scenario to an automation scenario.

5 Assumptions for the Simulation Model

To compare the current development to a possible world with automation two models are built, one baseline scenario and one automation scenario. For both scenarios specific assumptions had to be made.

The baseline scenario shows the current development without automation in Germany. It is assumed that the number of vehicle (188,481) is increasing with a rate of 4% each year. This development is based on data from 2010 to 2017 from the Federal Motor Transport Authority (Federal Motor Transport Authority 2018).

The average travelled kilometers per truck were 100,899 km in 2015. This number is decreasing with a rate of 1.1% per year (Federal Motor Transport Authority 2016). The current development shows that the number of vehicles increases while the travelled distances are getting shorter.

According to the Federal Employment Agency (2013), 546,479 truck drivers were employed in Germany in 2015, with a decreasing tendency of –1% per year. For the automation model a function is used to build up the number of employed drivers. The initial stock decreases over time as more automated vehicles enter the market.

To calculate the insurance costs, the net list price of a vehicle has to be determined. Therefore, the investment costs (102,000 Euro) as well as an impact factor (1.53%) have to be multiplied. The insurance costs are assumed as 1.5% from the net list price. The impact factor and the investment costs are also necessary to calculate the annuity costs for a vehicle (Wietschel et al. 2017).

Accident or safety costs also influence the total transportation costs. Trucks caused approximately 10,414 accidents in 2015 (Federal Statistical Office 2016). Per accident costs in the amount of around 33,000 Euro arise. The number is built from lost productivity and damage costs (Zaloshnja and Miller 2007). It is assumed that repair and maintenance costs are 0.143 Euro per kilometer travelled.

Fuel costs are built up by fuel consumption (0.311 L/kilometer) and the price per liter (1.17 Euro/liter) times the travelled kilometers of the truck. In Germany AdBlue is required for trucks, so additional costs (AdBlue: 0.006 Euro/kilometer) are calculated as well as taxes and toll. Taxes are assumed as 0.01 Euro per kilometer travelled, the toll rate is assumed as 0.2 Euro per kilometer (Wietschel et al. 2017).

For the automation scenario, it is assumed that all vehicles that enter the market are driverless. This is made for the reason that just driverless vehicles offer a realistic scenario for logistics companies. Logistics companies reach for a cost minimization in all operational costs. Just by eliminating the driver a visible change might occur. The implementation of the driverless vehicles is assumed according to Ginsberg (2017) in a S-shaped growth with less growth in the beginning and a strong increase after 25 years. Almost full market penetration (99%) occurs after 50 years as shown in Fig. 3.

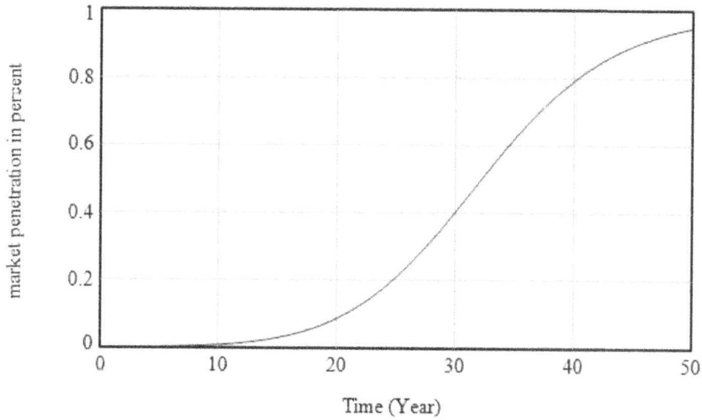

Fig. 3. Growth of market penetration of driverless trucks

The market penetration curve is impacting accidents per year, number of drivers, insurance costs, travelled kilometer per truck, fuel consumption as well as technology costs.

It is assumed that through automation the number of accidents is decreasing according to the development of market penetration of autonomous trucks. The same applies to the number of drivers. Insurance costs are as well changing according to the market penetration. All three variables are decreasing.

Furthermore, it is assumed that travelled kilometers of trucks are increasing through automation over time. This assumption is based on the fact that due to automation, rest periods are no longer necessary and therefore more kilometers can be covered in one trip and thus in total.

The fuel consumption is changing depending on the travelled kilometers. As the driven kilometers are increasing, the fuel consumption is increasing as well, although the consumption per liter is reduced from 0.311 to 0.276 L per kilometer due to a better fuel efficiency through automation.

Technology costs are the costs for automation technology. To implement automation technology in the truck, costs of approximately 14,200 Euro per truck arise. The technology costs are just included in the automation model.

6 Results of the Simulation

The results of the simulation have to be separated into a business and an economic view. The business view is just on one vehicle and strictly speaking on operational costs. As already mentioned in the assumptions, today's development is determined by a decreasing trend in truck drivers. This trend might be reinforced by autonomous driving and costs are almost completely reduced over time which would be a huge benefit for logistics companies. Although the results in Fig. 4 display the development for all drivers in Germany, the costs can be broken down to just one vehicle. The relation between the baseline and the automation scenario would be the same. The same effect can be shown with all other operational variables as fuel costs, insurance costs, taxes, toll, repair and maintenance and annuity costs. The focus on just one truck and the influencing costs show a positive development for the logistics companies.

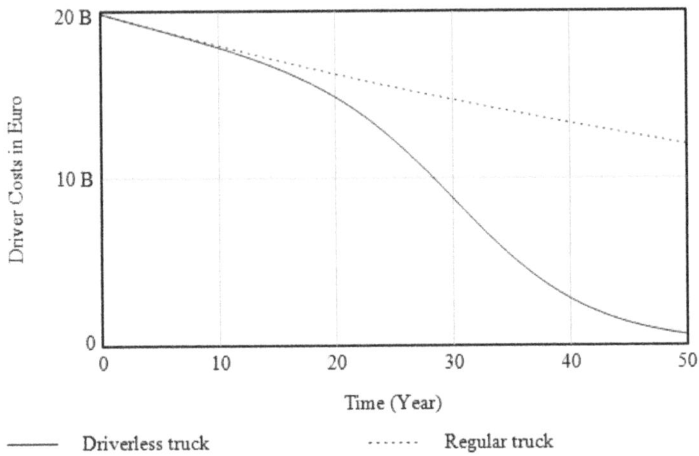

Fig. 4. Development of driver costs

In comparison, the overall economic view displays that the increase in travelled kilometers and the growing number of vehicles on the streets lead to the fact that the advantages are relativized and lead to rising overall costs. The positive effects of the decreasing driver costs are too small in comparison to the significant increase in fuel costs due to the rising travelled kilometers. Although the fuel consumption liter per kilometer is less than in the baseline scenario, the rising number of travelled kilometers per year is too high. Therefore, the biggest cost driver is still fuel costs. Figure 5 shows the total operational costs for the baseline and the automation scenario.

This result permits the consideration that automated driving only offers economic cost advantages if the type of drive of the vehicle is also reconsidered. In conclusion, driverless vehicles in road haulage have to be e.g. electrified to reduce fuel costs. This would not only result in cost advantages, but also in ecological benefits.

In comparison, if the assumption that driverless vehicles lead to more kilometers being driven each year is changed so that the same annual mileage is used as in the baseline scenario, driverless driving would lead to overall cost savings as seen in Fig. 6.

The simulation showed that by implementing driverless vehicles without changing the type of drive and by a further increase in travelled kilometers, the total operational costs are higher for all vehicles (total) in the automation scenario than the costs in the

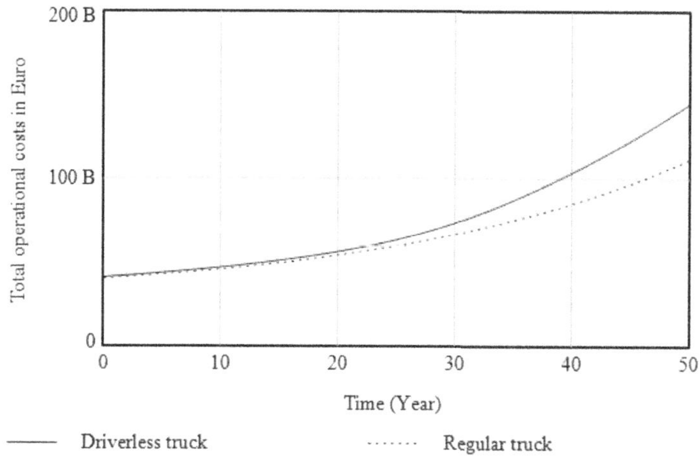

Fig. 5. Development of total operational costs with an increasing annual mileage

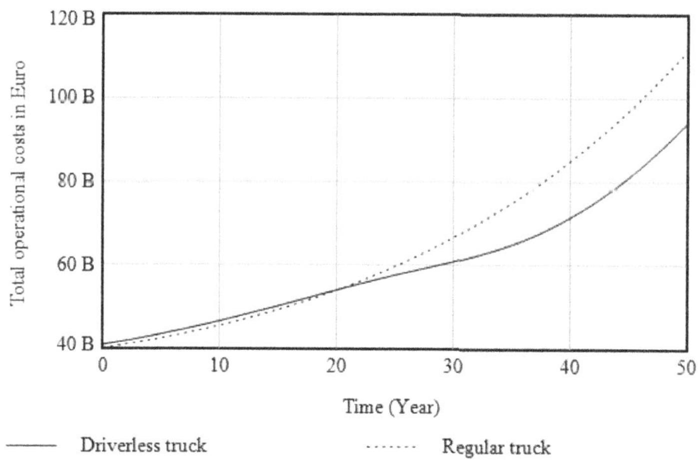

Fig. 6. Development of total operational costs with a constant annual mileage

baseline scenario after 50 years. This result is based on the increasing demand for road freight transportation which is expressed in the model by rising travelled kilometers. If only the individual vehicle (total cost per vehicle) is considered, however, costs are reduced.

To compare the simulation model to the theoretical one, the type of drive has to be included in the consideration. Furthermore, if transportation costs increase as shown in the simulation by using autonomous vehicles the demand for road freight transport might decrease and what's more, all depending variables are impacted. If transportation costs would decrease, the opposite effect would occur and the demand would rise. This could have a destabilizing effect on the system as a whole.

7 Conclusion

The analysis of the dependencies on automated driving and transport costs showed that cost savings could be expected, but would not necessarily have to be at a business and an economic level. At the economic level, however, it was discovered that driverless driving does not enable any savings to be made in the current development of transport volumes. The increasing transport volume and thus the travelled kilometers outweighs the advantages of driverless driving and leads to further increases in costs, which are even increased by driverless driving.

Overall, it should be noted that the results of the simulation depend strongly on the assumptions made. However, the simulation has shown a direction that can be caused by changing certain variables.

The system dynamical approach showed that driverless vehicles in road haulage might lead to a disadvantage in the transportation market which cannot be identified by just analyzing on a business level. The variables have to be considered in an overall context to understand the dependencies between them as the development of one variable can affect the whole system in different ways. In the end, the findings can be used for political measurement as the biggest problem is possibly the increasing volume of traffic. In order to continue to achieve economic advantages in the transport market while at the same time introducing automated driving in freight transport, special framework conditions must be created. On the one hand, this could lie in the introduction of alternative drives in road haulage or in the limitation of travel times or routes by politics. However, automated driving might be the only way to counteract the increasing shortage of drivers in the logistics industry in order to remain competitive.

References

Bossel, H.: Systeme, Dynamik, Simulation. Modellbildung, Analyse und Simulation komplexer Systeme. Books on Demand, Norderstedt, Germany (2004)

Federal Statistical Office: Verkehr. Verkehrsunfälle 2014. Fachserie 8 Reihe 7. Wiesbaden, Germany (2015)

Federal Statistical Office: Verkehrsunfälle. Unfälle von Güterkraftfahrzeugen im Straßenverkehr 2015. Wiesbaden, Germany (2016)

Engblom, J., Solakivi, T., Töyli, J., Ojala, L.: Multiple-method analysis of logistics costs. Int. J. Prod. Econ. **137**, 29–35 (2012)

Federal Employment Agency: Beschäftigte nach Berufen (KldB 2010) (Quartalszahlen). Nuremberg, Germany (2018)

Zaloshnja, E., Miller, T.R.: Unit Costs of Medium and Heavy Truck Crashes. U.S. Department of Transportation, Federal Motor Carrier Safety Administration (2007)

Flämig, H., Sjöstedt, L., Hertel, C.: Multimodal transport: an integrated element for last-mile-solutions? In: Proceedings of the International Congress on Freight Transport Automation and Multimodality: Organizational and Technological Innovations, Delft, pp. 1–17 (2002)

Ginsburg, R., Uygur, A.R.: Changing Technology in Transportation: Automated Vehicles in Freight Final Report. University of Illinois, IL (2017)

Federal Motor Transport Authority: Verkehr in Kilometern der deutschen Kraftfahrzeuge im Jahr 2015. Flensburg, Germany (2016)

Federal Motor Transport Authority: Bestand in den Jahren 1960 bis 2018 nach Fahrzeugklassen. Flensburg, Germany (2018)

Kühn, A.: Ökologische und ökonomische Effekte einer Regionalisierung der Automobilzulieferer – Ein systemdynamisches Wirkungsmodell. Ph.D. Thesis, Hamburg University of Technology. Hamburg, Germany (2016)

Meadows, D.H.: The unavoidable a prior. In: Randers, J. (ed) Elements of the System Dynamics Method. The MIT Press, Cambridge (1980)

Nowak, G., Viereckl, R., Kauschke, P., Starke, F.: The era of digitized trucking. Charting your transformation to a new business model. PwC Strategy (2018)

Stępień, M., Łęgowik-Świącik, S., Skibińska, W., Turek, I.: Identification and measurement of logistics cost parameters in the company. In: Transportation Research Procedia, vol. 16, pp. 490–497. Szczecin, Poland (2016)

Randers, J.: Elements of the System Dynamics Method. MIT Press, Cambridge (1980)

Vester, F.: Ausfahrt Zukunft Supplement. Studiengruppe für Biologie und Umwelt. Munich, Germany (1991)

Wietschel, M., Gnann, T., Kühn, A., Plötz, P., Moll, C., Speth, D., Buch, J., Boßmann, T., Stütz, S., Schellert, M., Rüdiger, D., Balz, W., Frik, H., Waßmuth, V., Rödl, A., Schade, W., Mader, S.: Machbarkeitsstudie zur Ermittlung der Potentiale des Hybrid-Oberleitungs-Lkw. Studie im Rahmen der Wissenschaftlichen Beratung des BMVI zur Mobilitäts- und Kraftstoffstrategie. Karlsruhe, Germany (2017)

Wittenbrink, P.: Transportkostenmanagement im Straßengüterverkehr. Grundlagen, Optimierungspotenzial, Green Logistics. Gabler, Wiesbaden, Germany (2011)

Standard Designs for Multifunctional Small and Medium-Sized Transhipment Sites

Dirk Bruckmann[1(✉)], Tobias Fumasoli[2], Stefan Schneider[3], and Martin Ruesch[4]

[1] Rhine-Waal University of Applied Sciences, Friedrich-Heinrich-Allee 25, 47475 Kamp-Lintfort, Germany
dirk.bruckmann@hsrw.eu
[2] Institute for Transport Planning and Systems, Swiss Federal Institute of Technology (ETH Zurich), Zürich, Switzerland
[3] Rapp Infra AG, Basel, Switzerland
[4] Rapp Trans AG, Zürich, Switzerland

Abstract. One of the major goals of the European transport policies is the modal shift from road to rail transportation. However, especially regarding the wagonload transport, this goal is jeopardised (hard to achieve). Due to a reduction in the number of private sidings and public loading points, the accessibility of the rail network is reasonably reduced. Thus, a new approach to counteract the reduction, especially for small and medium-sized transhipment sites is required. This paper develops a list of requirements for small and medium-sized transhipment sites based on shippers' demand. Further, based on these requirements, standard types of transhipment sites are defined. Additionally, for one of the standard type, a reference layout is also developed.

Keywords: Transhipment site · Rail freight · Single wagonload

1 Introduction

One of the major goals of the transportation policies of the European Union is a modal shift from road transportation to more environmentally friendly transport modes like railways or inland waterway shipping [1]. The freight transport volume in Europe exhibited an increase during the last decades due to the changing production strategies and the introduction of international free trade agreements. However, the transport volume of rail freight in Europe is showing stagnation during recent times. The market share for rail freight transport has been decreasing since the 1970s due to the changes in the types of goods transported (i.e. the shift from bulk goods towards manufactured goods) and new logistics strategies such as decrease in inventory, reductions in vertical integration of production and transport practices which place an emphasis on flexibility, reliability and smaller but more frequent deliveries. All these factors contributed to an increase in the modal share of road transportation. Within the rail freight sector in Europe, especially the wagonload based production schemes are largely at risk and are at the verge of vanishing completely [2].

U. Clausen et al. (Eds.): ICPLT 2019, LNLO, pp. 128–141, 2019.
https://doi.org/10.1007/978-3-030-13535-5_10

According to Guglielminetti et al. [3], the most critical issue for wagonload transport is the reduction of private sidings. Tracks in sidings and freight stations are removed resulting into the shippers' loose of the access to the rail freight. Thus, it becomes impossible for them to shift the transports from road to rail. Islam and Zunder [4] also argue that access points like terminals are a facilitator of a competitive rail service.

Thus, the crucial issues to keep rail freight competitive in future are new concepts for the design of transhipment sites. These new concepts need to fulfil the changing requirements of shippers as well as the operational requirements of the rail freight companies. Furthermore, the construction costs of these new transhipment sites should be cut down by modularisation and standardisation of their components because of the fact that other sectors of construction industry already recognise the benefits from those standardisations [5].

The development of standardised modules may help the shippers as well as the planners of railway loading and transhipment infrastructures to design and to build future proof facilities. From an academic point of view the research shall gain a more in-depth knowledge about the functional requirements for those transhipment sites. The functional requirements shall be standardised and transferred into standardised modules. The application of the modules in case studies shall approve the usability of standardised transhipment modules under real world conditions. From a practical perspective, guidelines for a catalogue of standardised transhipment modules shall be provided. The application of these standardised modules shall facilitate the design of new as well as the redesign of the existing transhipment sites.

According to the forementioned academic and practical objectives this study aims to answer the following research questions:

- What are the future functional requirements of the shippers and the rail freight companies for small and medium transhipment sites?
- Is it possible to transfer these functional requirements into a restricted number of standardised module designs?
- Is it possible to apply the modules to design a transhipment site under real world conditions?

2 Definitions

For further clarifications, the basic terms in this context need to be defined. These definitions are as follows:

- The term (freight) transhipment is defined as those facilities in which goods are transhipped between different transport modes - multimodal and/or intermodal (between road, rail and inland waterways).
- During transhipment, goods can either remain unpacked in the loading units of a combined transport (container, swap bodies, semi-trailer) or in other transport containers (pallets, sacks, wheeled boxes, skip/dumpster etc.), or they can be pumped in, blown in, transferred or poured out/dumped.

- Small transhipment sites have a capacity of up to 20 freight wagons per day, medium transhipment sites have a capacity of between 20 and 80 freight wagons per day.

3 Literature Review

The topic of the design of small and medium transhipment sites becomes more relevant as the structure of goods in Europe shift from mass and industrial products to consumer goods; the integration of rail freight in urban logistics processes becomes more and more relevant [6]. Furthermore, the integration of rail freight in urban logistic chains is also required. Present measures often address the terminal infrastructures, their capacities and locations. The remaining private sidings mainly focus on the supply of goods to large industrial sites.

New transhipment sites and terminals of combined transport are designed for an efficient transhipment process of long distance full-train loads, and primarily for combined transport. Based on this focus, most of the recent guidelines deal with larger transhipment sites. On an international level the guidelines of DB Netz [7] and Arendt [8] are implemented and on the Swiss level the studies about the new terminals in Switzerland by Ickert et al. [9] must be mentioned.

Nevertheless, the distribution of wagonload shipments in a distribution network can be competitive with road transportation. Bruckmann et al. [10] proved this for container distribution in Swiss sidings. However, to increase the number of wagonload networks, accessibility to the network must be ensured. This requires new approaches in the design of small and medium sized transhipment site. Yet, their design principles are only mentioned in the context of new transhipment technologies of combined transport [11, 12]. Very little literature is available on conventional transhipment technologies in the context of rail freight.

4 Structure and Methodology

The paper starts in Sect. 5 with a brief description of the todays' demand and supply structures of rail freight in Switzerland to show the current status of rail freight in Switzerland. Thus, Sect. 5 describes the starting point of the research. Section 6 analyses the future market requirements based on the analysis of the best practices from the current transhipment sites and a macro analysis of the future demand structures and expert interviews with shippers. Section 7 derives a set of different standardised transhipment modules from the functional requirements. In Sect. 8 the requirements are combined into six standard transhipment modules. Keeping the length restriction of this paper in mind only one example with the full design is integrated in this paper. In Sect. 9 the conclusions regarding the research questions are provided.

5 Current Situation of Rail Freight in Switzerland

The first step to assess the current Swiss situation was to analyse the current structure of private sidings and public loading platforms for rail-freight (in Switzerland named as "Freiverlad"). Public loading platforms are provided as a part of the open access network, which allow any shipper to use these access points. The current leading operator for SWL transport is SBB Cargo. The analysis of the status quo is based upon their transport and network data from the year 2015. In 2015 there were 386 stations, which were served within the single wagonload network of SBB Cargo. Most of the stations (242) are served regularly as part of the basic network. 86 of them were part of the customer specific network. They were served according to the specific demands of the shippers. 58 were in the networks of (narrow gauge) private railways, and 9 were outside Switzerland. 187 out of the 386 stations provided a public loading platform, and 145 of the stations provided the opportunity to tranship intermodal transport units. This provided a basic overview of the network structure and the relevance of public loading platforms in Switzerland.

According to the number of loaded and unloaded wagons, only 3% of the stations can be considered as large stations. 12% are medium sized stations, while most of the stations (85%) are small stations with a demand of fewer than 20 wagons per day.

6 Market Requirements

To find out more about the current demand structures and the future needs for small and medium-sized transhipment sites, the authors:

- Evaluated the best practices for transhipment sites in Switzerland and other European countries,
- Made a demand prognosis for the future demand structures regarding the freight types (based on the type of consignment) and
- Conducted expert interviews with larger shippers of wagonload consignments in Switzerland

6.1 Best Practices

In order to get some information about success factors for new small and medium transhipment sites, two of them (Gossau SG and Samedan in Switzerland) were analysed. The two sites have been chosen as they have been recently redesigned (during the last five years) and they show an increase of the amount of transhipped goods. Thus, the success factors from these two sites shall be integrated in the requirements for future transhipment sites.

The prime success factors were; factors which were supported by the local authorities (canton and local community) and synergies due to the availability of railway staff for other purposes as well as a high operational reliability and stability. The best practices from other European countries were not fully transferable to Switzerland since only larger transhipment sites are in operation in other European countries.

6.2 Aggregated Demand Structures

The market analysis was conducted in two ways; firstly, as an aggregated analysis of the future demand based on existing demand prognoses [12, 13] and secondly, the authors performed their own calculations based on existing road freight data surveys of the Federal Office for Statistics [14].

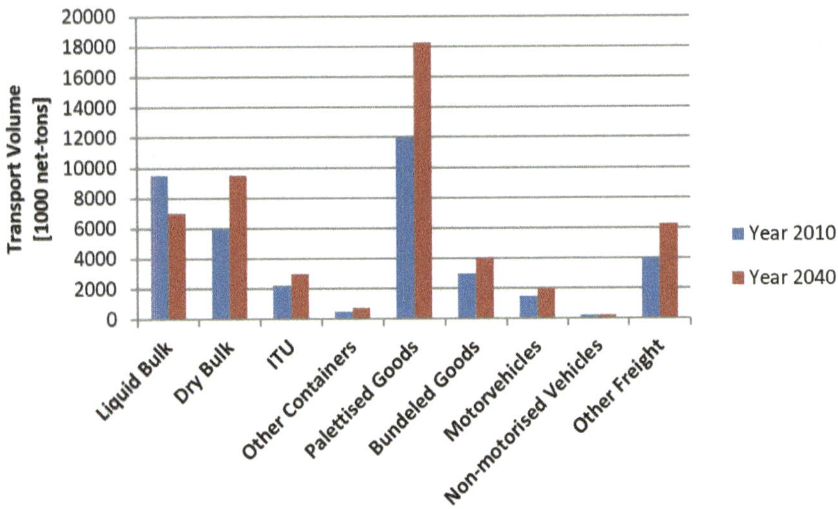

Fig. 1. Transport Volumes of rail freight differentiated by the type of goods for the years 2010 and 2040 [12, 14].

According to Fig. 1, in the year 2040, there will be significant changes in the type of the transported goods. The transport volume of liquid bulk is expected to decrease, whereas the volume of dry bulk and palletised goods are anticipated to increase. Prime drivers for this development are presumed to be the growing demands for mostly palletised consumer goods [13]. The growth in dry bulk results from expected changes in transport regulation. In the future, shippers will be forced to use rail freight for the supply of construction materials and industrial rocks and minerals to large construction sites. These changes in the characteristics of the consignments will influence the requirements for transhipment sites as well. The increasing amount of palletised goods need to be especially considered in the future design principles.

6.3 Expert Interviews with Shippers

The authors interviewed, in total, six experts from industrial sectors relevant for rail freight transport. Some of the interviewed shippers had their own intermodal transhipment sites. The interviewed shippers conduct the transports via public loading platforms, private sidings and/or container terminals. Thus, a range of possible requirements for transhipment sites could be assessed (Table 1).

Table 1. List of the interviewed shippers.

Shipper	Sector	Type of goods	Type of freight
1	Food and beverages	Food (frozen)	ITU
2	Retail	Food, other consumer goods	ITU
3	Waste and recycling	Waste, secondary raw material	ITU, bundled goods
4	Construction and construction materials	Construction material	Dry bulk
5	Construction and construction materials	Industrial rocks and minerals	Dry bulk, ITU
6	Courier, Express, Parcel	Consolidated Cargo, other goods	Other freight, ITU

The main reasons for those shippers to use rail freight are; the availability of own transhipment facilities in their production and logistics sites, lower costs of rail freight compared to the other modes of transportation, good service offers from the railways regarding reliability and punctuality and specific need for rail freight transport due to environmental reasons.

The shippers also explained their reasons to tranship goods on public loading platforms. The reasons are presented in Table 2.

Table 2. Reasons to tranship goods on public loading platforms.

Shipper	Reason(s)
1	Own private sidings
	Avoidance of congested road sections
2	Delivery times cannot be kept with road transportation (night driving ban)
3	Own small/medium transhipment site
4	Customers require rail transport
	Own private sidings
	Environmental strategy
5	Costs
	Own private sidings
	Environmental strategy
	Overnight transport (night driving ban)
6	Costs
	Own sidings
	Environmental aspects

The interviews indicate that for shippers who own private sidings and face fewer obstacles when integrating public loading platforms with their logistics network when compared to those who do not own private sidings. Thus, private sidings seem to be a facilitator for the use of public transhipment sites at the other end of the transport chain.

The requirements for loading platforms or transhipment sites themselves are mainly defined by the transported goods and the type of the consignments. Especially for shippers who use their private sidings and special transhipment equipment (e.g. warehouses with loading ramps, gravel quarries with chutes and bunkers, cement mills with equipment for dry bulk) the required transhipment equipment is already pre-defined. They require public loading platforms and the same other equipment that exist in their private sidings.

The transhipment of ITU[1] requires additional yard areas for the temporary storage of ITUs. This allows the decoupling of road and rail transport processes and delays and other disruptions in the transport processes may be compensated. The transhipment of palletised goods requires side ramps for the loading and unloading of the freight wagons. Depending on the type of transhipped goods, different requirements need to be fulfilled for the transhipment of dry bulk. The inter-tank transfer of cement re-quires only a paved loading quay beside the track and bunkers to facilitate the transhipment of industrial rocks and minerals.

Most shippers focus only on a quick and efficient transhipment process. Thus, most of them do not need additional services on the transhipment sites. Few of them use the sites for the stabling of freight wagons or for the weighing of the trucks and the loading units.

Additional areas for the consolidation and the sorting of partial loads provide additional logistics opportunities for shippers. Nevertheless, most do not use the sites for additional services since they focus only on a quick and efficient transhipment.

In general, the shippers mentioned these functional requirements (Table 3) for transhipment sites.

Table 3. Shippers' functional requirements for transhipment sites.

Shipper	Transhipment and storage area	Logistics and operational equipment	Service equipment
1	Highly stable pavement	Only transhipment	Electricity supply
2	Gantry crane, loading quay	Only transhipment	–
3	Housing (due to unpleasant odour)	Special equipment for the consolidation of waste and secondary raw materials	–
4	Simple loading quay	Only transhipment	Truck scale, stationary air compressor
5	Paved areal, bunker	Only transhipment	Truck scale
6	Side loading ramp	Preliminary sortation facilities	–

[1] ITU – Intermodal Transport Unit: Maritime container, semitrailer and swap bodies.

7 Requirements for Multifunctional Transhipment Sites

The results from the best-practices' analysis, the market analysis and the interviews with shippers were transferred into a catalogue of functional requirements for transhipment sites. The requirements were categorised as follows:

- Transhipment and storage area
- Infrastructures for rail and road vehicles
- Additional logistics services and operational equipment
- Equipment for measuring, weighing and safety.

The requirements are differentiated between small and medium sized transhipment sites. Furthermore, the stakeholder groups (users of the site, operators of the site and the administration) which require these characteristics are indicated.

7.1 Transhipment and Storage Area

As already mentioned, the transhipment sites are mostly used for palletised goods, (dry) bulk, general cargo, bundled goods and ITUs. Thus, Table 4 shows the functional requirements of the transhipment sites.

Table 4. Requirements for the transhipment and storage area. (X compulsory/(X) optional)

Requirement	Size		Stakeholder group		
	Small	Medium	User	Operator	Government
Transhipment of ACTS[a] containers	X	X			
Transhipment of ISO containers, swap bodies and semitrailers	(X)	X			
Transhipment of bulk goods	X	X			
Transhipment of overlong goods	X	X			
Transhipment of hazardous goods	X	X			
Transhipment of general cargo	X	X			
Transhipment of hazardous goods	(X)	X			
Yard for short term storage of ITUs	(X)	X			
Yard for long term storage of ITUs	(X)	(X)			
Storage of bulk goods	(X)	(X)			
Storage of palletised, bundled or other goods	X	(X)			
Storage of refer container	(X)	(X)			
Storage of hazardous goods	(X)	(X)			
Roof	(X)	(X)			
Housing	(X)	(X)			

[a]ACTS is a specific type of ITUs for rail and road transport.

7.2 Infrastructures for Rail and Road Vehicles

The infrastructures for road and rail are mainly defined by the amount of transhipped goods. Further requirements can be derived from the goods types and the consignments. Usually, the equipment for transhipments (e.g. Reach Stacker) defines the pavement's minimum stress resistance in the transhipment area.

The connection between transhipment site and public railway network shall be provided within a railway station and not as an additional siding on the open line (Table 5).

Table 5. Requirements for rail and road operations. (X compulsory/(X) optional)

Requirement	Size		Stakeholder group		
	Small	Medium	User	Operator	Government
Road – only one way roads	(X)	X			
Road – single lane roads used in both directions	X	(X)			
Road – unpaved surface	(X)	(X)			
Road – paved surface	X	X			
Road – extra strong paved surface	(X)	(X)			
Road – suitable for hazardous goods	X	X			
Railway – single sided connection to mainline	X	X			
Railway – double sided connection to mainline	(X)	X			
Railway – non electrified tracks	X	X			
Railway – electrified tracks sections towards the main line	(X)	(X)			
Railway – connected to station tracks	X				

7.3 Additional Logistics Services and Operational Equipment

The requirement of additional logistics services and operational equipment strongly depends on the specific demand structure of the transhipment site. In general, the demand of the single shipper's for equipment is quite low. Thus, these facilities should be shared between several shippers (Table 6).

Table 6. Additional logistics services and operational equipment. (X compulsory/(X) optional)

Requirement	Size		Stakeholder group		
	Small	Medium	User	Operator	Government
Service – stuffing/stripping of containers	(X)	(X)			
Service – maintenance of containers and other equipment	(X)	(X)			
Service – cross docking	(X)	(X)			
Service – customs clearance	(X)	(X)			
Waiting areas for trucks	(X)	(X)			
Truck parking facilities	(X)	X			
Gate/Offices	–	(X)			
Parking lots for employees	(X)	X			
Parking of equipment	–	(X)			

7.4 Equipment for Measuring, Weighing and Safety

The required equipment for measuring, weighing, control and safety are devised from the goods' types and consignments transhipped at the specific site. Generally, the only required equipment is the equipment which is used for weighing trucks, rail cars and/or consignments. This might be either demanded from the shippers themselves or as safety requirement from the rail freight or truck operators. Furthermore, the identification of vehicles or load units may be relevant. Due to security reasons the sites also may have fencing, gates and video surveillance. The security equipment will become more relevant in future (Table 7).

Table 7. Equipment for measuring, weighing and safety. (X as a rule/(X) optional)

Requirement	Size		Stakeholder group		
	Small	Medium	User	Operator	Government
Identification/Counting	(X)	(X)			
Load scanner (X-ray)	–	(X)			
Truck and/or freight wagon scale	(X)	(X)			
Container scale	(X)	(X)			
Fencing	(X)	(X)			
Gate	(X)	X			
Video survey	–	(X)			
Equipment for losses/Fire extinguishers	(X)	(X)			

8 Example Module

8.1 Definition of the Modules

Based on these defined requirements which are further based on the application of morphological boxes, three standardised transhipment modules for the supply of urban regions (Table 8) and three modules for rural regions (Table 9) have been developed.

Table 8. Standard modules for urban regions. (X compulsory/(X) optional)

Type of transhipment site	Logistics function	Type of freight				Overlong goods	Hazardous goods	Size
		Bulk	ITUs	Palletised	Bundled			
Urban consumer goods	Transhipment		X	X	(X)		(X)	Small or medium
	Storage		X	X	(X)		(X)	
	Additional services		X	X	(X)			
Urban construction materials	Transhipment	X	X			X		Small or medium
	Storage	X				X		
	Additional services							
Urban waste	Transhipment	(X)	X					Small or medium
	Storage		X					
	Additional services							

Table 9. Standard modules for rural regions. (X compulsory/(X) optional)

Type of transhipment site	Logistics function	Type of freight				Overlong goods	Hazardous goods	Size
		Bulk	ITUs	Palletised	Bundled			
Rural Construction materials	Transhipment	X	X			(X)		Small or medium
	Storage	X	(X)			(X)		
	Additional services							
Rural waste	Transhipment	X	X			(X)	(X)	Small or medium
	Storage		X			(X)	(X)	
	Additional services	(X)	(X)					
Rural agricultural products	Transhipment	X	X			X		Small or medium
	Storage	X	(X)			X		
	Additional services	(X)	(X)					

8.2 Example Design

For each of the above mentioned types of sites, a standard design has been developed. Here as an example, only the type rural construction materials are depicted (in Fig. 2). The standardised design includes the following characteristics:

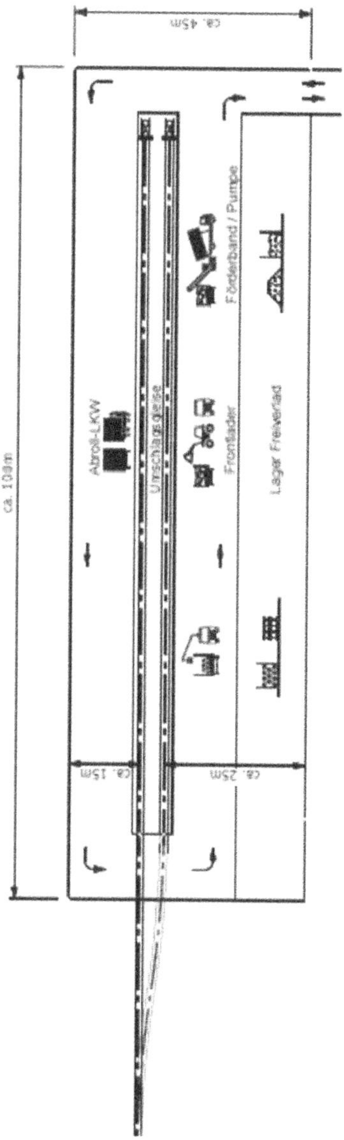

Fig. 2. Standard design of a transhipment site for construction material in rural areas

- Transhipment of ACTS container is possible.
- The available transhipment equipment is a truck-mounted crane, front loaders and conveyor belts for bulk materials.
- The roads are paved and used in one-way traffic.
- The siding is connected to the railway line is connected at only one side, the tracks are not electrified.
- There are additional storage facilities for bulk goods and overlong goods.
- The transhipment site is illuminated.

9 Results and Conclusion

According to the market analysis, there will be a future demand for small and medium sized transhipment sites for wagonload consignments. The required transhipment equipment will change according to the increasing demand for palletised goods and dry bulk goods. But there will be also a remaining demand for conventional liquid bulk goods.

For each type of the goods a different set of requirements towards transhipment site could be identified. Nevertheless, it was possible to derive a standardised transhipment module for each demanded type of goods. All of these modules were integrated in a catalogue of transhipment modules.

In a case study a detailed design for a transhipment site for construction materials in a rural region under usage of the standardised modules was developed. Thus, the real world applicability of the standard modules could be approved.

References

1. European Commission: White Paper, Roadmap to a Single European Transport Area – Towards a competitive and resource efficient transport system, Brussels (2011)
2. Galonske, N., Riebe, E., Toubol, A., et al.: The ViWaS project: future-proof solutions for wagonload transport. Transp. Res. Procedia **14**, 2850–2859 (2016). https://doi.org/10.1016/j.trpro.2016.05.402
3. Guglielminetti, P., Lagraulet, M.F., Artuso, D. et al.: Study on Single Wagonload Traffic in Europe – challenges, prospects and policy options: Final Report (2015)
4. Islam, D.M.Z., Zunder, T.H.: Experiences of rail intermodal freight transport for low-density high value (LDHV) goods in Europe. Eur. Transp. Res. Rev. **10**(2), 1723 (2018). https://doi.org/10.1186/s12544-018-0295-7
5. Gibb, A.G.F., Isack, F.: Client drivers for construction projects: implications for standardization. Eng. Constr. Architectural Manage. **8**(1), 46–58 (2001). https://doi.org/10.1108/eb021169
6. Fumasoli, T.: Estimations of the Potential and the Impact of Rail Freight in Urban Areas. ETH Zurich
7. DB Netz AG (2004) Richtlinien 800.06. Netzinfrastruktur Technik entwerfen. Bauliche Anlagen des Kombinierten Verkehrs. Accessed 1 June 2004
8. Arendt, T.: Mindestanforderungen an das Layout von Terminals. Schiffart, Hafen, Bahn und Technik 4/2014 (2014)

9. Ickert, L., Maibach, M., Bieler, C. et al.: Grossterminalstudie, Beurteilung der Terminal-projekte Gateway Limmattal und Basel-Nord (2012b)
10. Bruckmann, D., Dober, P., Mancera, A., et al.: Swiss split—a holistic approach to the distribution of containers to private sidings. Eur. Transp. Res. Rev. **8**(4), 169 (2016). https://doi.org/10.1007/s12544-016-0214-8
11. Ruesch, M., Wichser, J., Karrer, R., et al.: Ausgestaltung von Terminals für den (unbegleiteten) kombinierten Ladungsverkehr (2005)
12. Ruesch, M., Mohammad, T.: Freight Terminals. Challenges and Good Practices (2011)
13. Bundesamt für Raumentwicklung: Perspektiven des schweizerischen Personen- und Güterverkehrs bis 2040, Hauptbericht, Bern (2016)
14. Ickert, L., Maibach, M., Bertschmann, D., et al.: Marktanalyse und Marktprognose Schienengüterverkehr: Gesamtbericht, Bern/Basel/Zürich (2012a)
15. Bundesamt für Statistik: Gütertransporterhebung: (Data set) (2013)

A Survey on the Potentials of Indoor Localization Systems in Production

Carina Mieth[1,3(✉)], Philipp Humbeck[1,2], and Georg Herzwurm[2]

[1] TRUMPF Werkzeugmaschinen GmbH & Co. KG, Ditzingen, Germany
carina.mieth@trumpf.com
[2] University of Stuttgart, Stuttgart, Germany
[3] TU Dortmund University, Dortmund, Germany

Abstract. Indoor localization systems are primarily designed for the precise determination of the position of objects. In addition, the position data obtained can be used to create further potentials, so-called secondary benefits. So far, only a few of these are mentioned sporadically in the literature. In this survey, they are now collected and categorized. Further potentials are revealed in two consecutive expert workshops and transferred into a five-level model that contains a description of each use case and can be regarded as a recommended logical order for practical implementation for both developers and production managers.

Keywords: Indoor localization system · Cyber-physical systems · Data-based services · Production · Intralogistics · Potentials of data in industry 4.0

1 Introduction

Digital transformation increases both the availability and the significance of data [1, 2] and thus increasingly blurs the boundary between mechanical engineering and IT [3]. Within the scope of developing solutions for the Smart Factory, mechanical engineers are not only concerned with the further development of machines. The focus is now on the development of cyber-physical systems (CPS), since they can capture large amounts of data for use in data-based services over the entire product life cycle [4–6].

Acatech, the German Academy of Engineering Sciences, defines the term CPS as a system with "integrated software […] that can collect physical data using sensors and directly influence physical processes" [7]. The CPS considered here, is any real-time indoor localization system (RTILS), regardless the underlying technology. The located physical objects are part of intralogistics of production systems e.g. bins or vehicles. The position data from RTILSs hold great potentials [8] and will thus be analyzed to derive secondary benefits.

In view of the trend towards declining lot sizes and mass customization, the understanding of intralogistics plays a decisive role in the development of Industry 4.0 applications [9]. To better monitor the processes on the shop floor, one approach is to digitize them with the available location technologies. In addition, the position data from RTILSs are often enriched with context information, e.g. about the production

U. Clausen et al. (Eds.): ICPLT 2019, LNLO, pp. 142–154, 2019.
https://doi.org/10.1007/978-3-030-13535-5_11

order, its production sequence, layout information or process knowledge. This enables a wide range of application scenarios and optimization potentials. Using CPSs is therefore the next step towards autonomous production.

The challenge for the operators of a RTILS is now to use the collected data usefully. So far, there are only very generic descriptions of the potentials of RTILS data and the respective possible data-based services [10]. The focus of this work is therefore on the collection, analysis and description of promising use cases (see Sect. 5) to answer the central research question:

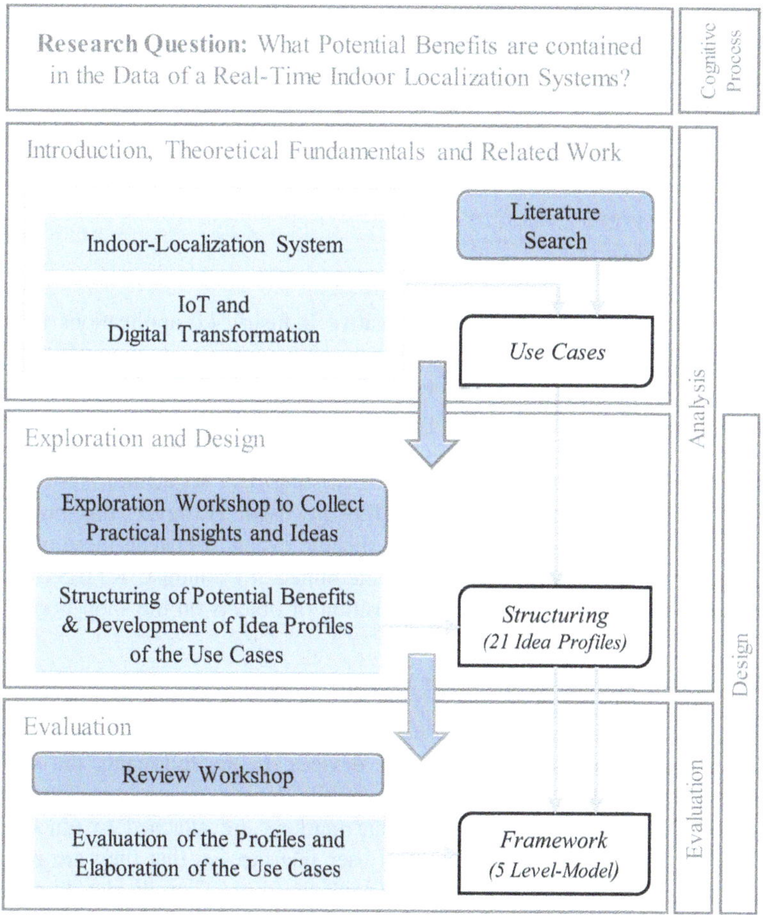

Fig. 1. Methodological approach and steps to derive the potentials of indoor localization data

The methodology and structure of the paper is illustrated in Fig. 1. Within this research work a literature analysis, an exploration workshop and a review workshop will be carried out. [11]. To ensure that the objectives are achieved, the work with the phase's analysis, design and evaluation is oriented towards the knowledge process of design-oriented business informatics. This can be assigned to the field of design science. [12] The analysis phase initially serves to work out the problem with subsequent derivation of the resulting research question. In addition, this phase serves to determine the research design in order to answer the research question.

This work examines the potential benefits of RTILSs based on a literature review and expert workshops. A distinction is made between the possible advantages for the user and for the system provider. Different ideas for the usage of an indoor localization system within a production system are derived from literature (see table in Fig. 3) and prepared in idea profiles. These profiles are assessed by interdisciplinary teams consisting of the RTILS's users and employees from different departments of the system's provider. In a first step, the idea profiles are classified by two groups according to their overall potential. Next, the classification from the two groups are compared and merged. In the last step, the profiles are evaluated against selected criteria: (a) benefits for the user of a RTILS, (b) benefits for the system provider i.e. developer, (c) requirements for implementation. For each criterion, a Likert scale is used, ranging from 1 "very high" to 5 "very low".

2 Fundamentals

An important challenge for enabling innovative Industry 4.0 applications in intralogistics is the integration of technologies that capture processes, material and information flows and thus create transparency. Production feedback data can usually be collected using production data acquisition systems, but this option is so far only used by about half of the companies. Especially small companies use these systems only sporadically [13]. As a basic technology for existing data acquisition systems, techniques for automatic identification (Auto-ID) are used. However, this only allows objects to be identified at certain places at specific events. Between these events, the exact position of the objects cannot be determined. In contrast, RTILSs have the advantage of continuously recording the position of objects on the shop floor.

2.1 Real-Time Indoor Localization System

A RTILS enables the acquisition of position data from different objects in real-time and typically consists of stationary and mobile devices. In the following, the stationary devices are referred to as satellites and the mobile devices as markers. The satellites make up the reference system, whereas the markers are attached to objects to be tracked. The markers are equipped with a user interface so that they are not just a sensor, but an actuator through interaction with the user. They display him relevant instructions on how to work on the object. According to Uckelmann and Wendeberg [8], these objects can be classified in the following five categories:

- vehicles such as forklifts or route trains
- load carriers
- goods, parts, etc.
- employees
- tools.

For manufacturing companies, the use of these RTILSs has until now not been widespread, mainly because of the heterogeneous production facilities, which imply different requirements and environmental conditions for RTILSs, but also because of difficulties calculating the return on investment. The underlying technology in RTILSs is typically radio-based. In the industry, the production environments are versatile with changing conditions and e.g. shiny materials like sheet metal pose a major challenge on RTILSs due to reflections. In this case, the usage of ultra-wide-band (UWB) is particularly promising, since it is not as sensitive to reflections in comparison to other technologies and thus comes up with high accuracies. Strong growth in the entire RTILS market can be expected, if technological challenges can be overcome. MarketsandMarkets [14] estimates annual average growth of 24.5% and a total volume of $ 8.09 billion till 2022.

2.2 Classification in the Value Creation Levels of the Internet of Things

Through the real-time localization of various objects in intralogistics of production systems, the indoor localization system offers the possibility of digitizing physical objects by supplying the information technology that bridges the gap between the physical and the digital world in the value chain. This can be seen in the level model of the value creation levels in the Internet of Things in Fig. 2.

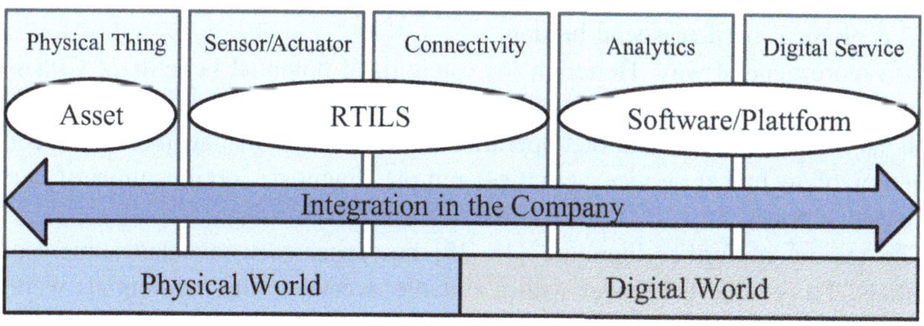

Fig. 2. Classification of real-time indoor localization systems (RTILSs) in the value creation levels of the Internet of Things - own illustration adopted from [15]

On the first level is the physical thing, in the case of RTILSs, this level corresponds to any physical object (see list in Subsect. 2.1) that can be tracked. On the second level, these objects are equipped with sensors and actuators, which corresponds to the markers for indoor localization. For global access, the satellites and the localization server provide a connection to the Internet on the third level. On the fourth level, the collected position data is stored and analyzed. Based on the data analysis, digital services are created on the fifth level. Depending on the application, a system designed according to this scheme offers not only the primary benefit of the physical local resource, but also an additional digital global benefit for the user. Fleisch et al. [15] claim that these levels cannot be created independently of each other. The greatest overall benefit of these applications can only be achieved through adequate integration from the physical level to the digital service level. The design of the digital service level has a direct influence on the development of the product at the physical level and vice versa.

3 Related Work

Most survey publications on RTILSs [6, 16, 17] focus only on technological characteristics. These are not discussed here, since the aim is to outline the cross-technological potentials of data from RTILSs. The literature dealing with the potential benefits of data from RTILSs or more generally with the benefits of data from CPS is manageable and summarized in Fig. 3. In this Section, two references [4, 8] are looked at in more detail, as they are the only ones, that contain a kind of overview of their mentioned potentials. All other references give only isolated potentials.

Uckelmann and Wendeberg [8] think that the use of location information opens opportunities for making production processes more flexible, as processes can now run on demand and according to rules, independently of predefined and rigid routes or storage locations. Their work identifies the chances of RTILSs to prevent waste according to the Muda concept. They point out that the RTILS data can only be used for minimizing waste, if the waste type has any spatial reference. In tabular form, they briefly related the spatial reference to the following types of waste: material movements, stocks, movements, waiting times, processing, over production, correction and faults, ecological burden, social burden.

In a more general way, Herterich [4] categorized potential benefits of CPS in the context of industrial services. They listed seven potential benefits in tabular form, which are: product optimization, operation optimization, management and control, prediction of technical service activities, remote diagnosis, optimization of service, data-based services.

The review of further literature [18–25] has repeatedly revealed some isolated potentials. To provide the reader with a suitable overview, these potentials were collected and categorized in Fig. 3. Five intuitive categories that group similar potential benefits together were introduced: process transparency, assistance systems, fault management, optimization, production control. In Sect. 5, these categories are also used in the structure of the presented five-level model in Fig. 4. It can be concluded that there is no clear demarcation and no detailed description of the potentials in literature. By using a RTILS the primary benefit is the localization of objects, but with additional context information and analyses of historical position data, secondary potentials can evolve. These are of interested in the further course of this paper and will therefore be analyzed and extended by experts in the following workshops.

	[1] Niehues 2016	[6] Zafari 2017	[17] Kashevnik & Shchekotov 2012	[8] Uckelmann & Wendeberg 2015	[18] Schürzinger 2014	[19] Schuhmacher & Hummel 2018	[20] Erol et al. 2016	[21] Bremer & Hummel 2017	[22] Fiedler 2014	[23] Fraunhofer IML 2012	[24] Ubisense 2014	[9] Lieberoth-Leden et al. 2017	[25] Vogel-Heuser et al. 2017	[3] Bauernhansel et al. 2015	[10] Herterich et al. 2015
primary benefit															
localization of objects	x	x	x	x	x	x	x	x	x	x	x	x			
secondary benefits															
process transparency															
time and cost advantages									x	x					
automatic posting	x								x						
record part times and movement logic	x		x	x	x										
assistance systems															
navigation and context information		x	x					x				x			
prioritization of rework											x				
monitoring of regulations				x											
fault management															
time prediction of technical support activities											x				x
optimization															
operational optimization				x	x						x				x
avoidance of waste / process optimization				x	x				x		x				
improvement of inventory management				x			x		x	x			x		
product optimization									x		x			x	x
production control															
localization as input for real-time control and management										x			x		x
adaptive workshop control through order status monitoring	x									x		x			

Fig. 3. Potential benefits of data from the real-time indoor localization system in production

4 Exploration and Design

An exploration workshop with experts from industry was conducted consecutively to the literature review from Sect. 3. The aim of the exploration workshop was to structure and evaluate the potential benefits and to gain practical insights into the industrial environment and their requirements. During the exploration workshop, potential benefits from the literature review were presented to a group of experts in production and mechanical engineering. They were familiar with RTILSs in production, but only concerning the primary purpose of localizing objects. They were asked about possible RTILS applications and based on their practical insight, new ideas for the usage of RTILSs were developed.

The instrument of the idea profile was chosen to prepare the ideas for the subsequent evaluation. In an idea profile, the holistic aspects of an idea, both positive and critical, are summarized. A total of 21 idea profiles were developed during the exploration workshop. Each profile contains the following elements according to Hartschen et al. [26]:

- *name of the idea:* is self-explanatory
- *user's problem:* describes the problem that is to be solved with this idea
- *actors:* persons that are affected by the problem or which benefit directly from this idea
- *description of the idea:* describes how the idea solves the user's problem
- *visualization:* a visual aid that contributes to the understanding of the idea
- *necessary context information:* tracking application level and geofences that are needed for the implementation of the idea
- *possible additional data:* e.g. ERP/MES/machine
- *user's benefit:* advantage of the idea for the user
- *discussion/problems:* list of disadvantages are questions to be discussed in the Review Workshop
- *evaluation scale for overall potential:* five-level Likert-type scale to evaluate the potential of the idea

5 Evaluation: Potential Analysis and Genesis

The idea profiles were evaluated with the help of an evaluation workshop. Experts from different departments e.g. production management, development and service were consulted for this purpose. The result is the five-level-model in Fig. 4.

5.1 Review Workshop

The aim of the workshop was to evaluate the 21 idea profiles that resulted from the exploration workshop. Besides the analysis of the benefits from the user's point of view, the internal benefits for the companies resulting from the utilization of the data were discussed. For this purpose, eight experts from the development, production management and service departments of mechanical engineering companies were consulted. The participants of the workshop were divided into two groups, taking care to ensure that they were as interdisciplinary as possible. In the first phase, each group assessed the idea profiles based on the evaluation scale on the profile and had to agree on a rating. In the second phase, the idea profiles were discussed in plenary and the evaluation results were combined. The six most promising ideas were selected. In the third phase, these six ideas were evaluated again by the participants. Three criteria were distinguished: (a) user benefit, (b) internal benefit, (c) requirements. For each criterion, a five-level Likert scale was used, ranging from one - "very high" to five - "very low".

5.2 Five-Level-Model

As the result of the workshop, the five-level model in Fig. 4 was set up. It can be understood as a ranking according to the difficulty of implementation starting with easier implementable use cases towards more difficult one's. The experts identified five application areas for the use of data of RTILSs. The names of the application areas were previously used to group the potentials from the literature in Fig. 3. For each level, a short description of the derived use cases is shown in Fig. 4 and explained in the following:

1. *Process Transparency*

(1a) *Automatic recording of processing times as an information tool for machine operators and production managers*:
A marker is assigned to each production order to automatically record the corresponding process times. The automatic booking of the process time takes place when a marker changes the geofences. The geofencing areas are divided in such a way that a clear differentiation of the individual production steps is possible. This first level enables real-time transparency of the current production progress. It should be noted that the division of the geofences directly influences the granularity of the process times and thus the possibilities of further analyses.

(1b). *Analysis of processing times and calculation of operative key figures as a basis for decision-making for the production manager:*
With the availability of process times in real time, various key figures that describe the current production status can be calculated and visualized automatically. These figures can refer, for example, to specific production orders, production steps, shifts or a period. This enables a quantitative comparison of the production's performance and can serve as real-time decision-support for production managers. In addition, benchmarking with the performance of similar production systems within a company or even with competitors is possible if values can be aggregated anonymously to form reference values.

	Application Areas	Short Description of Use Case	
1	Process Transparency	Automatic recording of process times as an information tool for machine operators and production managers	1a)
		Analysis of process times and calculation of operative key figures as a basis for decision-making for the production manager	1b)
		Comparison of actual with target process times as a basis for production planning and production controlling	1c)
2	Assistance Systems	Picking help in warehouse or buffer zone	2a)
		Prioritization in the buffer zone	2b)
3	Fault Management	Recording of disturbances in the production process	3a)
		Prediction of disturbances in the production process	3b)
4	Optimization	Acquisition of all material flow data for the optimization of processes	4a)
		Analysis of material flow data as decision basis for factory planning	4b)
5	Production Control	Real-time production control and simulation	5a)

Fig. 4. Five-Level-Model: Overview of the five identified application areas for the use of data from RTILSs and the respective description of the use cases

(1c). *Comparison of actual with target process times as a basis for production planning and production controlling:*
Process target times are defined in the production planning system. These are usually based on estimates and are rarely reviewed or adjusted. Due to learning effects of employees, process or technology improvements, deviations between target and actual process times commonly occur. A dynamic adjustment of the planned target process times to the recorded average process times from the RTILS helps to set up more reliable production plans and cost calculations.

2. *Assistance Systems*
(2a). *Picking help in warehouse or buffer zone*
The RTILS as picking aid simplifies the processes in which objects are searched for and picked, for example in the warehouse or during material supply in assembly. The

employee can look up the position on a computer, smartphone or watch and the marker can draw attention to itself through light or sound. In addition, the RTILS supports the employee in merging tracked objects. Merging work steps are process steps in which, for example, different individual parts are combined to form an assembly or a product.

(2b). *Prioritization in the buffer zone*
Nowadays, employees are usually free to choose which production order they want to process next from a buffer. Different prioritizations of orders are rarely taken into account, as this would involve increased effort on the side of the employee. A virtual representation of the order queue can be created from the position data of the RTILS. This requires a geofence that covers the buffer area of the waiting orders in front of the machine. The employee's decision for the next order can now be supported or replaced if the information about the queue from the RTILS is linked to the proposed prioritization in the production schedule.

3. *Fault Management*
(3a). *Recording of disturbances in the production process*
From the RTILS data, disturbances in the production process can be derived by detecting anomalies. For example, RTILS can locate an unusually high number of orders in a geofence assigned to a buffer in front of a machine. If no transport of finished orders is registered there, it can be concluded that processing is delayed. Adaptive threshold values, which result from the node balance (incoming minus completed orders), can be used to determine when the production manager is to be notified of the congestion. In any case, error detection requires additional context knowledge about the production area associated with the geofence.

(3b). *Prediction of disturbances in the production process*
Locating production orders in real time enables the company to learn from this data what a normal - i.e. undisturbed - production state is. For example, the average number of waiting objects in the buffers can be determined and set as default. If the actual number is continuously drifting further away from the estimated mean value during ongoing production, it will be possible to proactively avoid a malfunction.

4. *Optimization*
(4a). *Acquisition of all material flow data for the optimization of processes*
During the common value stream analysis, only one order is accompanied by process engineers through production. However, RTILSs create transparency by recording all material flows of the tracked objects. This opens up additional optimization potential that could not previously be exploited. The recorded positioning data from the RTILS can be visualized in spaghetti diagrams or heat maps that help the process engineers by identifying unnecessary movements. Especially for load carriers, the knowledge about the recorded routes can be used to improve transportation and avoid detours. In this way, the routes of vehicles and employees can also be optimized.

(4b). *Analysis of material flow data as decision basis for factory planning*
The RTILS data can be used for both the adaptation planning of existing and the planning of new production systems. For adaptation planning, the RTILS data is converted into a material flow graph. This allows the main routes in the material flow to

be identified. This knowledge can be used for layout optimization by rearranging the individual areas in the direction of the material flow. In addition, newly planned processes can be evaluated based on the material flow data collected after implementation. For new production systems, better planning can be made based on historical data from existing comparable systems.

5. *Production Control*

(5a). *Real-time production control and simulation*

The RTILS collects position data which, in combination with context knowledge, makes it possible to map the current situation in production digitally. This is the basis for a self-optimizing and adaptive production control in which adaptation decisions are no longer made solely by the production manager. In the future, decisions can be made more autonomously based on real-time data from production. The simulation of the production system with RTILS data will play a decisive role here, as it can be used to forecast the future from the current production situation. This allows adaptation decisions to be evaluated simulatively and to autonomously execute the decision for the best adaptation.

6 Conclusion

A detailed overview on the potentials of the usage of real-time indoor localization systems (RTILS) in production was worked out for users and developers where a distinction between primary and secondary benefits was made. The proposed ranking from the expert workshops, which has been incorporated into the five-level model, allows developers to easily decide which use case they should implement next. This paves the way for a versatile use of RTILS data and the achievement of quick wins in manufacturing companies. The five identified levels and the respective use cases show the relevance and the possibilities of the use of RTILS data in data-based services. As secondary benefits use cases were presented that can be realized already by using RTILSs and more challenging use cases that are built upon more sophisticated analyses of RTILS data in advance. For some of the use cases a simulation model of the production system will be necessary for the decision support and the prediction of future production situations. In future work, the location data should be considered as input for simulation models to simplify the creation and to improve the results' significance of simulation models. In addition, the real-time integration of localization data into a digital twin and the development of suitable data-based services and corresponding business models represent future research fields. Finally, we encourage the interested reader to add newly developed use cases to the application areas of the presented five-level model to obtain a model that is as comprehensive and complete as possible.

References

1. Niehues, M.R.: Adaptive Produktionssteuerung für Werkstattfertigungssysteme durch fertigungsbegleitende Reihenfolgebildung (engl.: Adaptive production control for shop floor production systems through production-associated sequencing), Ph.D. Thesis, Technical University Munich (2016)
2. Westkämper, E., Löffler, C.: Strategien der Produktion (engl.: Strategies in Production). Springer, Berlin Heidelberg (2016). https://doi.org/10.1007/978-3-662-48914-7
3. Bauernhansl, T., Emmrich, V., Paulus-Rohmer, D., Döbele, M., Schatz, A., Weskamp, M.: Geschäftsmodell-Innovation durch Industrie 4.0. Chancen und Risiken für den Maschinen- und Anlagenbau (engl.: Business model innovation through industry 4.0. Opportunities and risks for the mechanical industry). Dr. Wieselhuber & Partner GmbH and Fraunhofer-Institut für Produktionstechnik und Automatisierung IPA (2015). https://www.wieselhuber.de/migrate/attachments/Geschaeftsmodell_Industrie40-Studie_Wieselhuber.pdf
4. Herterich, M., Uebernickel, F., Brenner, W.: The impact of cyber-physical systems on industrial services in manufacturing. In: Fleisch, E., Weinberger, M., Wortmann, F. (Hrsg.), Geschäftsmodelle im Internet der Dinge (Bd. 67, Bd. 30, S. 323–328) (2015)
5. Knoll, T., Lautz, A., Deuß, N.: Machine-To-Machine Communication: From Data To Intelligence. In: Vogel-Heuser, B., Bauernhansl, T., ten Hompel, M. (Hrsg.), Handbuch Industrie 4.0. Bd. 3: Logistik (Springer Reference Technik, 2., erweiterte und bearbeitete Auflage, S. 347–356). Springer Vieweg, Berlin (2017)
6. Zafari, F., Glekias, A., Leung, K.K.: A Survey of Indoor Localization Systems and Technologies. arXiv: 1709.01015 (2017). https://arxiv.org/pdf/1709.01015.pdf
7. acatech – Deutsche Akademie der Technikwissenschaften: Cyber-Physical Systems. Innovationsmotor für Mobilität, Gesundheit, Energie und Produktion (engl.: Cyber-Physical Systems. Innovation engine for mobility, health, energy and production) acatech Position, vol. 11. Springer, Berlin (2011). https://doi.org/10.1007/978-3-642-27567-8
8. Uckelmann, D., Wendeberg, J.: Echtzeitortungssysteme für Industrie 4.0: Potentiale und Stand der Technik. In: Zsifkovits, H.E., Altendorfer-Kaiser, S. (Hrsg.), Management logistischer Informationsflüsse. 3. Wissenschaftlicher Industrielogistik-Dialog in Leoben (WiLD). Rainer Hampp Verlag, Mering (2015)
9. Lieberoth-Leden, C., Röschinger, M., Lechner, J., Günthner, W.A.: Logistik 4.0. In: Reinhart, G. (Hrsg.), Handbuch Industrie 4.0. Geschäftsmodelle, Prozesse, Technik, S. 451–606. Hanser, München (2017)
10. Herterich, M., Uebernickel, F., Brenner, W.: Nutzenpotentiale cyber-physischer Systeme für industrielle Dienstleistungen 4.0. (engl.: Potential benefits of cyber-physical systems for industrial services 4.0.). HMD Praxis der Wirtschaftsinformatik **52**(5), 665–680 (2015). https://doi.org/10.1365/s40702-015-0164-y
11. Mayer, H.O.: Interview und schriftliche Befragung: Grundlagen und Methoden empirischer Sozialforschung (engl.: Interview and written survey: fundamentals and methods of empirical social research), 6th edn. Oldenbourg, München (2013)
12. Österle, H., Becker, J., Frank, U., Hess, T., Karagiannis, D., Krcmar, H., Loos, P., Mertens, P., Oberweis, A., Sinz, E.J.: Memorandum zur gestaltungsorientierten Wirtschaftsinformatik (engl.: Memorandum on design-oriented business informatics). Schmalenbachs Z betriebswirtsch Forsch **62**(6), 664–672 (2010)
13. Schuh, G.: ProSense. Ergebnisbericht des BMBF-Verbundprojektes; hochauflösende Produktionssteuerung auf Basis kybernetischer Unterstützungssysteme und intelligenter Sensorik (engl.: result report of the BMBF joint project; high-resolution production control based on cybernetic support systems and intelligent sensor technology). Apprimus Verl., Aachen (2015)

14. MarketsandMarkets Research Private Ltd.: Real-Time Locating Systems (RTLS) Market by Offering (Hardware, Software, Services), Technology (RFID, W-Fi, UWB, BLE), Vertical (Healthcare, Manufacturing, Retail, Education, Govt., Sports), Application/Use case, and Geography - Global Forecast to 2023 (2018). https://www.marketsandmarkets.com/Market-Reports/real-time-location-systems-market-1322.html
15. Fleisch, E., Weinberger, M., Wortmann, F. (Hrsg.): Geschäftsmodelle im Internet der Dinge (engl.: business models in the Internet of Things)(Bd. 67) (2015)
16. Liu, J.: Survey of Wireless Based Indoor Localization Technologies. Department of Science & Engineering, Washington University (2014). https:// www.cse.wustl.edu/~jain/cse574-14/ftp/indoor.pdf
17. Kashevnik, A., Shchekotov, M.: Comparative analysis of indoor positioning systems based on communications supported by smartphones. In: Proceedings of The 12th Conference of Open Innovations Association FRUCT and Seminar on e-Tourism (S. 1–6) (2012)
18. Schürzinger, A.: Standortbestimmung in Echtzeit: Funkbasierte Ortung im Produktionsumfeld. (engl.: Positioning in real time: Radio-based positioning in the production environment) IT & Production (2014). https://www.it-production.com/allgemein/standortbestimmung-in-echtzeitfunkbasierte-ortung-im-produktionsumfeld/2/
19. Schuhmacher, J., Hummel, V.: Development of a descriptive model for intralogistics as a foundation for an autonomous control method for intralogistics systems. Procedia CIRP **23**, 225–230 (2018). https://doi.org/10.1016/j.promfg.2018.04.021
20. Erol, S., Jäger, A., Hold, P., Ott, K., Sihn, W.: Tangible industry 4.0: a scenario-based approach to learning for the future of production. Procedia CIRP **54**, 13–18 (2016). https://doi.org/10.1016/j.procir.2016.03.162
21. Brenner, B., Hummel, V.: Digital twin as enabler for an innovative digital shopfloor management system in the ESB logistics learning factory at Reutlingen - University. Procedia Manuf. **9**, 198–205 (2017). https://doi.org/10.1016/j.promfg.2017.04.039
22. Fiedler, M.: Abschlussbericht zum Verbundprojekt SMARTI (smart reusable transport items) im Leitthema: Wandelbare Logistiksysteme (engl:: Final report on the joint project SMARTI with the lead topic: adaptable logistics systems): Laufzeit vom 01.06.2010 bis 30.04.2014 (2014)
23. Fraunhofer-Institut für Materialfluss und Logistik IML (Hrsg.): Logistik entdecken (engl.: Fraunhofer Institute for material flow and logistics: explore logistics) Nr. 13 (2012). https://www.iml.fraunhofer.de/content/dam/iml/de/docments/OE%20983/Presse/Logistik%20entdecken/Logistik_entdecken_13_online_NEU.pdf
24. Ubisense (Hrsg.): 2014 Smart Manufacturing Technologies Survey (2014). http://go.ubisense.net/l/17692/2014-09-29/fny8m
25. Vogel-Heuser, B., Bauernhansl, T., Hompel, M. ten (Hrsg.): Handbuch Industrie 4.0. Bd. 3: Logistik (Springer Reference Technik, 2., erweiterte und bearbeitete Auflage). Springer Vieweg, Berlin (2017)
26. Hartschen, M., Scherer, J., Brügger, C. (Hrsg.): Innovationsmanagement. Die 6 Phasen von der Idee zur Umsetzung (Innovation management. The 6 phases from idea to implementation), 3rd edn. Gabal, Offenbach (2015)

Realization of ETA Predictions for Intermodal Logistics Networks Using Artificial Intelligence

Peter Poschmann[1], Manuel Weinke[1], Andreas Balster[2(✉)], Frank Straube[1], Hanno Friedrich[2], and André Ludwig[2]

[1] Chair of Logistics, Technische Universität Berlin, Straße des 17. Juni 135, 10623 Berlin, Germany
{poschmann, weinke, straube}@logistik.tu-berlin.de
[2] Kühne Logistics University, Großer Grasbrook 17, 20457 Hamburg, Germany
{andreas.balster, hanno.friedrich, andre.ludwig}@the-klu.org

1 Introduction

Intermodal logistics networks such as the maritime transport chain require a precise interaction of numerous actors. However, due to their complexity, the closely interlinked processes are highly susceptible to disruptions. Companies are constantly faced with the challenge of dealing effectively and efficiently with disruptions and resultant delays. At the same time, they are confronted with increasing logistical requirements related to higher quality and flexibility demands of customers (Straube et al. 2013). Supply chains are becoming increasingly vulnerable, due to the associated necessity to cope with increasing volatility while simultaneously reducing risk buffers in processes as a result of rising cost pressure. Combined with ongoing changes due to digitization, this situation contributes significantly to an increasing need for improved information transparency among companies and their customers.

To manage disruptions, transparency regarding the current status (actual time), the delay situation of a transport order, as well as the explanation for the delay, is required (Walter 2016). However, this descriptive information only detects an incident after it has occurred, so that only reactive interventions are possible (Fig. 1). At the same time, this descriptive information does not contain any derivations of the (temporal) effects on downstream processes and subsequent actors.

To address this challenge, reliable values for the Estimated Time of Arrival (ETA) of transport orders covering the entire transport chain must be predicted. With the ETA prediction, it would be possible to derive appropriate measures for disruptions and delays such as optimally reallocating company resources. According to conducted expert interviews, the significance of such an ETA will increase in the future.

Despite the usefulness of an ETA prediction for the entire maritime transport chain, only fractional and simplified solutions currently exist. This is partly due to the complexity of intermodal process chains, which are highly fractured according to a large number of actors with different processes, IT systems and means of transport. These actors are faced with very individual disruptions. As a result, many events have to be considered into an ETA calculation. Furthermore, the general availability of

U. Clausen et al. (Eds.): ICPLT 2019, LNLO, pp. 155–176, 2019.
https://doi.org/10.1007/978-3-030-13535-5_12

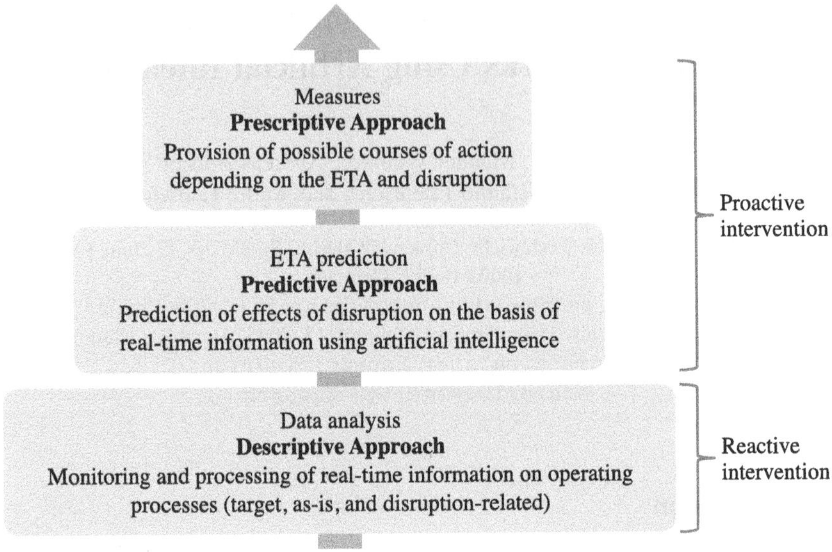

Fig. 1. Project approach: Using AI to make disruptions in logistics manageable

ETA-related information is often very limited. In addition, the transmission of existing information is not significantly prominent, so that no actor can trace the overall progress of a transport.

For maritime transport chains and intermodal networks in general, an integrated approach to delays is lacking. At the same time, within most processes, the tools for monitoring and dealing with disruptions have not matured. The implementation of utilizing data from Artificial Intelligence (AI) has so far only been carried out selectively. The research project SMECS (Smart Event Forecast for Seaports) (IHATEC 2017a), funded by the German Federal Ministry of Transport and Digital Infrastructure, aims to analyze and realize these AI-based potentials for a relevant use-case. In SMECS, the Chair of Logistics at the Technical University of Berlin, together with Kühne Logistics University, DB Cargo and numerous associated companies develop an IT system based on AI, which supports companies in the maritime transport chain with cross-actor ETA predictions and associated recommendations for action. Various methods of machine learning are tested and deployed in ETA models. Together with the actor-specific measures, these models are transferred into a knowledge-based AI system. The primary application case of SMECS is the pre leg of the maritime transport chain for sea freight containers that comprises the landside processes from the shipper or container depot, to the transshipment, to the sea transport in the deep-sea terminal. By combining the aspired ETA of SMECS with information of the main and subsequent leg, an overall view of the maritime chain could be achieved.

In the first phases of the project, a comprehensive system analysis and a demand analysis were carried out. Based on this work, the model development phase has already started. This paper presents the key results of these three phases. This includes the description of physical processes, associated data and disruptions. Based on this, the benefits and requirements of ETA prediction are shown and derived to prioritized application cases. In the final section, the approach of the model development according to the current status is described. In advance, the prevailing ETA-related methods used in companies and existing scientific solutions will be illuminated, based on a market and literature review.

2 Market and Literature Review

Initial business solutions for ETA predictions are available for all means of transport. In road transport, drivers usually choose their own routes. The existing ETA values are based on the calculations of the navigation system in the vehicle or other external services, combined with the experience of the drivers. These predictions are usually appropriate, but do not take into account random events such as road closures due to major accidents or weather influences. In contrast, ETA for rail transports are not predicted, but rather result from the forward projection of current delays. As a result, they are very unreliable and rarely used by other actors in the transport chain. More advanced business solutions and scientific concepts are currently being developed for rail transport, but primarily for passenger traffic. Rail freight transport is rarely considered. In addition, most scientific work is focused on the prediction of follow-up delays, but not on initial disruptive events.

Beyond these partial solutions, no ETA prediction tool for the entire maritime transport chain, or at least the pre leg with different means of transport, are already established and used in practice. Especially the nodes of the inland and deep-sea terminals as well as the marshaling yards are often neglected. Moreover, there is no discussion of the advantages of an overall-transport ETA or at least a consideration of the most significant disruptions within the transport chain. There is also still a need for research in the area of prescriptive tools, which provide actors with ETA-based recommendations for real-time disruption management. It was rather found that only descriptive tools (e.g. container tracking) cover the entire maritime transport chain. Examples of currently-developed platforms for integration of information from different actors of transport chains are KV 4.0 (BMVI 2017) and EMP 4.0 (IHATEC 2017b), which are also partner project of SMECS.

In scientific literature, a distinction can be made between two approaches to ETA and delay prediction: model-based approaches and data-based approaches. The model-based approaches can be categorized as simulation models and analytical models. The simulation models can be based on existing simulation software, such as OpenTrack (Nash and Huerlimann 2004) and Railsys (Radtke and Hauptmann 2004), or developed individually. The analytical models can be further divided into two categories, depending on whether they are based on graph theory or queuing theory. Examples for delay predictions based on graph theory use timed event graphs (Goverde 2010), activity graphs (Büker and Seybold 2012), and petri nets (Zegordi and Davarzani 2012). Queuing-theory models can be designed as Markov chains, as shown by Özekici and Sengör (1994). The model-based approaches are very precise, but not useful for real-time application. Data-based approaches are better suited for these applications.

Data-based approaches mostly use conventional statistical methods such as multiple linear regression, logistics regression, time series analysis, or Bayesian networks (Zhang et al. 2016). More modern methods, such as machine learning, are considered only selectively. In the field of machine learning, various approaches to delay predictions are available: Artificial neuronal networks (Yaghini et al. 2013), support vector machines (Markovic et al. 2015), decision trees (Van Riessen et al. 2016), and k-nearest neighbor algorithms (Chang et al. 2010), among other methods. Although they can be applied for real-time analyses, data-based approaches still represent the minority of scientific literature compared to model-based approaches.

3 Methodology

The system and demand analysis with the associated steps were performed to gain a deeper understanding of the maritime transport chain and identify benefits as well as requirements of an ETA prediction. This comprises a comprehensive analysis of operational processes, IT processes and disruptions by literature research and strong involvement of market representatives. Partner companies were involved in varied ways. On one hand, they have contributed important technical information and provided the possibilities to inspect various facilities, while on the other hand, they have provided the necessary operational data.

In order to verify and expand the elaborated results, semi-structured interviews with various companies for every actor role along the maritime transport chain were conducted (Fig. 2). To quantify the interview statements of the expert interviews, a survey based on standardized questionnaires was carried out afterwards. This two-stage approach ensures comparability of the results, and at the same time, enables the possibility to take into account the specific requirements of the actor. On this basis, the benefits and the preferred processes for the use of an ETA as well as the technical implementation possibilities with regard to data availability were derived and transferred into prioritized application cases, from both perspectives – overall and actor-related. Altogether, the views of 13 actor roles in the maritime transport chain are taken into account in the analysis by conducting interviews with 16 companies with one to five representatives each, lasting about two to four hours.

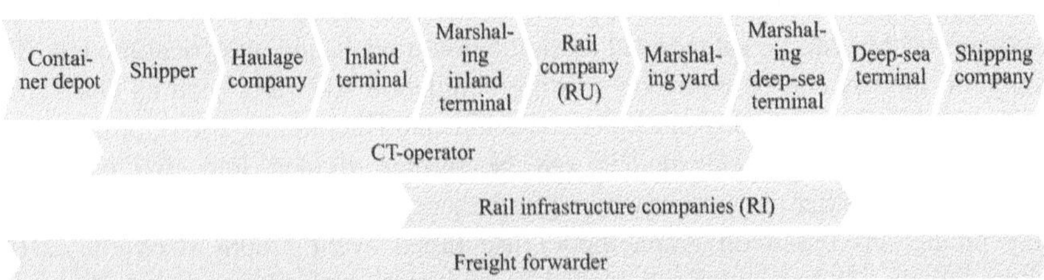

Fig. 2. Actor roles in the maritime transport chain

Workshops with several actors were also considered as an alternative to the bilateral expert interviews. Within the workshops more attention could have been paid to the interactions between actors. Bilateral expert interviews were ultimately preferred, as the competitive situation of various actors would have reduced the likelihood of participation in a workshop and biased the answers. However, afterwards a meeting for discussion and validation of the analysis results took place together with all interview partners.

4 Results

Within SMECS, a multistage approach to develop the model was conducted to meet the complexity of the maritime transport chain (Fig. 3).

In order to gain the necessary knowledge of the system, the first step involves the identification and systematization of the key operating processes as well as the accompanying information flows. For each process, during an expert- and data-based analysis the most important disruptions and their causes are identified and modeled. In the next step, specific data is identified that is relevant for predicting the observed disruption causes and influencing factors. After acquiring the data, this step can be used to uncover other previously unknown correlations in the occurrence of delays.

In parallel to the system analysis, a demand analysis was conducted to gain information about specific actor-related benefits and preferred application cases and process sections for the use of an ETA. For these cases, functional and non-functional requirements were elaborated.

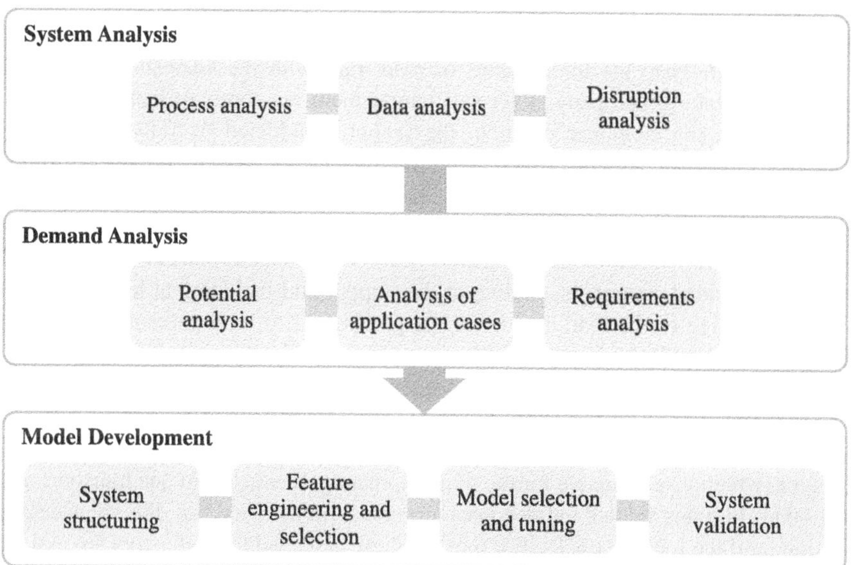

Fig. 3. Key activities of the project

In the next step of the model development, the complex problem of the ETA prediction is divided into several independent, but interacting, sub-systems. By considering and testing various approaches, the models for the sub-systems will be developed, trained and validated by methods of machine learning and the use of sub-system-specific data. The results of the different phases are presented in the following sections.

4.1 Processes Analysis

The pre leg of the maritime transport chain comprises various operational processes, which includes the physical processes of the container handling as well as associated administrative and planning activities. Whereas the physical processes start with the first movement of the empty container at the container depot or of the loaded container at the shipper (consignor), the administrative and planning processes are proceeded either in parallel to handling processes (e.g. customs, quality checks) or in advance (e.g. commissioning, document preparation). Involved companies in the transport chain cover different operational activities, strongly depending on the individual profile and service portfolio of each company. Accordingly, for an analysis and systemization it is rather useful to separate the transport chain into various actor roles regardless of any specific company characteristics. For this generalized picture of the transport chain, in industry and science there is a common understanding about the assignment of physical, administrative and planning processes to the various actors.

The process variants of the pre leg differ according to the involved actors. There are actors and associated processes, which are part of the execution of each transport order. This applies to the aforementioned container depot and the shipper, where the empty and loaded containers are collected, as well as to the seaport (deep-sea terminal), where the containers are transshipped to the sea transport by loading them to the sea ship. The involvement of other actors depends on the selected means of transport of a specific order. Accordingly to the three means of road, rail (with the associated actor of the railway undertaking company, RU) and barge, there are three variants of an entire transport from the container depot to the seaport conducted by only one mean of transport. Two other variants also include the transport of loaded containers from the shipper to the seaport by rail or barge only, but imply the collection of empty containers by truck. Those variants of an involvement of more than one transport mean do not yet fulfill the criteria for a combined transport (CT). This rather applies to variants for the transport of loaded containers, starting at the shipper and including at least two means without a change of the load carrier for the goods (in this case container) (Holderied 2005). The prevailing variants of CT are a combination of barge and truck as well as rail and truck, whereas the latter, in the sequence of rail and truck is the most common configuration of CT (Arnold et al. 2008).

In any combined transport, an inland or CT terminal as an additional actor is involved, which is responsible for the transshipment processes from one transport mean to another. In many inland and deep-sea terminals, trains could not directly access the location until the loading tracks. For most of these cases, additional actors are assigned to marshal the trains or wagons from the main rail infrastructure to the terminal tracks. Due to different processes, it is useful to distinguish between the marshaling roles for both types of terminals. Besides the inland terminal, in many combined or entirely rail transports there is a further node point, and consequently actor, for an additional train composition during the transport process. From an export perspective, trains arriving at these marshaling yards will be reassembled according to specific sequences for the supply of deep-sea terminals. Whereas most of the maritime actors process all types of operational activities, there are also a few overarching actors, which only cover administrative and planning processes, e.g. freight forwarders, the CT-operators and railway infrastructure companies (RI).

In order to gain the necessary understanding of the maritime transport chain in the SMECS project, the main operational processes including physical, administrative and planning activities are recorded and systematized. The physical processes of the loaded containers from shipper to sea transport and vice versa for the two prevailing variants of the landside maritime transports in Germany according to cargo volume (entirely truck transport and combined truck-train transport) are shown in the appendix.

4.2 Data Analysis

Data Requirements

The availability of data is an essential prerequisite for the realization of an ETA prediction using artificial intelligence. Required data for the approach could be categorized into different types, whereas a primary separation into process time data and data for qualifying delays (explanatory data) is useful.

For the project phase of system development, especially for the training of the prediction models, most important data are about historical transport orders with process information on the planned time (target values) and associated actual time values to identify occurred deviations (process delays). All time values must be linked with definite locations (addresses or coordinates). While planned times are provided discretely for specific measuring points like important milestones of the process (events), actual times are captured either as discrete timestamps or as continuous values by GPS tracking.

To understand the deviations of planned and actual time, further data are needed to explain the associated causes. These explanatory variables could be process-related disruption information, which is mostly directly recorded by the respective actor for reporting, quality or liability reasons. This disruption information comprises either container-related causes, that directly qualify a specific process delay of a container, or generalized information, which are not systematically linked with a specific container delay but reference disruptions concerning activities, resources or infrastructure of the affected process, e.g. information about congestions or construction sites on used routes of the delayed container.

Besides the process-related disruption information, there are many other data sources with information about environmental conditions and other factors, which have (potential) influence on the process time. These explanatory data with influence factors could be either further operational data (internal data) or external data. For operational data major types include information about:

- Infrastructure in terms of general characteristics (e.g. route length, number of switches), capacity (e.g. number of cranes, slots or tracks in terminal) and utilization (e.g. traffic situation on road and rail, number of containers in terminal),
- Resources like used transport vehicles (e.g. type, mass, length, traction) and assigned personnel (e.g. like sickness rate, planned changes of locomotive personnel),
- Order characteristics (e.g. load type, priority).

Examples for external data are weather data (e.g. temperature, wind, precipitation, warnings), geo-data (e.g. geo-positions to rail and road network, railway stations) and calendar data (e.g. vacation times, holidays).

The container handling in most processes is executed by using transport vehicles like trucks, trains or wagons. Depending on the respective actors, the used operational systems do not focus on containers, but rather on the mentioned vehicles or the associated transport order. Accordingly, the aforementioned information types are linked with different reference objects. However, a unique reference along the entire transport chain is required to ensure an end-to-end consistency of the prediction. This is only feasible with the container number, which leads to the necessity of gaining data for the respective assignment of other objects to this number. To retrace the consequences of delays in terms of missed subsequent processes planned and actual assignments are needed.

Data Availability and Maturity

The entire maritime pre leg was examined considering the availability and maturity of mentioned required data types, focusing on planned and actual process times as well as on disruption information at first. The analysis comprises the status of acquisition of the data, the associated data transmission between involved actors (e.g. information flows, source and target systems, data format and content) and the data usage with regard to already existing ETA prediction. Considering these three perspectives, a substantial potential for improvement in the transport chain was determined. Besides a lack of available digital information, there is a very limited cross-actor transparency of information since many data are not transmitted or at least not on the systemic way. Thereby, the analysis revealed different scenarios of data availability and maturity according to the type of data. Whereas the availably of planned times (e.g. timetable data) is ensured at almost all actors, the corresponding actual time stamps and delay causes are more often missing. Beyond that, the existing data are often not suitable for implementation into the prediction model due to a lack of volume (e.g. only available for a short period of time or in low granularity), validity, time of receipt, further processing possibility like unstructured in free text format or regularity.

As Fig. 4 illustrates, required process information shows little maturity concerning the mentioned criteria. This applies especially for disruption information, which is very often barely or manually acquired. The difference between values of planned and actual times results from the absence of actual timestamps for some associated set points. Both times are mostly discretely determined and captured along the maritime pre leg.

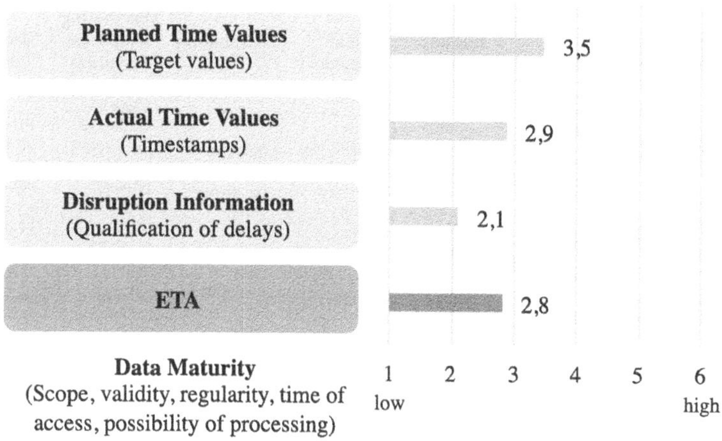

Fig. 4. Maturity of data for process information

Continuous GPS tracking for entire processes is only launched as pilot projects without sufficient data scope. Besides that ETA information is not available for several processes yet, the already existing values also show little maturity since they are not dynamically provided or are only calculated simply by focusing on expert knowledge or by linear updating prevailing delay times.

Analyzing each process, the best data is available for rail transport (Fig. 5). For all operating trains, there are comparatively dense time and location data as well as associated train-related and generalized disruption information. The inland and deep-sea terminals show the second best data. Compared with the rail transport, they acquire a smaller amount of process times and disruption information. In contrast to this, for the marshaling processes in the terminals as well as at the separate marshaling yard, there are only a few usable planned values available. Due to the prevailing disposition principle with its process execution according to the current order situation, planned times are only determined in very short-term. Moreover, the causes of delays in the marshaling yard are not systematically acquired. The same applies to road transport, for which disruption information is mostly manually recorded and does not reference a traceable place of the occurred delay. At the same time, beyond the events of departure and arrival, more planned and actual values for the road transport do not exist. However, road transport already shows better ETA solutions due to the use of external services like navigation systems and Google Maps. Comparing the landside transport means, the barge presents the worst data scenario. Among the overarching actors, the availability of information is significantly limited due to the lack of transmission. Accordingly, the scope of information decreases from the CT-operator over the forwarder to the shipper.

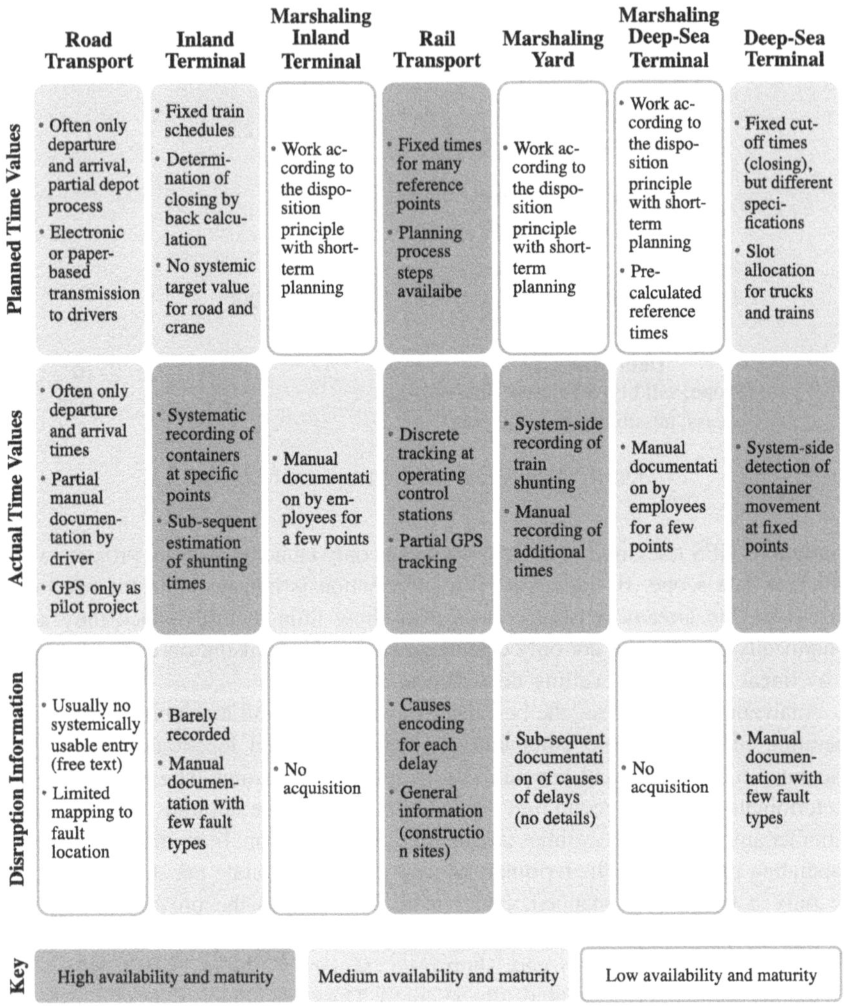

Fig. 5. Actor-related data availability and maturity

Data Relevance

As emphasized in Fig. 6, the importance of all ETA-related process information increases. Today, the planned and actual values show the slightly highest significance. In the future, it will still apply to the planned time, but together with the ETA, whereas the ETA will contain the highest increase of significance. Even though the disruption information is also getting more important, it is comparatively less today and in the future. This evaluation is reflected in the aforementioned low data availability.

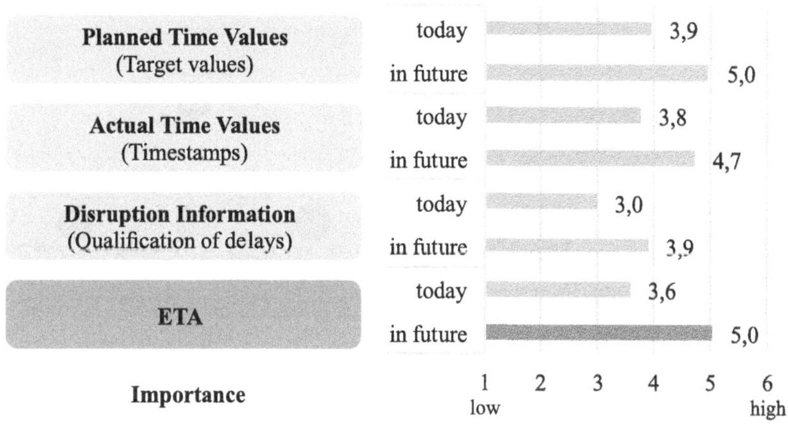

Fig. 6. Importance of process information

4.3 Disruption Analysis

In order to identify the most relevant disruptions to be considered in the prediction, a three-stage expert-based and a data-based disruption analysis was carried out. As the first step, an expert based assessment of the relevance of certain types of disruptions for all processes of maritime pre leg transport was carried out. The relevance of certain specified disruption categories was assessed on a scale of 1 (low relevance) to 5 (high relevance) on the basis of a questionnaire. In addition, the essential detailed disruptions as well as the causes and effects of disruptions were documented on the basis of expert discussions following a Failure Mode and Effect Analysis (FMEA).

In the second step, a data-based fault analysis was carried out for selected sections of the pre leg transport on the basis of historical disruption data from 10,000 train runs over 3 years. The frequency of the disruptions, spatial and temporal influences as well as the disruption effects in the form of delay minutes were evaluated.

Finally, in the third step, the relationships between causes and effects of most relevant disruptions were investigated in depth and qualitatively modeled to identify the necessary data for prediction.

The expert-based evaluation of the relevance of given failure categories shows a heterogeneous picture of possible failures, i.e. a variety of different disruption causes exists, which can lead to delays and have to be taken into account in the prediction (Fig. 7). Nevertheless, some categories are assessed as more relevant than others. The experts assessed operational (internal) causes for disruption with a higher relevance than external reasons. Among the operational causes, the limited availability of necessary process resources (i.e. process resources are not available or not usable) is considered to be the most relevant. Especially, the lack of human resources, vehicles and handling equipment (e.g. cranes) as well as the limited availability of IT systems are of particular importance for the occurrence of delays. Errors in process execution are also rated as important.

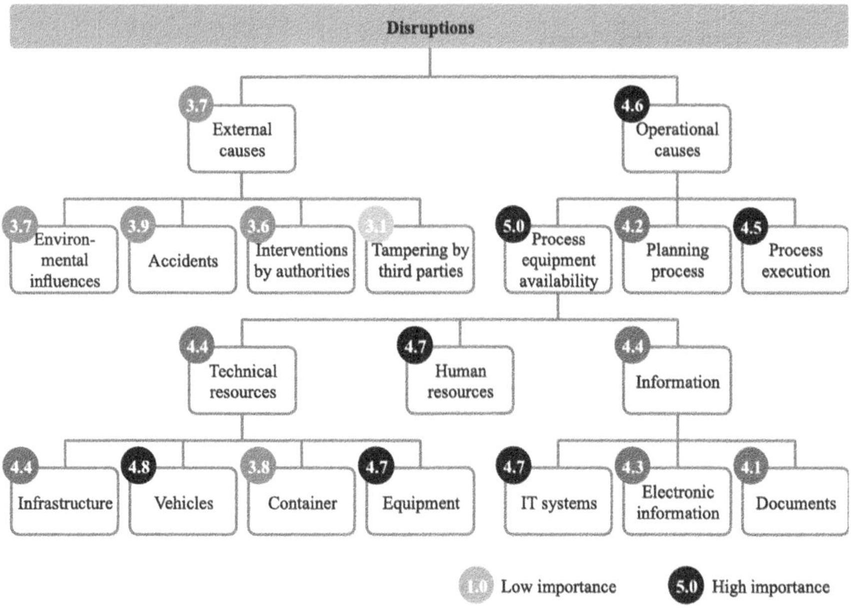

Fig. 7. Expert-based evaluation of the importance of disruption types in maritime pre leg

A data-based fault analysis was carried out for rail freight transport (Fig. 8). It has been found that delays in rail freight transport are strongly related to a small number of disruption categories. Thus, the listed disruptions represent about 15% of the documented disruption categories, but cause approximately 90% of the cumulated delay minutes of the evaluated trains.

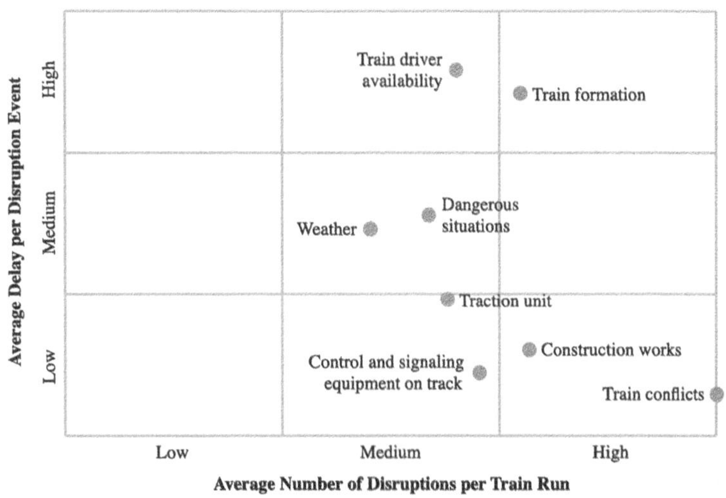

Fig. 8. Data-based evaluation of most relevant disruption types in rail transport

In an ETA prediction for rail transport, traffic effects in particular in the form of train path conflicts with passenger and freight traffic, lack of locomotive personnel, delays caused by construction sites and delays in train formation should be taken into account as causes of delays. In addition, integrating technical disruptions in the control and signaling equipment on track as well as in the traction unit and by incorporating weather effects could improve the quality of prediction. The consideration of dangerous situations is expected to be difficult due to their heterogeneity.

Finally, it should be noted that the disruption categories listed may contain several different disruptions, and that manual coding by train dispatchers may be subject to human error.

4.4 Potential Analysis

The actors of the maritime transport chain associate the provision of a cross-actor ETA with various functional benefits, which are summarized in Fig. 9. In principle, the ETA represents the possibility for better process transparency over all processes. The cross-actor approach enables users not only to quantify local disruptions effects with regard to their delays, but also to evaluate the consequences for upstream and downstream processes in particular. Using the ETA, appropriate measures could be taken by the actors, which keep the negative effects of disruptions as low as possible and still allow a fulfillment of the requirements for each order. Therefore, the ETA is primarily seen as an instrument for early support in operational decision problems in terms of process control and disposition of resources (e.g. personnel, vehicles, tools, infrastructure). This results in an improved demand and capacity management, which leads to an increase in asset utilization and a reduction in risk buffers in the transport chain. In addition to its

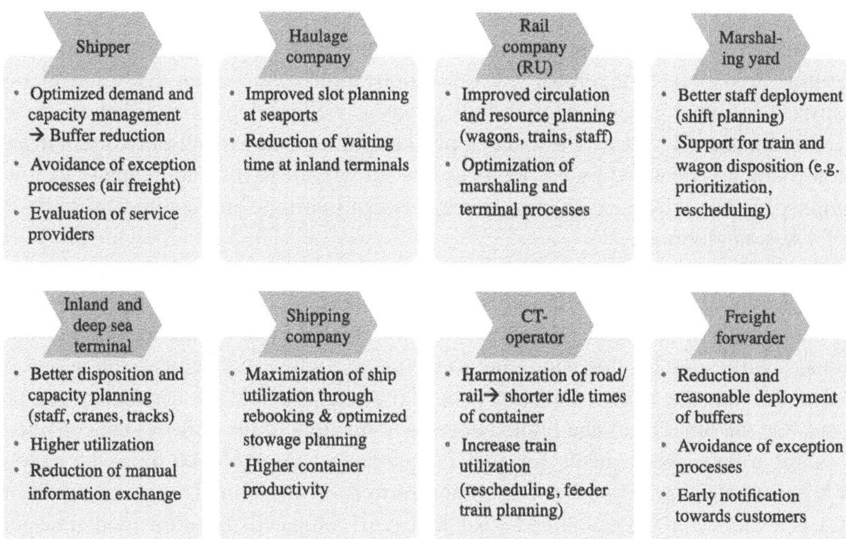

Fig. 9. Summary of actor-specific ETA benefits

use for the own operations management, the ETA is also important information for adjoining actors or customers. The enabled early communication of disruptions and delays together with improved punctuality makes an important contribution to increasing customer satisfaction. The overarching actors see the ETA also as a promising mean for quality management to evaluate the assigned service providers, as it provides improved transparency on delays and their causes. The systemic and dynamic provision of the ETA also includes the benefit of reducing the manual exchange of information with other actors.

4.5 Analysis of Application Cases

Due to individual susceptibilities of disruptions and a varying flexibility of downstream processes, the operational relevance of an ETA for the individual processes of the maritime transport chain differs. In the aggregated perspective, the involved actors attach greater relevance to landside process chains of the pre and subsequent leg than to the main leg. The major reason is the temporal decoupling of the main leg by the timetable of the sea ships. A missed ship closing due to delays in the pre leg could lead to high additional delays. At the same time, sea transport appropriate solutions for determining arrival times are already available. However, from the perspective of the overarching actor and the shipping companies, an integrated view of the legs is important to be able to evaluate the effects of delays and measures of the pre leg on the entire chain, especially on the arrival time at the consignee. In the comparison of pre and subsequent leg or export and import transport, there are no differences in relevance for the ETA. Rather, both transport directions must be considered together due to high process-related interactions, such as the circulation planning of trains between the inland and deep-sea terminal. Accordingly, a delay of an import train in the inland terminal could lead to limited crane, track and storage space capacities for the export ones.

Considering different transport variants, the process of load containers in combined maritime transport is the most relevant application case for an ETA calculation. Transports of other general cargo or bulk goods as well as empty containers are clearly of subordinated relevance. The same applies to pre leg chains with only one transport means, which is regarded by the actors as less susceptible to disruption due to its lower complexity. The difference in importance between maritime and continental traffic for an ETA is negligible.

Comparing the transport means and their combinations, rail transport represents the most relevant application, followed by road transport in combined transport (Fig. 10). On the one hand, the higher relevance of a rail-related ETA results in a higher operational significance for the actors due to the cargo volume per vehicle, the associated load peaks and the higher planning complexity, e.g. for loading and unloading processes. On the other hand, the limited degrees of freedom of the railways are considered to enable a better prerequisite for an ETA prediction. At the same time, the existing ETA possibilities are less advanced compared to road transport. The reason for higher ETA relevance for the combined road transport compared to entire road transports could be explained by the already introduced truck slot management of some seaports, such as Hamburg, which enables improved transparency of truck arrival times. Due to the comparatively smaller volume of transported goods, the barge is of the least operational relevance for an ETA.

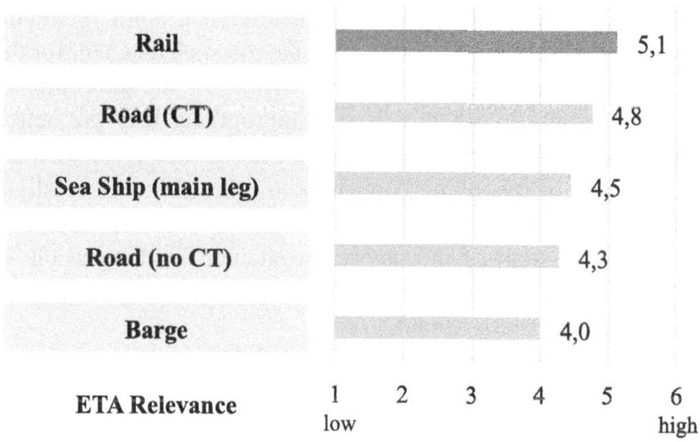

Fig. 10. Relevance of ETA for transport means

4.6 Requirement Analysis

As a major result of the data and relevance analysis, the combined road-rail-container transport from the shipper to the seaport was derived as an initial application case of the project. For the realization of the associated ETA prediction, information about functional and non-functional requirements of potential users must be considered. This comprises required reference points and objects of the prediction, the associated timing and accuracy, the prediction content as well as the way of provision.

The reference points of the ETA prediction are always process events with high operational significance, strongly depending on each actor. This comprises especially interfaces of actors, e.g. the ETA for arrival and departure of containers in terminals or for the transition to different rail infrastructures. However, reference points within processes are also required, e.g. places with locomotive personnel changes. In summary, for the combined road-rail-transport there are following important reference points:

- Departure from the shipper
- Road arrival at the inland terminal
- Rail departure at the inland terminal track
- Rail arrival at the main rail infrastructure (e.g. DB Netz)
- Rail arrival at the marshaling yard (e.g. Maschen Marshaling Yard)
- Rail departure at the marshaling yard
- Rail arrival at the railway station of the seaport (start of marshaling)
- Marshaling arrival at the terminal track of the seaport
- Container arrival at closing point (e.g. container yard or ship loading).

Although the container number is the only reference object, which ensures a consistently view on the entire transport chain, some actors like the inland terminals and marshaling yards primary need a rail-related ETA at wagon and train level. Therefore, the container-based ETA will transform and provide for other reference objects accordingly to each actor's need.

The timing of providing the ETA for each reference point (prediction horizon) depends on the feasible prediction accuracy and the associated benefit for the respective actor by an early communication of the ETA. While for the allocating of personnel within a shift a few hours in advance is still appropriate, they are activities, which requires a considerable long-term provision of the associated ETA such as registering late arrivals in seaports or rebooking containers to other ships. Generally, the ETA is requested as early as possible, whereas for the majority of the actors, a reasonable initial deployment is the start of the physical container handling at the depot or the shipper. In contrast, very short-term provisions are considered similar to the current operative planning without less benefit.

The accuracy of the ETA depends on the needed timing, the reference point and the associated benefit as well as on the arrival time of the assigned sea ship. Accordingly, for long-term prediction or in case of delayed ships the acceptance for a less precise prediction is higher. However, for some purposes a very accurate ETA is needed, e.g. for personnel planning in general and for the arrival of a truck at an inland terminal due to restricted opening hours. In general, there will be a continuous increasing of the accuracy towards the particular reference point.

Beside the ETA itself, the accuracy of each predicted value is important information for the users. In addition, there is a high need of transparency of the delay results on adjacent processes to be able to initiate appropriate measures. A clear majority of the actors prefer the calculation and provision of the ETA and the additional information by their own IT-systems due to data security reasons. Advantages for a provision via an external platform are considered in the integration of various actors and the avoidance of individual agreements or interfaces between them. However, this way of implementation requires well-established authorization and roles concepts.

4.7 Model Development

Based on the requirements analysis, a prototype solution of the ETA prediction for combined road-rail freight transport from the shipper to ship loading is implemented and tested through the example of certain maritime pre leg relations in Germany.

Different from most existing concepts for ETA predictions, a data-based approach using various methods of machine learning is used for model implementation. A key objective is to assess these AI-based approaches in terms of their superiority to existing approaches for ETA prediction.

The model development process essentially comprises four steps:

1. System Structuring,
2. Feature Engineering & Feature Selection,
3. Model Selection & Model Tuning,
4. System Validation.

Different modeling possibilities are available to build the overall system and each of its defined sub-systems. To identify the best modeling approach, a morphological box with six layers was compiled describing suitable variants, which can be combined to build ETA prediction algorithms (Fig. 11).

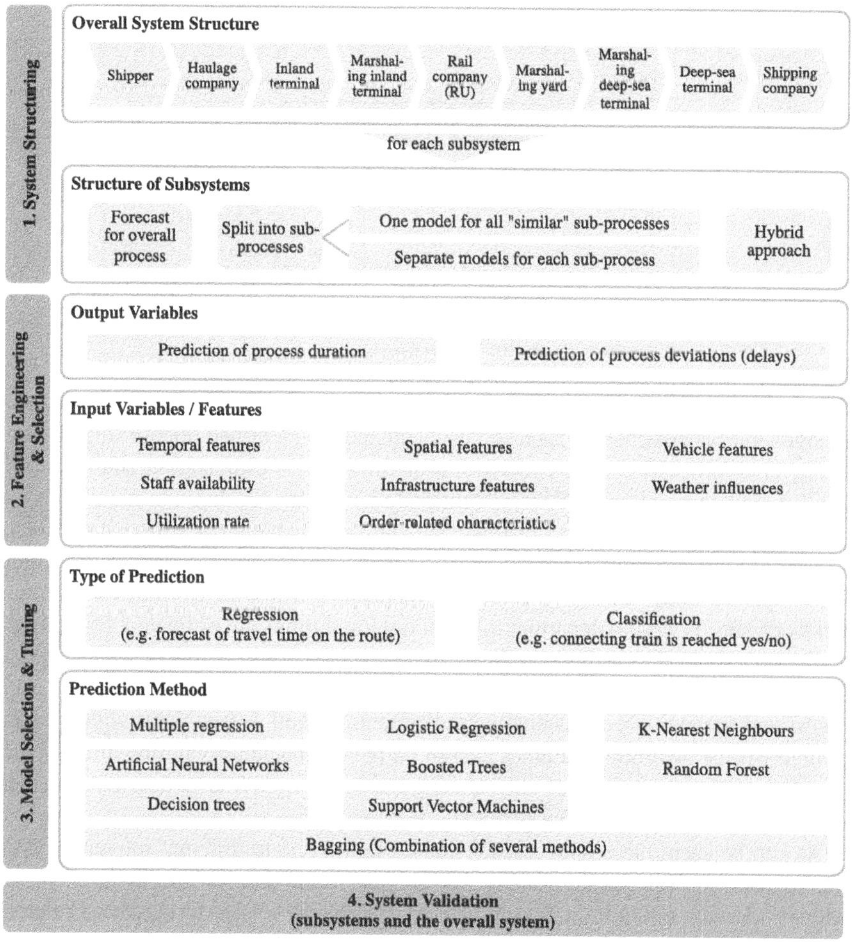

Fig. 11. Morphological box of possible modeling approaches

System Structuring

Based on the processes identified in the process analysis and the requirements determined with regards to the necessary reference points in maritime pre leg transport for which an ETA should be provided, the overall system structure is specified by defining suitable sub-systems which can be implemented separately (Fig. 12). The breakdown into sub-systems is necessary because the individual processes of the maritime transport chain differ with regards to modeling, data availability and relevant input data as well as the relationships between input and output parameters. The definition is made in

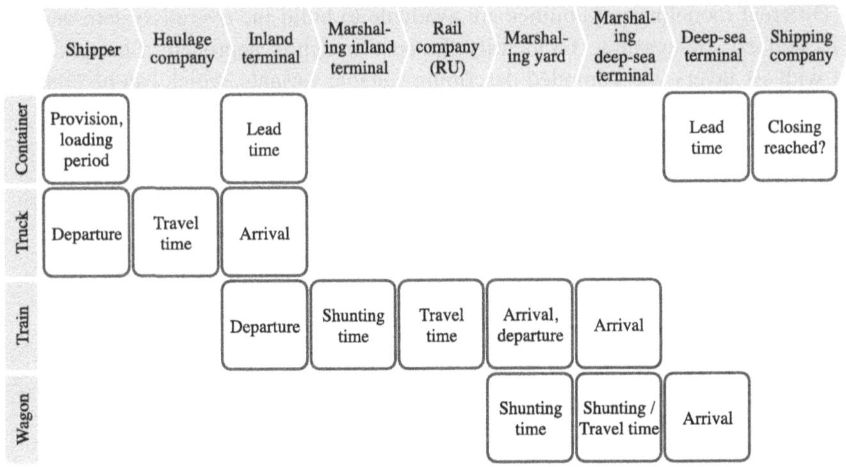

Fig. 12. Required sub-systems s for ETA prediction

such a way that the relationship between input features and model output are as homogeneous as possible for each sub problem and the identified relationships between the operating processes are considered. Furthermore, ETA estimates should be possible for all reference points which were identified as relevant for decision-making.

Depending on the modeled process, the sub-systems generate different outputs. In the case of transport processes of trucks and trains, the travel time will be determined. In the logistics nodes inland terminal and marshaling yard, arrival times of incoming trucks or trains as well as departure times of outgoing trains are estimated. Furthermore, the minimum lead-time will be predicted to decide if a certain shuttle or feeder train can be reached. In the area of port authority and port terminal arrival times of trains, process and travel times of marshaling movements as well as lead times within the port terminal will be determined. Finally, based on the estimated arrival time of a container and information about planned ship closing the system provides a decision whether a certain container will reach the planned container ship.

As can be seen, the output of each sub-system refers to different reference objects (e.g. trains, wagons, trucks, containers). By means of data on the assignment of a container or wagon to a train, all predictions can also be related to these two reference objects. This enables the ETA prediction to be used for different users and planning problems, as the ETA can be determined for each reference object that is relevant for the decision to be made on the basis of the ETA.

In order to generate the overall ETA, the sub-systems are integrated into an overall model logic. In this way it is possible both to map different transport chain configurations through the models and to determine ETAs for different intermediate points in the chain.

For each sub-system, various model structures with different granularities are possible. Specifically, either one prediction can be done for the overall process duration or the process can be split into further sub processes, whose duration will be predicted separately. In case of multiple sub processes, either an independent model for each sub process can be determined or one prediction model can be applied to multiple sub processes (e.g. one model for all lead time predictions at different inland terminals). As this approach allows increasing the amount of training data per model but requires a high similarity of the considered sub-processes, a balance must be achieved between an accurate modeling of process-, spatial and time-specific characteristics and the amount of training data.

Feature Engineering and Selection

Regarding the output of each sub-system, either the process duration (actual time) or the process deviation from the planned value (delay) can be predicted. To predict the process durations or deviations as best as possible, appropriate input parameters must be found which highly correlates with the model output and represents all crucial influences on ETA. Various types of input features are tested, comprising temporal (e.g. time of day, holidays) and spatial features (e.g. planned route), vehicle features (e.g. length and mass of a train), infrastructure features (e.g. number of cranes in inland terminal), staff-related features (e.g. planned staff changes), weather features (e.g. snow height at track), utilization rate-related features (e.g. current numbers of containers in inland terminal) and order-related features (e.g. dangerous goods container). As different input parameters can be optimal for each process, a feature engineering and feature selection process is done for each sub-system to identity the features, which lead to the highest possible prediction accuracy.

Model Selection and Tuning

Besides feature engineering, the choice of appropriate prediction methods and model configuration are important steps. With regard to the nature of the problem, a distinction can be made between a regression problem and a classification problem. For example, predicting the running time of a train can be seen as a regression problem as a quasi-continuous value should be estimated. A prediction about the probability that a certain train connection will be reached can be modeled as a classification problem as the model output will be a binary variable (1: train reached, 0: train not reached).

Several AI methods are usually possible for each sub problem, depending on the type of problem (regression or classification), the amount of data available, the number and type of input and output variables (continuous, categorical) as well as computation time. Furthermore, different tuning parameter configurations can be used for each model type and thus will be tested.

System Validation

Finally, each sub-system, as well as the overall solution, is validated regarding its prediction quality, using cross validation technique and historical data. In order to assess the suitability for a real-time use, the validation results are compared to the required prediction accuracies derived from requirements analysis.

5 Conclusion

For the maritime transport chain, there is a high demand for cross-actor systems which enable greater transparency for delays and their consequences, and which support logistics companies in carrying out appropriate measures in case of disruptions. Reasons for the lack of such solutions are primarily the heterogeneous objectives of the various actors and the diverse disruptions along the maritime transport chain. In addition, most of the monitoring and risk management tools used by the actors are not yet fully developed. Potentials resulting from processing and evaluation of existing data have so far only been utilized selectively. The SMECS project addresses this research gap by exploiting the high potentials of AI and developing a data-based cross-actor ETA prediction for the landside processes of the maritime transport chain. This allows not only an assessment of the potential of AI methods to improve complex transport chains, but also promotes in general a greater exploitation of AI in the logistics industry.

In cooperation with representatives of all actors in the maritime transport chain, process-related characteristics in terms of benefits, applications and feasibility for ETA development were investigated within the framework of a structured requirements analysis. Furthermore, the most relevant disruptions and corresponding data sources for an ETA prediction were identified. In an initial application, a prototype for the combined road-rail-container export from the shipper to the seaport is currently being developed using different AI methods like artificial neural networks and random forest.

After successful validation, this prototypically developed system for ETA prediction can be transferred to other means of transport. The resulting tools will allow the prediction and proactive communication of disruption effects to all actors involved. This leads to increased process robustness as well as process efficiency for all actors in the realization of complex transport orders and has therefore a positive effect on the reliability and cost-effectiveness of transport chains.

Acknowledgments. The authors would like to thank all dedicated experts who were available for interviews; especially the professionals from DB Cargo AG, who are also partners in the project alongside the Technical University of Berlin and Kühne Logistics University. The research project SMECS is funded by the German Federal Ministry of Transport and Digital Infrastructure as part of the IHATEC program for the development of innovative port technologies.

Appendix

See Fig. 13.

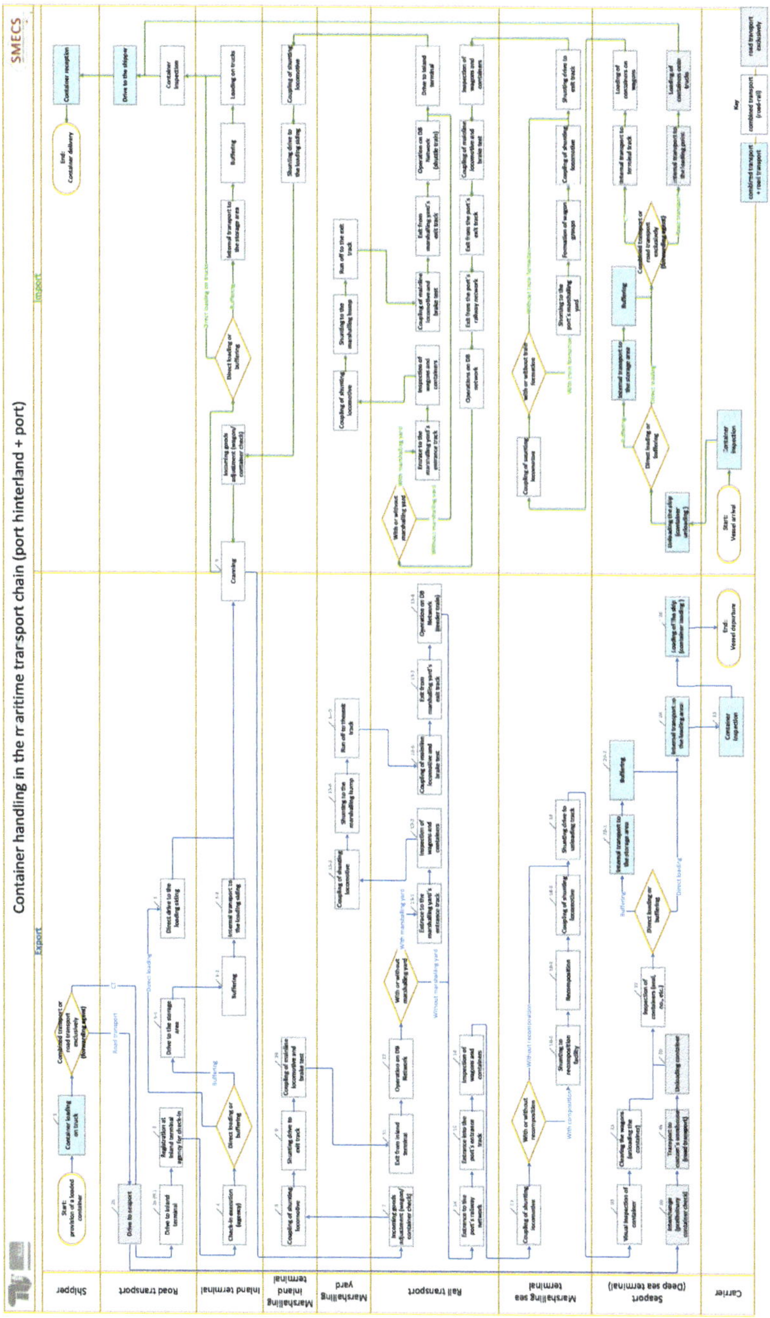

Fig. 13. Physical processes of landside container handling (export and import)

References

Arnold, D., Isermann, H., Kuhn, A., Tempelmeier, H., Furmans, K. (eds.): Handbuch Logistik, 3rd edn. Springer, Heidelberg (2008). https://doi.org/10.1007/978-3-540-72929-7

BMVI: Digitalisierung intermodaler Lieferketten – KV 4.0 (2017). https://www.bmvi.de/SharedDocs/DE/Artikel/DG/mfund-projekte/digitalisierung-intermodaler-lieferketten-kv40.html

Büker, T., Seybold, B.: Stochastic modelling of delay propagation in large networks. J. Rail Transp. Plan. Manag. **2**(1–2), 34–50 (2012)

Chang, H., Park, D., Lee, S., Lee, H., Baek, S.: Dynamic multi-interval bus travel time prediction using bus transit data. Transportmetrica **6**(1), 19–38 (2010)

Goverde, R.M.: A delay propagation algorithm for large-scale railway traffic networks. Transp. Res. Part C Emerg. Technol. **18**(3), 269–287 (2010)

Holderied, C.: Güterverkehr, Spedition und Logistik. Oldenbourg, München (2005)

IHATEC: SMECS – Smart Event Forecast for Seaports – Realisierung von akteursübergreifenden Prognose- und Risikomanagementmodellen für ein echtzeitfähiges Zuverlässigkeitsmanagement (2017a). https://www.innovativehafentechnologien.de/wp-content/uploads/2017/09/IHATEC_Projektsteckbrief_SMECS_formatiert.pdf

IHATEC: EMP 4.0 – Export Management Platform 4.0 – Ganzheitliche, digitaliserte und vernetzte Abwicklung des Exportprozesses (2017b). https://www.innovativehafentechnologien.de/wp-content/uploads/2017/08/IHATEC_Steckbrief_EMP4.0.pdf

Markovic, N., Milinkovic, S., Tikhonov, K.S., Schonfeld, P.: Analyzing passenger train arrival delays with support vector regression. Transp. Res. Part C Emerg. Technol. **56**, 251–262 (2015)

Nash, A., Huerlimann, D.: Railroad simulation using OpenTrack. Comput. Railw. **9**, 45–59 (2004)

Özekici, S., Sengör, S.: On a rail transportation model with scheduled services. Transp. Sci. **28** (3), 246–255 (1994)

Radtke, A., Hauptmann, D.: Automated planning of timetables in large railway networks using a microscopic data basis and railway simulation techniques. Comput. Railw. **9**, 615–625 (2004)

Straube, F., Handfield, R., Pfohl, H.C., Wieland, A.: Trends und Strategien in Logistik und Supply Chain Management. DVV Media Group, Hamburg (2013)

Walter, F.: Informationsaustausch in der maritimen Transportkette. Springer Gabler (2016)

Van Riessen, B., Negenborn, R.R., Dekker, R.: Real-time container transport planning with decision trees based on offline obtained optimal solutions. Decis. Support Syst. **89**, 1–16 (2016)

Yaghini, M., Khoshraftar, M.M., Seyedabadi, M.: Railway passenger train delay prediction via neural network model. J. Adv. Transp. **47**(3), 355–368 (2013)

Zegordi, S.H., Davarzani, H.: Developing a supply chain disruption analysis model: application of colored Petri-nets. Expert Syst. Appl. **39**(2), 2101–2111 (2012)

Zhang, J., Teixeira, A.P., Soares, C.G., Yan, X., Liu, K.: Maritime transportation risk assessment of Tianjin port with Bayesian belief networks. Risk Anal. **36**(6), 1171–1187 (2016)

Assessment of the Topology and Efficiency of a Railway Network: The Case of Deutsche Bahn

Arunava Putatunda(✉), Dirk Bruckmann, and Michael Schwind

Hochschule Rhein Waal, Friedrich-Heinrich-Allee 25, 47475 Kamp-Lintfort, Germany
arunava.putatunda@hochschule-rhein-waal.de

Abstract. Under the light of recent studies, it seems that not only the structure and topology of networks but also their growth follows some laws of nature. It appears that this statement is also true for transport and logistics networks. Moreover, specific attributes of a given network like small-world and scale-free properties have additional influences on the efficiency of a network. The principal concern of this study is to assess the topology and structure of the different types of weighted and unweighted networks of "Deutsche Bahn (DB)" or German Railway and to evaluate their local and global efficiencies. Additionally, the different types of efficiencies of a network are used to develop an algorithm that will further help in optimising the network in the case of some planned addition or alteration of edges to the underlying or basic topology of the network. The assessment processes are divided into two parts and carried out on three different networks (both weighted and unweighted) namely, the core infrastructure network of DB and the two service networks, i.e. the long-distance InterCity (IC) service network and the high-speed InterCity Express (ICE) service network. These networks are evaluated under the lens of six different indicators; the *average shortest path Length (L), clustering Coefficient (C), network Density (d), power law Coefficient* (γ)*, local Efficiency* (E_{Loc}) and *global Efficiency* (E_{Glob}). Finally, the results from the analyses are compared to the results available from the other studies on different railway networks of different sizes (as the transportation and logistics networks show similar properties irrespective of their sizes). The study reveals that the service networks are more efficient than the infrastructure network. Additionally, the ICE service network turns out to be more efficient than the IC service network. Further, during the development process of the algorithm, this study finds certain deficiencies in the existing quantification of the efficiency of a network and tries to form a basis to address these deficiencies.

Keywords: Weighted and unweighted network · Clustering coefficient · Transitivity · Average shortest path · Local and global efficiency · Flow in the network

U. Clausen et al. (Eds.): ICPLT 2019, LNLO, pp. 177–192, 2019.
https://doi.org/10.1007/978-3-030-13535-5_13

1 Introduction and Motivation

Transportation and logistics systems play a critical role in the modern world. For a given region, they may be considered as one of the most important indicators of that region's economic growth and development. They provide mobility which in turn allows the movements of goods and individuals from one place to another. A transportation and logistics system of a given region incorporates three most essential components, namely the network or the fixed infrastructure, the mobile infrastructure or the rolling stocks also known as the vehicles and the plans and policies of operation. A transportation and logistics system of a region is formed by a complex interaction among these three components. Out of these three, perhaps the network infrastructure of such a system is the most critical construct of the system, as it requires a considerable amount of investment and caters us for an extended period. Thus, it becomes an absolute necessity to carry out more innovative and ingenious research to understand the transport and logistics networks comprehensively, which in turn may directly influence the quality of life and may optimise the number of initial investments to build and modify them.

Recent studies prove the fact that networks could be used to model puzzling natural or artificial systems having complicated internal interactions within their constituting elements. The elements of such a system can be represented as the nodes of a network and the interactions between them can be characterised as the edges of that network [1]. Till now, a considerable number of studies have been carried out to understand the patterns and complexities of social networks [2], computer and communication networks [3, 4], neural networks [5], biological networks [6] and so on. In recent years the spatial networks such as transportation and logistics networks are also receiving much-required attention from the researcher communities worldwide. Examples of such research include studies on shipping networks [7], air transport networks [8], urban street networks [9], railway networks [10, 11] and so on. These scholarly works predominantly focus on the properties of spatial networks that are related to their topology and structure. However, the number of works that deal with the robustness [12] and efficiency [1, 13, 14] of such spatial networks are quite limited in number.

This study solely deals with transport and logistics networks. In this research, the topology and the structure of the different types of the networks (weighted and unweighted) of "Deutsche Bahn (DB)" or German Railway and their local and global efficiencies have been evaluated. The unweighted network is considered because all the edges in this network weigh unity, thus representing a topological or relational graph of the actual network. However, the weighted network with the direct geographical distances between the nodes as edge weights represents a more real-life situation [14]. The evaluation is carried out in two parts, namely the assessments of the chosen indicators of both the weighted and unweighted networks (Average shortest path length (L), Clustering coefficient (C), Network density (d), Power law coefficient (γ), Local efficiency (E_{Loc}) and Global efficiency (E_{Glob})) and a comparison of the results with available data. The study considers the core infrastructure network of DB, the regular long-distance InterCity (IC) service network and the high-speed InterCity Express (ICE) service network. Additionally, this study deals with the development of an algorithm to optimise a network utilising its efficiency after adding a new edge between the nodes. This work further tries to develop the quantification of the notion of efficiency [1, 14] of a network.

1.1 Transport and Logistics Networks as Small-World and Scale-Free Networks

For a long time, the topology of connectivity of nodes in networks is assumed to be either completely random or completely regular [15]. However, recent path-breaking research works of Watts and Strogatz (1998) show that many networks such as biological networks and social networks are not at all wholly random or disordered or completely regular. Instead, they fall somewhere in between these two extreme ends of the scale [15]. A majority of real-world networks fall in this category. On the one hand, they have small characteristic path lengths which resemble the random networks while, on the other hand, they show high clustering [15, 16]. The fuzziness of such networks is explained by the following figure (Fig. 1). Such networks are called "Small-World Networks" [15–17].

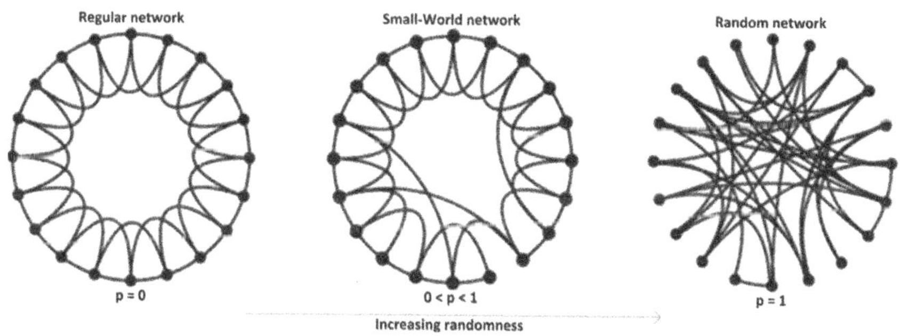

Fig. 1. At $p = 0$, the network is connected in complete order, and at $p = 1$ the connections are entirely random. The small-world networks are denoted by the fuzzy state, i.e., $0 < p < 1$ [15, 16].

In 1999 the ground-breaking work of Barabási and Albert showed that the large-scale networks show the properties of self-organisation [16]. Quantitatively, this meant that the probability of a connection of a new node to an existing network is not uniform, but somehow preferential, i.e. a new node connects (or tends to connect) to the nodes with high degree (hubs). This phenomenon gives rise to the hubs which are the highly connected nodes in the network [16] ultimately giving rise to the property of power-law. Such networks are also called scale-free networks [16]. One class of network that shows such properties is the class of transportation and logistics network. Numerous research works [7, 10, 18] have been carried out on different types of transportation networks which support this fact. Thus, it can be stated that the hub and spoke structure and scale-free properties of transportation and logistics network is a naturally grown.

1.2 The Case of Deutsche Bahn (DB)

DB is the biggest rail based mover of Germany. Additionally, DB claims to provide services to 130 different countries in the world. In 2017, it served about 2,075 million passengers in Germany alone through its rail-based services. With an overall upward trending revenue (42.693 billion Euro in 2017) it contributed a substantial percentage of revenue and service to the German economy. Regarding passenger services, DB sold 95,854 million person-KM in 2017 [19]. However, unfortunately, there are not sufficient studies available which evaluate the topology, structure and efficiency of such a big network. A look at the points mentioned above may enhance the ease of network maintenance and customer satisfaction dramatically.

2 The Analyses

2.1 Data Acquisition

The first step of the analysis includes data collection. The data is obtained from "Register of Infrastructure" of DB [20]. From this data source, a list of nodes (railway stations, halts, switches, diversions, turns and important service points) and connections between them along with their geographical longitude and latitude are acquired. These data are transformed into an infrastructure network of DB. Any node that falls on an edge, i.e. the nodes where a train can pass through is omitted, unless such nodes fall in particular places, e.g. border areas or have some special significance, e.g. serving a remote area in the form of a loop. Moreover, the network is assigned with the direct geographical distances between the nodes as edge weights, which are obtained from their longitudes and latitudes. Subsequently, two service-network are established which are the sub-network or the subgraph of the greater infrastructure network. The infrastructure network is topologically more comprehensive and exhaustive than the two service networks. The following figure (Fig. 2) distinguishes between these three networks and the abstractions. The data collected from DB sources are then modelled with the help of software R to build the required networks for analyses. Following figures (Figs. 3 and 4) represent the modelled networks. For the infrastructure network of DB, a total of 1766 nodes and 2452 links or edges relevant to this study are identified and considered. While for the long-distance IC and high-speed ICE service networks, 94 nodes with 143 links and 60 nodes with 89 links are identified and considered respectively.

2.2 Abstraction of the Networks

To carry out the analyses efficiently on the developed networks, an abstraction process is carried out during the data acquisition and network development stages. The infrastructure network is comprised of many nodes which are necessary for smooth operations but are seldom perceived during a train service run (IC or ICE), e.g. switches. There are specific nodes and stops where a train can pass through. Such nodes are omitted. In this study, these abstractions are termed as "Network-level abstractions".

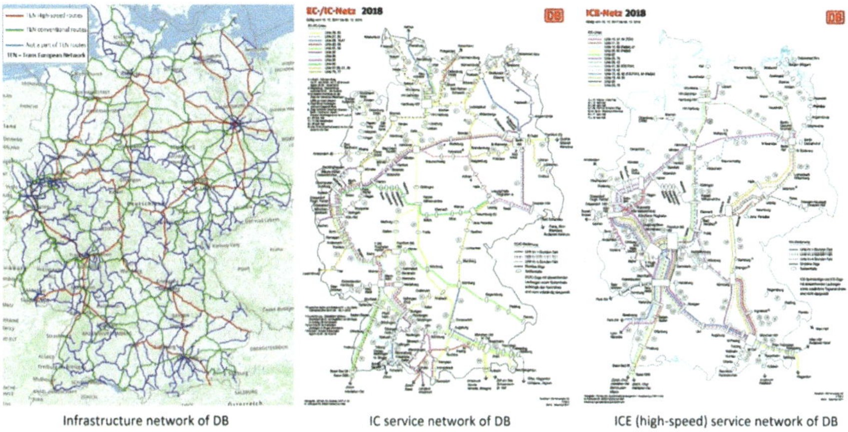

Fig. 2. Critical differences between DB's infrastructure network (the route colouring follows TEN classification), long-distance IC service net and high-speed ICE network [19, 20].

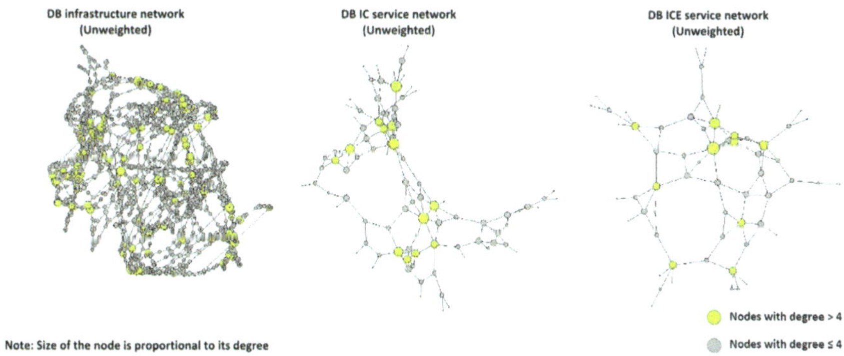

Fig. 3. The modelled unweighted infrastructure network, long-distance IC service network and high-speed ICE service network.

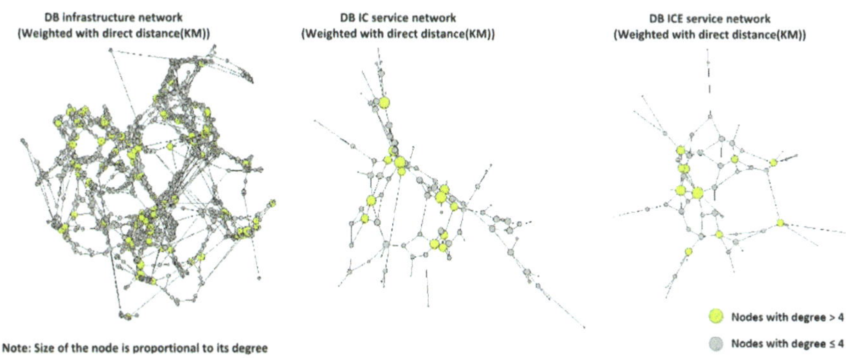

Fig. 4. The modelled weighted (considered weights are the direct geographical distances between the nodes) infrastructure, long-distance IC service network and high-speed ICE service network.

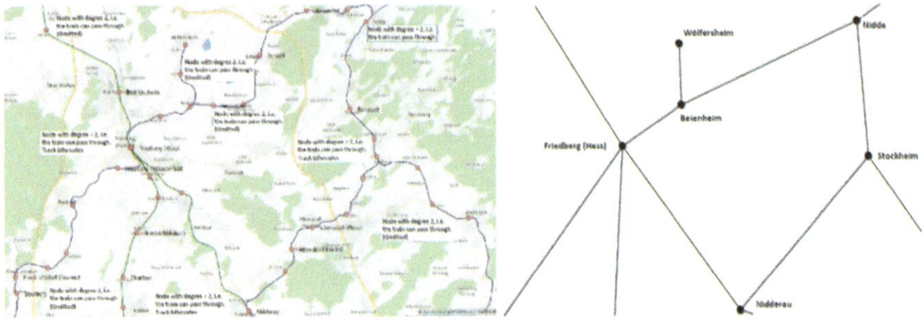

Fig. 5. Example of network-level abstraction.

Additionally, stops (considered in this study) which have multiple switches, turns and diversions within and around them are considered more comprehensively. Such stops (e.g. München Hbf) consisting of multiple nodes are considered as one node (Fig. 6). For the local services like S-Bahn, a range of nodes are perceived during the service runs around such stops; additionally, such nodes consist of switches, low hierarchy halts and so on. These points are omitted while developing the service networks. Such abstractions are termed as "Stop-level abstractions. Additionally, any node where the lines bifurcate is taken into consideration in both the cases. The figures (Figs. 5 and 6) describes the abstraction process.

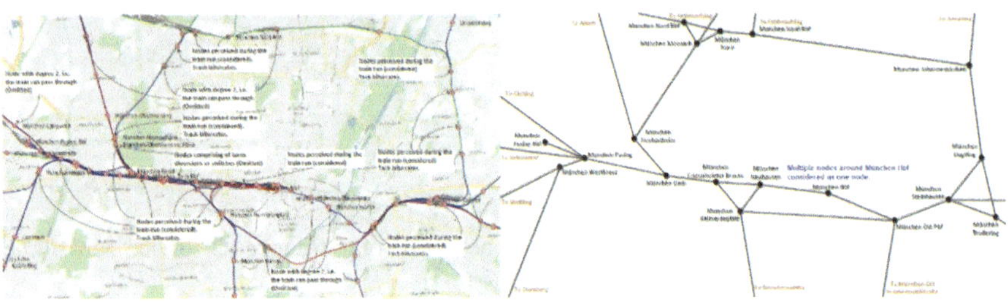

Fig. 6. Example of stop-level abstraction.

2.3 Indicators for Analysing the Networks

To assess the topology and structure of the networks, the fundamental intrinsic properties of the networks are considered as indicators. These indicators are evaluated and assessed for all the three networks in both weighted as well as unweighted form. The following paragraphs describe the considered indicators.

2.3.1 Average Shortest Path Length (L)

The Average shortest path length (L) between two vertices of a network (graph) is a significant intrinsic property to ascertain the efficiency of that network. In the subject of social networks, L is also commonly known as degrees of separation. In addition to this, L also assists us in determining the topology of the network [14, 17]. Since this research deals with the spatial network, thus Eucleadian Distances (direct geographical distances in KM) between the vertices are considered. The following formula has been considered to calculate the value of L.

$$L_G = \frac{1}{n(n-1)} \sum_{i \neq j \in G} d_{ij} \tag{1}$$

Where L_G is the average shortest path length in a graph G, n is the number of vertices or nodes and d_{ij} is the shortest distance between vertices i and j such that $i \neq j$.

2.3.2 Clustering Coefficients (C) Here Transitivity

The phenomenon of clustering (transitivity) is vital in understanding and analysing a network. In social networks as well as other networks including spatial networks, transitivity relates two nodes with the enhanced probability of connection between them. In other words, if vertex i is connected to vertex j and at the same time, vertex i is connected to vertex k, then high transitivity implies that there is an enhanced probability of an existing connection between vertex j and vertex k [14]. Mathematically, transitivity is represented as:

$$C_G = \frac{3 \times \text{number } of\ triangles\ in\ G}{\text{number } of\ connected\ triples\ of\ vertices\ in\ G} \tag{2}$$

where C_G is the transitivity of graph G.

2.3.3 Network Density (d)

Sparsity is a well-known property of small-world networks [15]. The spatial networks (transportation and logistics networks) also show sparsity of connection, i.e. the actual number of connections between the vertices of such networks are much smaller than the total number of possible connections within them, i.e. if K is the total number of edges present in a network which is sparse, then $K \ll N(N-1)/2$ [14]. Thus, to understand and analyse the topology of such a network, the phenomenon of sparsity must be quantified. It can be achieved by the following formula:

$$d_g = \frac{\text{number } of\ edges\ in\ G}{Total\ \text{number } of\ edges\ possible\ between\ all\ the\ nodes\ in\ G} \tag{3}$$

where d_g is the density of graph G, N is the number of vertices in G.

2.3.4 Power Law Coefficient (γ)

In 1999, a path-breaking work of Barabási and Albert showed that the mechanism of network growth is not random but preferential. Due to this growth mechanism, the property of scaling emerges in a network, i.e. the probability of interaction $P(k)$ of a vertex with k other vertices follows $P(k) \sim k^{-\gamma}$. The power law coefficient (γ) indicates the degree of self-organisation of large networks into a scale-free state [16]. The value of γ is obtained from a log-log plot of the degree distribution of the nodes of the network.

2.3.5 The Efficiency of a Network

Two significant scholarly works of Latora and Marchiori paved the path for determining the efficiency of networks. Qualitatively, the efficiency of a network may be defined as the ease of information exchange between the nodes of a network [1, 14]. The efficiency of a network may be calculated for the whole of the network (global efficiency) or for local sub-networks (subgraphs) around every individual node of the network (local efficiency). As described above, the phenomena of clustering may also be regarded as the primary estimation for efficiency [1]. The efficiency or ease of information exchange between vertices $\boldsymbol{i}$ and $\boldsymbol{j}$ in a network is inversely proportional to the shortest distance $\boldsymbol{d_{ij}}$ between them [1, 14]. These assumptions allow us to quantify the notion of the efficiency for a network.

The average efficiency of a graph $\boldsymbol{G}$ (network) may be represented as [1]

$$E(G) = \frac{1}{N(N-1)} \sum_{i \neq j \in G} \frac{1}{d_{ij}} \tag{4}$$

The average efficiency or $\boldsymbol{E(G)}$ is normalised with the efficiency of a complete variant of the graph $\boldsymbol{G}$ also termed as $\boldsymbol{G_{id}}$ with $\boldsymbol{N(N-1)/2}$ edges. Thus, the value of $\boldsymbol{E(G)}$ always lies in the range of [0, 1], i.e. $\boldsymbol{0 \leq E(G) \leq 1}$. This quantity is also termed as the *Global efficiency* (E_{Glob}) of graph $\boldsymbol{G}$ [1, 14].

The *Local efficiency* (E_{loc}) of the same graph $\boldsymbol{G}$ is defined as the average efficiency of the subgraphs of each nodes in $\boldsymbol{G}$. Mathematically it is represented as:

$$E(G_{Loc}) = \frac{1}{N} \sum_{i \in G} E(G_i) \tag{5}$$

Where $\boldsymbol{N}$ is the number of nodes in $\boldsymbol{G}$ and $\boldsymbol{E(G_i)}$ is the average or global efficiency of the immediate subgraph of node $\boldsymbol{i}$ [1, 14]. $\boldsymbol{E(G_{Loc})}$ represents the local tolerance of a network against faults i.e. how efficiently the network works locally if a node $\boldsymbol{i}$ is removed from it [1].

3 The Experiments and Their Results

In this section, the results of the analyses are presented. The first part of the analyses is carried out to assess the topology and structure of the three selected networks. This part explicitly refers to the indicators namely, average shortest path Length (L), clustering Coefficient (C), network density (d) and power-law coefficient (γ) obtained from the analyses on the three selected networks. The second part of the analyses is carried out to determine the efficiencies of the networks. This part refers to the Global and Local efficiencies of both the weighted (with direct geographical distances) and the unweighted (edge weights as unity) form of the selected three networks.

3.1 Assessment of the Topology and Structure of the Networks

3.1.1 Average Shortest Path Length (L)

The Average shortest path length (L) for the selected networks namely, DB's infrastructure network, DB's IC (long-distance) service network and DB's ICE (high-speed) service network are presented in a tabulated form below (Table 1):

Table 1. Average shortest path length (L) of the three selected networks in both weighted and unweighted form.

L	DB Infrastructure Network	IC Service Network	ICE Service Network
Unweighted	22.439	6.215	4.938
Weighted	363.152	429.747	379.471

The units for L are "Units" and "KM" for the unweighted and the weighted networks.

3.1.2 Clustering Coefficient (C) Here Transitivity

The Clustering coefficients (C) (in this case transitivity) of the three selected networks are presented below (Table 2):

Table 2. Clustering coefficients (C) (in this case transitivity)

C	DB infrastructure network	IC service network	ICE service network
Unweighted	0.142	0.241	0.22
Weighted	0.142	0.241	0.22

3.1.3 Network Density (D) and the Power-Law Coefficient (γ)

The Network density (d) and the power-law coefficient (γ) of the three selected networks are presented below in Tables 3 and 4 respectively. Additionally, figure (Fig. 7) represents the log-log plot of the degree distribution of nodes of the DB's infrastructure network.

Table 3. Network density (d)

d	DB infrastructure network	IC service network	ICE service network
Unweighted	0.001595	0.032715	0.050282
Weighted	0.001595	0.032715	0.050282

Table 4. Power law coefficients (γ) of the selected networks

γ	DB infrastructure network	IC service network	ICE service network
Unweighted	2.2741	0.8589	1.3111
Weighted	2.2741	0.8589	1.3111

The values of C, d and γ remain unchanged for both the weighted and unweighted network variants because the connection topology of the networks remains the same for both the variants.

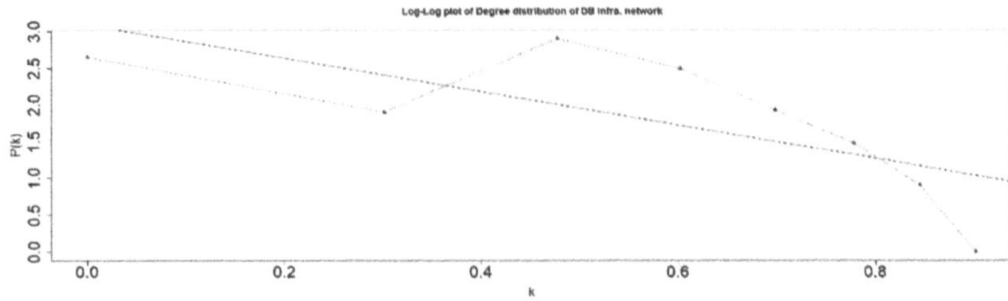

Fig. 7. Log-Log plot of the degree distribution of the DB infrastructure networks.

3.2 A Brief Discussion

The first part of the analyses yields results that unveil the true nature of the three selected subject networks (weighted and unweighted). All the three networks show sparsity and scale-free properties, whereas the two service networks additionally exhibit the properties of the small-world network type. The reason can be traced by studying the precise geometry, abstraction and structural differences between the DB's infrastructure and the service networks. Geometrically and structurally, the infrastructure network is more complicated than the service network. The former contains more nodes (bifurcations, switches and so on) than the latter. These extra nodes are not directly perceived by the service runs. Due to these reasons, the infrastructure network which seems to be a small-world network intuitively has a much longer actual average shortest path length (L) (here the L of the unweighted infrastructure network has been considered, as it provides the absolute degree of separation) than the small world networks. For the weighted variants of all the networks, the average length of an average edge is 16.184 km, 69.147 km and 76,847 km for infrastructure, IC and ICE service networks respectively (these values are obtained from $L_{Weighted}/L_{Unweighted}$). This assessment proves that the average edge length of the infrastructure network is much smaller than the same of the service networks, supporting the fact that the former contains more nodes that are not directly perceived by the service runs. The IC and ICE service networks have longer average edge lengths supporting the fact that there is a lesser number of redundant nodes in these networks. The infrastructure network along with the service networks show the scale-free properties because there exist multiple hubs instigating the preferential attachment mechanism (of the nodes with low degrees) with them [16]. The indicator, network density (d) supports the fact that the ICE service network is the densest followed by IC service network and the infrastructure network. Moreover, the indicator transitivity (C) supports the fact that the IC service network is much widespread (to the remotest corner) and connected than the ICE service network because ICE high-speed services are available to the nodes of a higher order[1]. These differences are comprehensively described by the preceding figure (Fig. 2).

[1] The ranges or benchmarks namely C (<1), L (≤ 6), γ (example dependent, usually <3) and d (<10%) are derived from the various example (Internet, Power-grid, various social networks and so on) of small-world and scale-free networks described in the works of Watts and Strogatz [15], Barabási and Albert [16], Gaure (1992) and Sohn et al. (2017).

A comparison between these indicators of the Indian railway (IR) network and the selected networks of DB are given in the table (Table 5) [10]. Although of completely different sizes, but these railway (transportation) networks exhibit similar properties like Hub and Spoke character, sparsity, low L and so on and thus are compared together.

Table 5. Indicator comparison between Indian railway network and the selected DB networks [10]

	Transitivity (C)	Average shortest path (L)	Power law coeff. (γ)
IR Net	0.69	2.16 Units	1.38
DB Infra. Net (weighted)	0.142	22.439 Units	2.2741
DB Infra. Net (unweighted)	0.142	363.152 km	2.2741
DB IC Net (weighted)	0.241	6.215 Units	0.8589
DB IC Net (unweighted)	0.241	429.747 km	0.8589
DB ICE Net (weighted)	0.22	4.938 Units	1.3111
DB ICE Net (unweighted)	0.22	379.471 km	1.3111

3.3 Assessment Based on Efficiency

3.3.1 Global Efficiency (E_{Glob}) and Local Efficiency (E_{Loc})

The global and local efficiency (E_{Glob} and E_{Loc}) are presented below in a tabulated form (Table 6):

Table 6. Global and local efficiencies of the selected networks

	E_{Glob} (Weighted)	E_{Glob} (Unweighted)	E_{Loc} (Weighted)	ELoc (Unweighted)
DB Infra Network	0.00495	0.06034	0.05205	0.10887
DB IC Network	0.00451	0.22073	0.00697	0.21493
DB ICE Network	0.00605	0.27283	0.01496	0.18007

3.3.2 A Brief Discussion

The analyses by efficiency yield results as per predictions. Both the unweighted service networks are globally more efficient than the infrastructure network. It shows the presence of redundant nodes in the latter thus increasing L. However, locally the unweighted IC service network is the most efficient followed by the ICE and infrastructure networks because of the widespread and more connected nature of the IC network and low degree of separation. Although, the ICE network has the lowest degree of separation, but the service limitation to the selected specific higher and lower order nodes lowers the local efficiency of this network.

In the case of weighted networks, the ICE network turns out to be the most efficient followed by the infrastructure and IC network. It may be explained by considering the combination of the degree of separation and the shortest average distance between the nodes. Although, the infrastructure network has a high degree of separation, but the distance between the nodes are much lower than IC and ICE networks. Further, locally the weighted ICE network is most efficient followed by the IC and infrastructure network. Following table (Table 7) presents a comparison of different efficiencies of the selected networks and Boston transport network (MBTA underground and Boston Transport Network) system [1]. All the logistics and transportation networks show similar properties of sparsity, low L and so on and thus the selected DB networks are compared with the MBTA transportation network.

Table 7. Efficiency comparison [1]

	E_{Glob}	E_{Loc}
MBTA underground (unweighted)	0.10	0.006
MBTA underground (weighted)	0.63	0.03
Boston City Transport Network (Road + Underground)	0.72	0.46
DB Infra Net (unweighted)	0.06034	0.10887
DB Infra Net (weighted)	0.00495	0.05205
DB IC Net (unweighted)	0.22073	0.21493
DB IC Net (weighted)	0.00451	0.00697
DB ICE Net (unweighted)	0.27283	0.18007
DB ICE Net (weighted)	0.00605	0.014961

4 An Algorithm to Optimise a Network by Its Efficiencies

Apart from assessing the topology and structure of the selected networks of DB, this research develops an algorithm to develop a network in the case of planned addition or alteration of the edges to the basic structure of the network. It does so by providing the most optimised edge set placement in the network based on the optimised (maximised) efficiencies of the final structure of the network. This way the final (planned) network remains efficient. Efficiency is a parameter that can represent and quantify the robustness of a network [1]. The algorithm calculates the efficiencies (E_{Glob} and E_{Loc}) of a network after addition (removal is also possible but not considered in this research) of an edge (between two vertices) to the basic layout of the network. The algorithm can calculate the values for all such combinations possible throughout the network according to the hierarchy (number of degrees) of the nodes. During the development of the algorithm, the existing quantifications of efficiencies are used, and [1, 14] are highly scrutinised. After analysing the results, this research recommends inclusions of certain critical phenomena of the network to the existing quantification of the efficiency. These recommendations will be analysed in future research. The following steps describe the algorithm in detail:

- **Step 1:** The nodes of a network are identified and classified according to their degrees. In this study, the nodes are identified and classified into three groups namely, High Order Nodes (HON with $k > 4$) degree nodes, Medium Order Nodes (MON with $5 > k > 1$) and Low Order Nodes (LON with $k<2$). The nodes are sorted solely on the basis of their degrees and not according to the theory of central places. Thus a node which is of lower order according to the central place theory may have higher order here.

- **Step 2:** All possible combinations of new edge connection are created. In this study, the combinations are HON-HON, HON-MON, HON-LON, MON-MON, MON-LON and LON-LON.
- **Step 3:** A set of individual edges is created according to these combinations. The edge sets are same as the above six combination sets.
- **Step 4:** One new edge at a time is added to the network with its respective weight.
- **Step 5:** The efficiencies are calculated for the new network.
- **Step 6:** A plot is generated to show the deviation of the new efficiencies from the base values (the efficiencies of the original network).
- **Step 7:** The most efficient network topology may then be chosen from this set (according to the need) thus optimising the final network regarding efficiency.

The algorithm is applied to the selected networks, and the results are explained below.

4.1 Results from the Algorithm

The following figure (Fig. 8) represents the node classification, and one combination (HON-HON) from the infrastructure network, while Fig. 8 represents a plot of the E_{Loc} and E_{Glob} on the addition of one HON-HON edge to the weighted infrastructure network at a time. The numbers of HON, MON and LON in the infrastructure, IC service and ICE service networks are 124, 1187, 443 (a total of 1754 due to the presence of some redundant nodes in the system); 13, 56, 25 and 10, 33, 17 respectively.

Node_ID	Node_Name	Degree	High_Order_Node	Medium_Order_Node	Low_Order_Node
1293	Muehldorf_Oberbay	5	HON	Not_MON	Not_LON
1232	Muenchen_Westkreuz	3	Not_HON	MON	Not_LON
1236	Olching	3	Not_HON	MON	Not_LON
1237	Mering	3	Not_HON	MON	Not_LON
1238	Augsburg_Hochzoll_Abzweig	3	Not_HON	MON	Not_LON
1240	Augsburg_Morellstrasse	3	Not_HON	MON	Not_LON
1243	Augsburg_Oberhausen	3	Not_HON	MON	Not_LON
1244	Gessertshausen	3	Not_HON	MON	Not_LON

Nodes ordered by the algorithm according to their degrees.

HON_FROM	HON_TO	Dist_KM
2644	2644	0
1536	2644	359.0454525
1568	2644	353.6799847
1625	2644	314.6396613
1759	2644	395.5982191
1777	2644	393.1225047
2166	2644	306.760996

New HON-HON edge generated by the algorithm with direct distance in KM between the HON nodes.

Fig. 8. The left part of this figure shows the sorting of the nodes according to their degrees by the algorithm whereas the right part shows the new HON-HON edges (that will be added to the network in stages by the algorithm) with direct distance in KM between the constituting nodes.

In the case of a weighted network, the E_{Glob} shows no change after adding a new HON-HON edge, whereas E_{Loc} shows substantial changes. This observation holds for both the weighted service networks. Whereas, in the case of the unweighted networks, the E_{Glob} shows only increasing changes and the E_{loc} shows both increasing and decreasing changes from the base E_{Golb} and E_{Loc} values (Fig. 9).

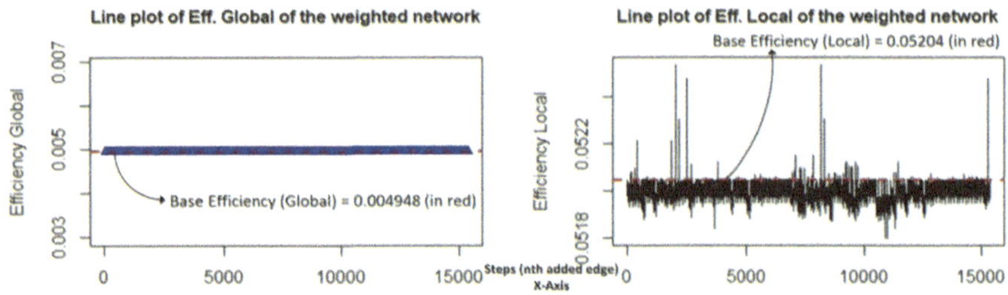

Fig. 9. The plots of E_{Glob} and E_{Loc} for every new HON-HON edge addition.

4.2 A Brief Discussion

The fluctuations of E_{Loc} are on both positive and negative sides from the base E_{Loc} value in the case of a weighted network. This is because some new connections (generated by the algorithm) are long-range connections (edges with long direct distances between them). When such an edge is added to the base network, it turns the (local) subgraph around the considered nodes (nodes considered for calculating the E_{Loc}) much larger thus reducing the efficiency (as efficiency is indirectly proportional to the shortest distance) of that particular subgraph (E_{Loc}) which is further propagated to the whole system through the process averaging. This phenomenon lowers the E_{Loc} of the whole network substantially. The following figure (Fig. 10) explain this phenomenon.

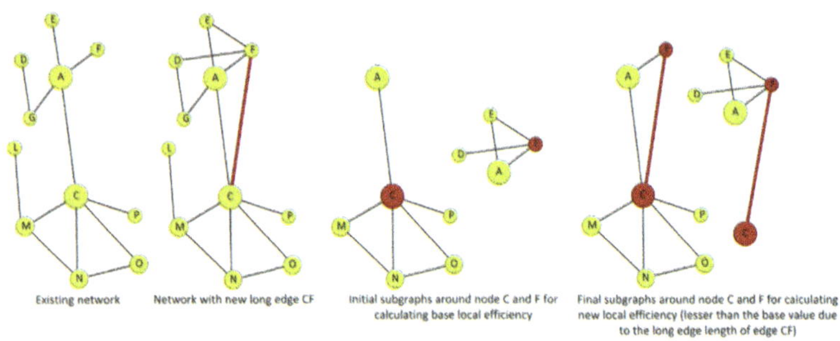

Fig. 10. Change in E_{Loc} on the addition of a long edge (edge with substantial weight).

However, the E_{Glob} does not show any changes. Although, it seems plausible and intuitive, that the addition of a new HON-HON edge will increase the E_{Glob} of a given network. This observation cannot be explained entirely by the existing quantification of network efficiency. It may be explained and quantified precisely by considering the flows in a network within the quantification of the efficiency of a network. This research shall be succeeded by a new set of studies in which the flows in a network shall be considered as a significant factor for the efficiency quantification and calculation so that a more sensitive and precise instrument can be obtained which may further help in optimising a network in the real-time frame. The directions in which this study would proceed are:

- Drawing the relationship between the route choice procedure of 4-Step demand modelling (flow) and network efficiency [21].
- Drawing the relationship between flows inside a directed network with its efficiency and robustness.
- Agent-based modelling to ascertain the relationship between the flow and efficiency of a network [22].
- Dealing with the dynamicity of flow in the network and relating it to the efficiency of a network.
- The relationship between the cost of laying a new edge within a network and its efficiency.

5 Conclusions and Outlook

This study presents a new perspective of optimising and developing a network on the basis of its efficiency. The algorithm may be used to develop an optimised network in terms of its efficiency which is rapidly growing like Delhi Metro Rail Corporation (DMRC) network or some other transport networks of the developing world which add several kilometres of new edges (as an extension and also within the network) every year. The algorithm may also be used to detect the impact of the addition of new edges in the case of an already developed network like the network of DB, e.g. the newly constructed high-speed railway line between Munich to Berlin. Additionally, this algorithm may also be of great help while developing and optimising service networks (e.g. post delivery network) in a given region. However, it may be argued that guidelines like RIN 2008, FGSV provide a well-elaborated method to design and develop efficient transportation (road and rail) networks, but such guidelines do not offer any method of network development and optimisation which is based on the efficiency and robustness of the network.

Thus, from this research, it becomes clear that the DB's infrastructure, IC service and ICE service networks are scale-free and sparse. At the same time, these service networks also exhibit the properties of the small-world network. Additionally, it becomes clear from the output results of the newly developed algorithm that the flows in a network may act as an essential phenomenon for quantifying the efficiency of a network. This study shall be succeeded by a new set of studies which will consider the flows and costs of laying a new edges in a network as significant factors including some other factors (described in preceding section) while quantifying the efficiency of a network to obtain a more precise and sensitive instrument which may further optimise networks in the real-time frame.

References

1. Latora, V., Marchiori, M.: Efficient behavior of small-world networks. Phys. Rev. Lett. **87**(19), 198701 (2001). https://doi.org/10.1103/PhysRevLett.87.198701
2. Jackson, M.O.: Social and Economic Networks. Princeton University Press, Princeton (2011)
3. Pastor-Satorras, R., Vespignani, A.: Evolution and Structure of the Internet: A Statistical Physics Approach, Digitally, printed edn. Cambridge University Press, Cambridge (2007)
4. Hein, O., Schwind, M., König, W.: Scale-free networks. Wirtsch. Inform. **48**(4), 267–275 (2006). https://doi.org/10.1007/s11576-006-0058-2
5. Abbott, L.F., van Vreeswijk, C.: Asynchronous states in networks of pulse-coupled oscillators. Phys. Rev. E **48**(2), 1483–1490 (1993). https://doi.org/10.1103/PhysRevE.48.1483
6. Alm, E.: Biological networks. Curr. Opin. Struct. Biol. **13**(2), 193–202 (2003). https://doi.org/10.1016/S0959-440X(03)00031-9
7. Xu, X., Hu, J., Liu, F.: Empirical analysis of the ship-transport network of China. Chaos **17**(2), 23129 (2007). https://doi.org/10.1063/1.2740564
8. Li, W., Cai, X.: Statistical analysis of airport network of China. Phys. Rev. E: Stat., Nonlin, Soft Matter Phys. **69**(4 Pt 2), 46106 (2004). https://doi.org/10.1103/PhysRevE.69.046106
9. Crucitti, P., Latora, V., Porta, S.: Centrality in networks of urban streets. Chaos **16**(1), 15113 (2006). https://doi.org/10.1063/1.2150162
10. Sen, P., Dasgupta, S., Chatterjee, A., et al.: Small-world properties of the Indian railway network. Phys. Rev. E: Stat., Nonlin, Soft Matter Phys. **67**(3 Pt 2), 36106 (2003). https://doi.org/10.1103/PhysRevE.67.036106
11. Álvarez, A.G., Cañizares, M.P.M., Pita, A.L.: Structure and topology of high-speed rail networks. Proc. ICE – Trans. **168**(5), 415–424 (2015). https://doi.org/10.1680/tran.13.00015
12. Dorbritz, R.: Methodology for assessing the structural and operational robustness of railway networks. ETH Zurich (2012)
13. Erath, A., Löchl, M., Axhausen, K.W.: Graph-theoretical analysis of the swiss road and railway networks over time. Netw. Spat. Econ. **9**(3), 379–400 (2009). https://doi.org/10.1007/s11067-008-9074-7
14. Latora, V., Marchiori, M.: Economic small-world behavior in weighted networks. Eur. Phys. J. B **32**(2), 249–263 (2003). https://doi.org/10.1140/epjb/e2003-00095-5
15. Watts, D.J., Strogatz, S.H.: Collective dynamics of 'small-world' networks. Nature **393**(6684), 440–442 (1998). https://doi.org/10.1038/30918
16. Barabási, A.: Emergence of scaling in random networks. Science **286**(5439), 509–512 (1999). https://doi.org/10.1126/science.286.5439.509
17. Guare, J.: Six Degrees of Separation. Dramatists Play Service, New York (1992)
18. Guimerà, R., Mossa, S., Turtschi, A., et al.: The worldwide air transportation network: anomalous centrality, community structure, and cities' global roles. Proc. Natl. Acad. Sci. U.S.A. **102**(22), 7794–7799 (2005). https://doi.org/10.1073/pnas.0407994102
19. Deutsche Bahn Official Website. https://www.deutschebahn.com/de. Accessed 24 Oct 2018
20. DB Netz AG: Interaktive Karte: Infrastrukturregister (2018). Accessed 07 Jul 2018
21. JdD, Ortúzar S., Willumsen, L.G.: Modelling Transport, 4th edn. Wiley, Chichester (2011)
22. Mancera, A., Bruckman, D., Weidmann, U.: Single wagonload production schemes improvements using GüterSim (Agent-based Simulation Tool). Trans. Res. Procedia **10**, 615–624 (2015). https://doi.org/10.1016/j.trpro.2015.09.015

Forecast of Transport Demand Effects of Longer Trucks in Germany

Robert Burg[1(✉)], Elisabeth Neumann[1], Jan-André Bühne[2], and Marco Irzik[2]

[1] SSP Consult Beratende Ingenieure GmbH, Waldkirch, Germany
burg@fr.ssp-consult.de
[2] Federal Highway Research Institute (BASt), Bergisch Gladbach, Germany

Abstract. In 2012 the German Federal Government started a five-year field trial with longer trucks. One scientific project focused on empirical data of the logistic and transport processes as well as the modelling of future mileage and market shares of longer trucks. Based on predefined scenarios, potential areas of application and accessible routes for longer trucks were determined using a likelihood approach with particular attention to the different definitions of a 'longer truck suitable road network'. In a second step, empirical data covering logistic characteristics collected from participating carriers and forwarders were integrated into a sophisticated transport model in order to estimate transport demand effects and emissions of air pollutants as well as greenhouse gases stemming from longer trucks in normal business operations. Analysis of the traffic demand modelling for the scenarios shows for the reference years 2014 and 2030 that due to logistical constraints and requirements, only a small part of all German heavy good vehicle-km (from 0.03% up to 0.25%) and even a smaller part of all rail/inland waterway transport performance (from 0.05‰ up to 0.3‰) are estimated to be actually shifted to longer trucks.

Keywords: Longer trucks · Freight transport demand · Transport modelling · Energy efficiency

1 Overview

Despite of progress in the reduction of the transport sector's environmental impact in the last couple of years, transport is still responsible for 25% of European Union (EU) greenhouse gas emissions and contributes significantly to air and noise pollution and habitat fragmentation [9]. In order to achieve the European Commission's ambitious goal of a 60% reduction in greenhouse gas emissions by 2050 while guaranteeing economic growth, new and effective technologies and measures have to be used and implemented [8].

As one possible solution to address this challenge, the introduction of heavy goods vehicles with a higher capacity for freight volume was discussed in many European countries (e.g. [1, 11, 14, 15]). With the adoption of the Council Directive 96/53/EC of 25th July 1996 [19] the EU opened up the possibility for the use of overlength vehicles and vehicle combinations (here referred to as "longer trucks") on certain roads and

U. Clausen et al. (Eds.): ICPLT 2019, LNLO, pp. 193–204, 2019.
https://doi.org/10.1007/978-3-030-13535-5_14

routes based on exemption regulations in the member states (The Council of the European Union 1996). In 2011, the German Federal Ministry of Transport, Building and Urban Development (BMVBS, currently BMVI) commissioned the Federal Highway Research Institute (BASt) to conduct a supporting scientific study on a nationwide field trial of longer trucks [4].

The field trial was launched on 1st January 2012 and was scheduled to run for five years. It is part of the BMVBS's Freight Transport and Logistics Action Plan [3]. The legal basis for the conduction of the field trial is constituted by regulations issued by the Federal Minister of Transport on "Exemptions from Road Traffic Law Provisions governing Overlength Vehicles and Vehicle Combinations" ("the Exemption Regulations") of 19th December 2011 and the relevant amending regulations. A key requirement for the field trial participants was the participation in the supporting scientific research.

Moreover, while longer trucks may be designed with a length up to 25.25 m, the Federal Government insisted on the current weight limits of 40 tonnes (or 44 tonnes on the initial and terminal hauls in combined transport) for the field trial due to infrastructural and road safety concerns examined in earlier studies by the Federal Highway Research Institute [10].

Finally, although longer trucks also have to meet the turning circle requirements set out in Section 32d of the German Road Vehicles Registration and Licensing Regulations [17], experience so far suggests that it might not be possible for some vehicle combinations, especially those with a length of up to 25.25 m, to operate on all highway facilities. Therefore, longer trucks may operate only on suitable origin-destination pairs, i.e. only on roads that have been declared suitable by the competent ministries of the federal states concerned for the operation of longer trucks, that have been notified to the BMVI and that have subsequently been published in the Exemption Regulations (authorized or positive network). Based on the 6th exemption regulation the authorized network had a total length of almost 11,600 km with around 70% federal motorways. This was equivalent to just over 60% of all federal motorways, and a few per mille of all local roads in Germany. Currently, the positive network covers partly 15 federal states.

In general, the operation of longer trucks was criticized by several stakeholders and organizations. Concerns included the increased length per se, as well as a potential shift in freight transport demand ultimately resulting in more road freight traffic. Thus, one important aim of the field trial's supporting scientific research was to provide data and detailed information for the debate on "longer trucks" [13].

Out of over twenty scientific projects within the field test, one project aimed at estimating the market potential of longer trucks in Germany. The "Freight Transport Demand Study" focused on empirical data of the logistic and transport processes as well as the modelling of future mileage and market shares of longer trucks and included the estimation of the scope and dimension of potential demand shifts.

This paper is based on the freight transport demand study and is organized as follows: In Sect. (2) the study design and methods are outlined, including a detailed description of the model approach. Section (3) highlights important findings, and Sect. (4) concludes.

2 Methods

2.1 Study Design

The freight transport demand study [5] was carried out in three stages. Observational empirical data of transport processes collected in a baseline study during the first phase of the field trial served as basic input for the study [16]. A first model of the market potential of longer trucks was drafted. Subsequently, to verify the results and to possibly gain new insights, an additional empirical survey of transport processes of the members of the test trial was conducted building on the results and methodology of the baseline study. Again, the main objective was the observation and collection of actual transport data of longer trucks within the test run. These quantitative data were supplemented by qualitative interviews in a parallel survey with participants and also with non-participants of the field trial, e.g. about their assessments of transport demand effects and market opportunities of longer trucks.

Two network scenarios were created in order to assess a wide range of impacts regarding the use of longer trucks with different road networks restrictions (positive network): (1) The network access according to the 6th network regulation, and (2) an extended network, consisting of a core network including an extended catchment area.

In a third step potential areas of application and the accessible routes of longer trucks were determined for predefined scenarios using a likelihood approach with particular attention to the different definitions of a 'longer truck suitable road network' (positive network). The empirical data from the participating carriers and forwarders were integrated into a transport model in order to estimate transport demand effects and emissions of air pollutants as well as greenhouse gases stemming from longer trucks in normal business operations. Based on the defined positive network, the transport demand effects of longer trucks were calculated for the reference years 2014 and 2030 via an integrated transport model with special focus on an enhanced mode choice model for longer trucks. A given traffic demand based on the national freight traffic forecast of the BMVI served as the foundation of the analysis.

As a main result, current national intra-modal market potentials for longer trucks were derived based on national statistical data of heavy good vehicles (see Fig. 1). As a basis for modelling, the medium transport demand version (15 Mio. truck-trips p.a.) was used since it was similar to the characteristics of the empirical findings. Flatbed trucks, tip trucks, tank trucks, as well as bulk transport were excluded as potential business cases for longer trucks. Furthermore, collecting trips and delivery trips were also neglected for the estimation of the market potential of longer trucks. In version 2 only dedicated trips with general cargo and over 70% volume utilization form the basis for the market potential calculations. In total the intra-modal market potential of longer trucks in 2014 amounts to 869 million veh-km (see also Sect. 3).

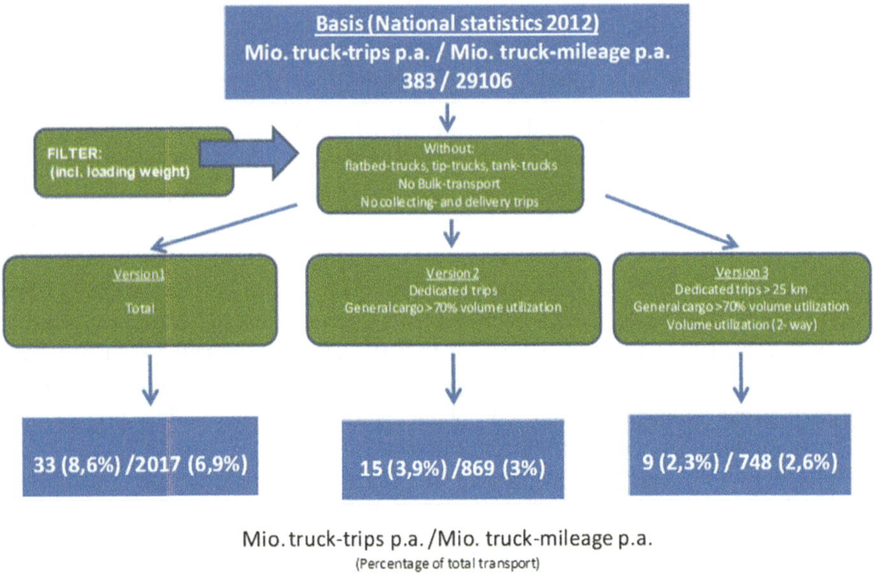

Fig. 1. Potentials for longer trucks

2.2 Model Approach

The basic model approach was designed on two levels:

- Integrated transport model with special focus on enhanced mode choice model for longer trucks.
- Spatial pattern model: Likelihood approach for the determination of potential areas of application and accessible routes of longer trucks with particular attention to the different definitions of a *positive network*.

Transport Model

Based on the defined suitable road network, the transport demand effects of longer trucks were calculated for the reference years 2014 and 2030 via an integrated transport model with special focus on an enhanced mode choice. A given traffic demand, based on the national freight traffic forecast of the Federal Ministry of Transport and Digital Infrastructure [12], served as the foundation of the analysis. Modelling tools from SSP-Freight were used as a basis for modelling the traffic post-fractions, particularly the effects in the investigated scenarios.

The SSP-Freight model, which is the general basis for the modelling approach, is a five-step approach with feedback iterations. SSP-Freight is a collection of specific modules, which are combined by well-defined interfaces. It can be used for the estimation of current transport flows as well as for the forecast of future transport patterns applying different scenarios. Based on the integrated vehicle model, the approach provides an assignment of traffic flows for all modes within the network and is able to analyse route choices and bottlenecks. Mainly the following modules have been used:

- Network assignment model for trucks
- Cost models
- Vehicle and potential analysis model for longer trucks
- Mode-choice model.

The network assignment model provides the routes and the characteristics for transport time and distance for normal trucks as well as longer trucks. These transport characteristics serve as input for cost models that calculate transport costs for all relations, distinguished between commodity groups and transport types. This part of the model responds sensitively to scenario-specific network configurations (*positive network* for longer trucks) and provides scenario-specific costs of longer trucks per relation as input for the mode-choice model.

As the state of the art for calculating costs and services for transport and logistics companies, business costs for road and rail freight traffic are used as a basis for the reference year 2014. Neither economic cost elements, such as external environmental costs, nor market prices are integrated in the cost calculation. In addition to the pure transport costs all elements specific to the transport chain are taken into account. The cost elements are generally subdivided into fixed (time based) and variable (performance based) costs components.

A vehicle model is applied based on the origin/destination freight flows of the traffic forecast 2030 [12]. Depending on the type and distance-specific average utilization rates from the (annual) freight flows, the daily vehicle flows are determined. The acquisition of back loading has been taken into account. In addition, a potential analysis model is used for longer trucks based on special logistical patterns derived from the field survey. In some cases, these patterns can be assigned to the commodity groups (for example, no bulk goods); some of them are patterns based on transport type, e.g. collection/distribution transport versus point-to-point transport, which requires a new segmentation. This segmentation is part of the potential model for longer trucks that was determined in the first transport demand study [16].

The aggregated mode-choice model is one of the main elements of the modelling approach. It regards all transport means within a commodity group and a distance class as logistically equivalent. In consideration of the results of the potential analysis model, the original mode-choice model differentiates between truck, rail, and inland waterway transport. An aggregated Logit Model is applied for the estimation of mode shift effects from rail – differentiated according to the production modes single wagon load and combined transport – and the inland waterway vessel (container) to longer trucks. This model approach supports a detailed, group-specific comparison of individual modes of transport, which is related to individual transport relations and thus also reflects the international state of the art in freight transport modelling (for example: [2, 6, 7, 18]).

The Logit Model has the following formulation:

$$p_v = \frac{exp(U_v)}{\sum_k exp(U_v)} \tag{1}$$

with

$$U_{v;i,j} = \alpha_{v;i,j} + \beta_k BC(K_{v;i,j}, \gamma_k) + \beta_T BC(T_{v;i,j}, \gamma_T) \tag{2}$$

In which:

pv	Likelihood to select the transport mode v
i, j	Relation i->j
α	Parameter to adjust the as-is modal split
β_K, β_T	Weighting parameter
γ	Parameter for the Box-Cox Transformation
K	Transport costs
T	Transport times
BC	Box-Cox Transformation.

This approach is based on the assumption that mode choice is determined by the relative supply characteristics of the competing carriers taking into account the different weighted decisions based on commodity segments and type of transport. The required transport characteristics in the main haulage of the alternative modes are provided as a result of the network assignment models distinguished by commodity groups and transport type (e.g. container transport). To ensure the integration of longer trucks as a separate transport mode, the original logit approach is enhanced by a "nested" stage. Within this stage, the transport mean "truck transport" distinguishes between the subsegments of conventional trucks and longer trucks.

Within the intramodal mode choice approach, which describes the potential shift from conventional to longer trucks, generalized costs (joint consideration of transport costs and travel time) between the alternatives are compared. Potential areas of application and accessible routes of longer trucks with particular attention to the different definitions of a *positive network* are taken into account in a later model stage. In comparison to that, mode choice models based on generalized price elasticities of demand are not appropriate for a determination of mode shift effects on this level of detail.

Another key input of mode choice modelling is the quantification of the increased efficiency of longer trucks, expressed as an efficiency factor. In the mode choice model, two efficiency components are distinguished: Firstly, the efficiency influences mode choice directly by decreasing cost components, and secondly efficiency affects the mileage (in truck-km) replaced by longer trucks. The efficiency factor is derived based on the smallest used loading unit (Euro Pallet). In contrast to the economic view, which represents the substituted number of journeys without taking into account the actual capacity utilization, this macroscopic approach builds on the relational substitution of the actually transported volumes of goods in Germany.

Spatial Pattern Model

The spatial pattern and the road network model are important elements of the general model design. For the modelling of the spatial pattern, a grid model developed by SSP Consult was used, which is based on an intersection of industrial real estate areas with a 500 m × 500 m grid in Germany. In order to keep the number of grid elements manageable, these grid elements were combined into a 1.5 km × 1.5 km grid model. Furthermore, a limit value of 5 ha was defined for industrial real estate per grid element. This resulted in approximately 31,000 grid elements (see Fig. 2). Only with this elaborate model approach it is possible to model transport demand on a detailed level.

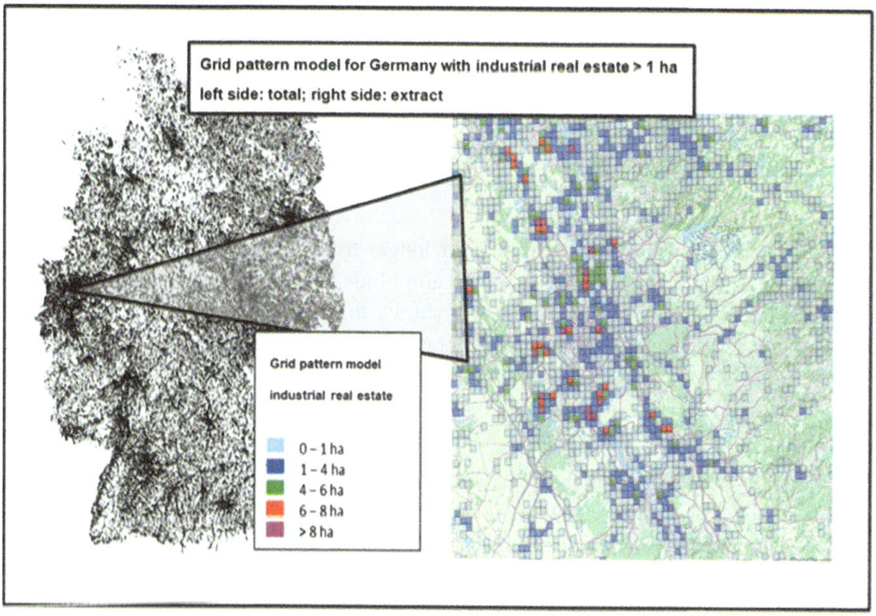

Fig. 2. Grid pattern model for industrial real estate in Germany as basis for the spatial pattern of the model

The road network model used here covers the German main road network from motorways down to the municipal roads. In addition, the route information was supplemented in a very elaborate procedure to match the 6th positive network regulation. Network models for rail and inland waterway transport have been used based on the national freight traffic forecast of the Federal Ministry of Transport and Digital Infrastructure [12]. This is the basic input for modelling the potential areas of application and the accessible routes for longer trucks. Therefore, different scenarios have been defined (see Sect. 2) and analysed based on a suitable likelihood and approximation approach that takes into account that a wide definition of a maximum positive network for longer trucks in Germany is far beyond the current situation. This approach defines the accessibility of the individual grid elements for longer trucks as follows: For individual routes or road categories, likelihood values will be gathered if these routes can be driven by longer trucks. Criteria for this are road category, number of lanes, control speed, and adjacent residential areas. Since the likelihood will be multiplied for sequent stretches, the likelihood approach for homogeneous stretches can be defined as:

$$w(l) = exp^{-\alpha * l} \quad (3)$$

In which:

w Likelihood that a longer truck could drive on an homogenous road path
l The length of the path (per km)
α Path specific model parameter.

With that the parameter α can be construed as "half amplitude length" L(1/2). With each half-length of a stretch

$$L_{1/2} = \frac{In(2)}{\alpha} \sim \frac{0,7}{\alpha} \quad (4)$$

the likelihood of a route's accessibility for longer trucks is cut in half. The parameter selection of $\alpha = 1$ corresponds to a half amplitude length of approximately 700 m, $\alpha = 0.1$ corresponds to 7 km. The table shows the likelihood-based half amplitude length dedicated to the relevant road categories. The outcome of this is a core network for longer trucks, which consists of highways and comparable roads with at least two lanes per direction and without direct crossings of residential areas that could be accessible without any restrictions. For all other categories and combinations a likelihood-gradation of the accessibility for longer trucks will be defined (Table 1).

Table 1. Parameters of the likelihood approach (positive network)

Category of road	Number of lanes	Control speed (km/h)	Adjacent residential areas	Half amplitude length
A (motorway)	All	–	All	∞
B (federal road), L (rural road) or S (state road)	≥ 2	–	No	∞
		–	Yes	20 km
B (federal road)	1	–	No	20 km
		–	Yes	8 km
L (rural road) or S (state road)	1	–	No	15 km
		–	Yes	5 km
K/G (municipal road)	≥ 2	–	No	20 km
		–	Yes	5 km
	1	–	No	10 km
		–	Yes	1 km
F (ferry)	All	–	All	0
Other	≥ 2	≥ 60	No	20 km
			Yes	5 km
		<60	No	5 km
			Yes	1 km
	1	–	No	2 km
		–	Yes	0.5 km

3 Modelling Results

The analysis of the freight transport demand modelling for the scenarios proves that generally only a small part of all German heavy good vehicles runs (about 3.0% (2014; 869 m. veh-km) to 3.2% (2030; 1,179 m. veh-km)) and even a smaller part of all rail and inland waterway transport performance (1.7 and 2.9%) can be seen as shiftable intra-modal or inter-modal potentials to longer trucks in both scenarios due to logistical constraints and requirements (see Table 2). These shiftable potentials do not mean an actual shift of intra-modal or inter-modal transport demand. The decision about the scheduling of transport routes is rather determined by transport cost differences as well as accessibility considerations. Therefore, a detailed and scenario-based analysis of the actual transport demand shifts has to be conducted. In scenario A, estimations of the actual vehicle mileage of longer trucks amount to approximately 9.6 m. veh-km in 2014 and 15.1 m. veh-km in 2030 respectively. These numbers account for about 1% of the sum of the intra-modal and inter-modal longer truck potential. In scenario B, approximately 70.2 m. veh-km (2014) and 97.8 m. veh-km (2030) account for 7% (2014) and 6% (2030) respectively of the total longer truck potential.

Table 2. Results of the transport modelling

Year 2014 vehicle mileage millions (veh-km)	Scenario A	Scenario B
Overall road freight transport	29,450	29,450
Overall rail transport/inland water vessels*	10,118	10,118
Intra-modal potential	869	869
Inter-modal potential	176	176
Longer trucks intra-modal	9.1	68.4
Longer trucks inter-modal	0.5	1.8
Longer trucks total	9.6	70.2
Conventional trucks	856.7	769.0
Trucks total	866.4	839.2
Year 2030 vehicle mileage millions (veh-km)	5	6
Overall road freight transport	37,402	37,402
Overall rail transport/inland water vessels*	13,541	13,541
Intra-modal potential	1,179	1,179
Inter-modal potential	390	390
Longer trucks intra-modal	13.9	93.9
Longer trucks inter-modal	1.2	3.9
Longer trucks total	15.1	97.8
Conventional trucks	1,159.4	1,040.9
Trucks total	1,174.5	1,138.7

*as longer truck vehicle-km equivalent

As mentioned above, a maximum of 1.7% (2014) and 2.9% (2030) of the railway and inland waterway transport performance can potentially be shifted to longer trucks. Actually, in scenario A 0.5 m. veh-km (2014) and 1.2 m. veh-km (2030) are estimated to be shifted intermodally. This freight transport demand shift amounts to 0.3% of the total inter-modal mileage potential only. For Scenario B the results are similar. In comparison to the total German transport amount of rail and inland waterway transport performance, model-based derived intermodal shifts to longer trucks are nearly non-existent and thus negligible (Scenario A: 0.05% in 2014; Scenario B: 0.3% in 2030). The comparison of the results of both scenarios shows that the vehicle mileage (veh-km) of all trucks would in total be about 3.1% lower in 2014 and 3.0% in 2030 respectively due to the expanded network in scenario B. An additional reason for this decrease besides the growing routes of longer trucks lies within the possibility to avoid detours due to the previous networks constraints. Associated with the changes of the freight transport demand, the emissions of CO_2 and air pollutants (NO_X, particles) decreases around 0.4% in 2014/2030 for scenario A and around 3.2% (2014)/3.0% (2030) for scenario B compared to the longer trucks emission potential.

4 Conclusion

In general it can be summarized that the use of longer trucks in normal operation currently generates a positive overall transport demand effect regarding the reduction of driven truck mileage and accordingly a reduction of greenhouse gas emissions and air pollutants. This positive effect will be enhanced by a possible extension of the road infrastructure that is allowed to be used (*positive network*). Intermodal shifts from the railways and inland waterways to longer trucks were not apparent in the empirical observations nor are they considered likely given the logistical and freight structures and characteristics observed in the deployment of longer trucks. This assessment is backed up by the transport demand modelling which also shows that modal shift effects from rail and inland waterway transport are very small and almost negligible. On the other hand, the results also clearly show that the operation of longer trucks is not solely the solution for reducing freight traffic growth and the associated negative environmental effects. But, in certain market segments there are economic and traffic demand-side benefits. Finally, some questions cannot be answered by either field trials or by experimental or theoretical model studies. To forego the limits of the methodology used in this paper, future research should evaluate the topic by the observation and analysis of real-world operation data from several years.

Acknowledgements. The research work presented in this paper was carried out during the scientific monitoring of the German field test on longer trucks between 2012 and 2016. This paper is inspired by two R&D projects (FE-Nr. 89.0273/2012; 89.0315/2015) financed by the Federal Ministry of Transport and Digital Infrastructure (BMVI) and commissioned by the Federal Highway Research Institute (BASt). Responsibility for the content of this paper lies with the authors.

References

1. Aarts, L., Honer, M.: Längere und Schwerere Lastkraftwagen (LZVs) in den Niederlanden – Einblicke und Erfahrungen im Zeitraum 1995–2010. Rijkswaterstaat – Dienst Verkeer en Scheepvaart, Herausgegeben vom Ministerie van Verkeer en Waterstaat, Den Haag (2010)
2. Ben-Akiva, M., Meersman, H., Van de Voorde, E.: Recent developments in freight transport modelling. In: Ben-Akiva, M., Meersman, H., Van de Voorde, E. (eds.) Freight Transport Modelling. Emerald Group Publishing, Bingley (2013)
3. BMVBS – Federal Ministry of Transport, Building and Urban Development 2010. Aktionsplan Güterverkehr und Logistik – Logistikinitiative für Deutschland, Berlin (2010)
4. BMVBS – Federal Ministry of Transport, Building and Urban Development 2011. Verordnung über Ausnahmen von straßenverkehrsrechtlichen Vorschriften für Fahrzeuge und Fahrzeugkombinationen mit Überlänge (LKWÜberlStVAusnV). https://www.ebundesanzeiger.de in the "Official Section" of the Federal Gazette, date of publication 21 December 2011
5. Burg, R., Schrempp, S., Röhling, W., Klaas-Wissing, T., Schreiner, S.: Verkehrsnachfragewirkungen von Lang-Lkw. Final Report on R&D project 89.0315/2015. St. Gallen, Waldkirch (2016)
6. BVU: Entwicklung eines Modells zur Berechnung von modalen Verlagerungen im Güterverkehr für die Ableitung konsistenter Bewertungsansätze für die Bundesverkehrswegeplanung, Freiburg (2014)
7. de Jong, G., Vierth, I., Tavasszy, L., Ben-Akiva, M.: Recent developments in national and international freight transport models within Europe. Transportation **40**(2), 347–371 (2012)
8. EC – European Commission 2011. White Paper - Roadmap to a Single European Transport Area – Towards a Competitive and Resource efficient Transport System, COM(2011) 144 final, European Commission, Brussels, 28 March 2011
9. EEA – European Environment Agency 2015. SOER 2015 European Briefings – Transport, Copenhagen
10. Glaeser, K.-P., Zander, U., Lerner, M., Roder, K., Weber, R., Wolf, A., Zander, U.: Auswirkungen von neuen Fahrzeugkonzepten auf die Infrastruktur des Bundesfernstraßennetzes. Federal Highway Research Institute, Bergisch Gladbach (2006)
11. Hellung-Larsen, M.: Evaluation of trial with European modular system – Danish experience. In: Conference paper, 12th International Symposium on Heavy Vehicle Transportation Technology, The Danish Road Directorate (2011)
12. Intraplan et al.: Verkehrsverflechtungsprognose 2030. Los 3: Erstellung der Prognose der deutschlandweiten Verkehrsverflechtungen unter Berücksichtigung des Luftverkehrs, München (2014)
13. Irzik, M., Kranz, T., Bühne, J.-A., Glaeser, K.-P., Limbeck, S., Gail, J., Bartolomaeus, W., Wolf, A., Sistenich, C., Kaundinya, I., Jungfeld, I., Ellmers, U., Kübler, J., Holte, H., Kaschner, R.: Feldversuch mit Lang-Lkw, Bergisch Gladbach (2016)
14. OECD: Moving Freight with Better Trucks: Improving Safety, Productivity, and Sustainability. International Transport Forum, Paris (2011)
15. Ramberg, K.: Three Short become Two Long, if the EU Follows the Example Set by Sweden and Finland. Fewer Trucks Improve the Environment. Confederation of Swedish Enterprise (2004)
16. Röhling, W., Burg, R., Klaas-Wissing, Th.: Verkehrsnachfragewirkungen von Lang-Lkw – Grundlagenermittlung. St. Gallen, Waldkirch (2014)

17. StVZO 2012 – German Road Vehicles Registration and Licensing Regulations in the Version Promulgated on 28 September 1988 (Federal Law Gazette I, p. 1793), as amended by Article 6 of the Regulations of 13 January 2012 (Federal Law Gazette I, p. 103)
18. Tavasszy, L., De Jong, G.: Modelling Freight Transport. Elsevier, Oxford (2013)
19. The Council of the European Union 1996. Council Directive 96/53/EC of 25 July 1996, laying down for certain road vehicles circulating within the Community the maximum authorized dimensions in national and international traffic and the maximum authorized weights in international traffic, Brussels

Delivery and Shipping Behaviour of Commercial Clients of the CEP Service Providers

A Descriptive Analysis

Carina Thaller(✉), Lara Papendorf, Peiman Dabidian, Uwe Clausen, and Gernot Liedtke

German Aerospace Center (DLR), Institute of Transport Research, Rutherfordstrasse 2, 12489 Berlin, Germany
carina.thaller@dlr.de

1 Motivation

The courier, express and parcel (CEP) service continues its growing over the last decade. There are almost three billion deliveries shipped within Germany – with rising tendency. One reason for a further growing is the booming of e-commerce. As the importance of online shopping behaviour is growing, so far a lot of research has been carried out with main focus on delivery and ordering behaviour of private households in the e-commerce segment. At the moment, several household models and firm-specific datasets are available to analyse and to forecast parcel deliveries of private households. But businesses are still responsible for nearly half of all CEP services. There is still a lack of knowledge with regard to the delivery and shipping behaviour of commercial clients and their use of CEP service providers. However, the way shops are delivered has undergone an enormous change during the past decade. Based on an extensive literature analysis, we have identified the sketched gap and need for market research.

2 Research Objectives

In this paper, we analyse the delivery and shipping behaviour of commercial clients with a focus on parcel deliveries. This research is part of an overall project "Behaviour based Multi-Agent Simulation of CEP markets at urban level to evaluate transport policy and logistics strategies and measures", which is funded by the German Research Foundation (DFG) under contracts LI 1729/3-1 and CL 318/18-1. Whilst the identified delivery and shipping behaviour patterns are used to parametrise a simulation model of CEP transport for a whole city, the present paper shows the descriptive analysis of the empirical survey of commercial parcel recipients and shippers.

U. Clausen et al. (Eds.): ICPLT 2019, LNLO, pp. 205–220, 2019.
https://doi.org/10.1007/978-3-030-13535-5_15

3 State of the Art

In the following chapter, the state of the art in freight demand in CEP market is presented to derive the research gap for this approach.

In this context studies are discussed, which deal with the characterisation of the CEP demand segments. In this case, the demand segments of the CEP service providers are clients, which can be differentiated in commercial clients, the so-called business-to-business (B2B) segment, and private clients. Private clients are defined as the business-to-consumer (B2C) and consumer-to-consumer (C2C) segments.

There are numerous studies focussing on the private client segment (B2C, C2C) of the CEP service provider dealing with e-commerce. These studies analyse the specific structures, markets and transactions in the electronic trading of private households. These investigations focus on e-commerce structures at national and international level. In addition, they also examine the impacts of e-commerce on transport, the economy and society (including Esser and Kurte 2016; Bogdanski 2015; Manner-Romberg 2015).

Furthermore, further studies focus on spatially and temporally differentiated demand structures of the CEP market. These analyse location factors in electronic retail as well as interactions between clients, producers, retailers and logistics service providers. Henschel (2001), Popp and Rauh (2003) and Schellenberg (2005) reflect the impact of retail-related e-commerce on the supply and location structures of retail and retail-related services. Farag et al. (2005) examine the correlation between the frequency of internet use, online shopping and shopping trips. Farag (2006) derives the factors that influence shopping behaviour in e-commerce and in stores.

Nerlich et al. (2010) focus their research approach on the end-consumer perspective and investigate the implications of online shopping for mobility and shopping behaviour as well as transport and spatial development. Various contributions in the anthology by Lenz et al. (2010) examine the transport and spatial effects of e-commerce on the adaptation requirements of logistics systems and process chains (including CEP service providers). Esser and Kurte (2016) concentrate on concrete relationships between the segments of the CEP sector in inner cities. The study provides empirical data and profiles on CEP-related economic sectors as well as private households. Visser and Lanzendorf (2004) focus on the influence of e-commerce on individual activity patterns and mobility behaviour. Thereby, they derive interdependencies between commercial transport, logistical decision-makers and private clients. Schliephake et al. (2004) examine mobility behaviour, infrastructure development and logistical decisions, which are in the area of tension between supply and demand. The Federal Association of E-Commerce and Mail Order (BEVH) carries out primary data surveys at annual intervals. Thereby, 40,000 private households in Germany are surveyed on their consumption behaviour in interactive retailing. The NHTS (2009) is a representative data base containing parameters, which describe consumption and mobility behaviour of private households in the USA. The sample consists of 150,147 households and 308,901 persons.

Sánchez-Díaz et al. (2016) pursue the goal of modelling freight transport generation. In this context, they examine its impact on the urban environment.

Jaller et al. (2015) identify and quantify freight transport generation. They focus on companies, which generate a large share of the truck transport. Zhou and Wang (2014) investigate the correlation between online shopping and shopping trips. Ren and Kwan (2009) analyse the impact of geographic conditions on consumers' online shopping behaviour. In this context, they examine the influence of geographical proximity and accessibility of stores on online shopping. They also investigate the research question, which parameters affect the order frequency and amount of expenses for online shopping. Weltevreden and van Rietbergen (2007) focus on the attraction force of the city center and its impact on online shopping and shopping in retail stores. Krizek et al. (2005) aim to shed light on the propensity of metropolitan residents for online shopping and online banking. Gould and Golob (1997) deal with the prediction probability of purchasing activities. In addition, they concentrate their research on the purchasing behaviour of people working at home compared to those with a permanent job. In this work, the effects of online shopping and shopping trips are analysed in detail.

The review presented of the state of art shows that the primary focus is on the investigation of the consumption behaviour of private households in e-commerce and in stationary retailing. The role of business sectors as receivers and shippers of parcels or commercial clients of CEP service providers was only occasionally considered. However, a comprehensive description of the delivery and shipping behaviour of commercial clients of the CEP service providers is still pending.

In the present empirical study, the focus is on the CEP affinity of businesses (including retail, gastronomy, services) and administration (e.g. research institutions, associations and public administration units). The authors carry out a primary data collection with a sample of 431 cases for the investigation area Berlin, Germany (see Dahidian et al. 2016).

4 Research Design of the Empirical Study

In the following, the research design of the primary data collection will be presented. The objective of this survey is to describe and understand the delivery and shipping behaviour of companies and administration as commercial clients of the CEP service providers. For this reason, the focus is on companies and administration, which represent the commercial freight demand. First, the potential commercial clients of the CEP service providers had to be identified. In this context, the authors specified the economic sectors, which receive and ship parcels. The economic sectors retail, gastronomy and service have been identified as CEP affine businesses. In addition, associations, public administrative bodies as well as research and educational institutions were classified as CEP affine. The investigation area is Berlin, Germany.

Therefore, the questionnaire design had to be developed. The questionnaire includes questions to company-specific information and demand behaviour (e.g. delivery and shipping frequency) of CEP affine companies and administration. Table 1 shows the characteristics of the commercial clients of the CEP service providers, which are considered in the questionnaire.

Table 1. Design of the questionnaire for the primary data collection (Dabidian et al. 2016)

Characteristics of the companies and administration
Company-specific information
Sectors
Sales revenue
Number of employees
Operating area
Storage area
Location
Demand behaviour differentiated according to delivery and shipping
Type of delivery or shipment
Number of logistics service providers
Delivery and shipping frequency
Number of deliveries or shipments

According to the commercial register of the Chamber of Commerce and Industry (IHK) Berlin, a total of 11,477 retail businesses (excluding food sector), 10,978 gastronomy companies, 58,504 service companies and 473 administrations were registered in Berlin.

From this total population of 81,432 companies, a sample was taken for the primary data collection. In this context, sampling is a proportionally stratified sample. It determines the minimum required sample size for a finite population to meet the minimum representativeness requirements. As a result, at least 383 companies (including 54 retail, 52 restaurants, 275 services and 2 administrations) had to be interviewed to achieve representativeness. The sample size could be achieved within the scope of the project.

These stakeholders are interviewed within the framework of a quantitative primary data collection. The freight demand behaviour of these companies and administrations was collected by questionnaire-based surveys of business owners and stores. The provisional questionnaire was tested by means of a pre-test. The questionnaire was adapted to personal face-to-face interviews, telephone interviews and online-surveys.

For this purpose, addresses from the IHK Berlin Commercial Register, public telephone directory and various online platforms were collected in order to use them as contacts for the survey.

Experiences from the survey phase have shown that the response rate of face-to-face interviews and online-surveys was very low. For this reason, the responses were mainly collected by telephone interviews. The companies were informed by e-mail and also by post about the survey in advance. During the survey phase, two survey waves were performed. After the first survey wave, the responses of the participants and the interviews were evaluated. The questionnaire was for the first wave too long for conducting an interview with busy employers or employees during business hours. To counter this time trial, the questionnaire has been shortened for the second survey wave. Nevertheless, it was ensured that the questions and collected content of the two waves are comparable.

This data set was collected as part of the DFG research project (GZ: LI 1729/3-1 and CL 318/18-1) by means of a quantitative survey of 431 companies.

This data set has been prepared to be available for statistical analysis as data input. This includes a comprehensive description of the contained features with their respective scale levels and feature values. This will enable an examination of the commercial freight demand of CEP service providers.

5 Descriptive Analysis of the Empirical Study

Based on the prepared primary data collection, a descriptive analysis will be carried out in the following. With regard to the structure of the questionnaire design, the authors will investigate the delivery and shipping behaviour of commercial clients of the CEP service providers. The main focus is on choices with regard to number of logistics service providers, delivery and shipping frequency as well as number of deliveries and shipments. Finally, the business-specific characteristics of the observed commercial clients are analysed (i. a. branches, specification of the branches, number of employees, revenue of the business unit, total area, storage area, geographical location of the commercial clients).

Based on the primary data collection, the following questions will be analysed in detail.

- Which branches receive or ship goods?
- Which delivery and shipping method is chosen by which branch?
- How many logistics service providers or suppliers are used in average per commercial client in the respective branch?
- How often do these commercial clients receive and ship deliveries?
- How many parcels per delivery or shipment do they usually receive and ship?

In total, 431 companies and administrations were interviewed with focus on their delivery and shipping behaviour. Of these, 65 cases did not provide any information and 59 cases did not generate either delivery or shipping transport. For further analyses, 307 cases could be used.

In Fig. 1 the share of receivers and shippers of parcels are shown. 57% of commercial clients only receive parcels from CEP service providers, while 1% ship exclusively. 42% of commercial clients receive and ship parcels from their business location. In total, there are 305 companies receiving goods and 131 shippers.

Commercial clients who only receive are predominantly 69.4% from the service sector. Approximately one fifth (21.4%) of the receivers are from the gastronomy and 8.7% from the retail sector. Since only two commercial clients ship exclusively, the evaluation is not significant enough. In this case, a company from the service sector and an administrative institution exclusively ship. Even commercial clients, both receiving and shipping, account for just over two thirds (66.1%) of the service sector. Gastronomy (12.1%) and retail (17.7%) together make up about one third. Again, the administration is the least represented with 4%.

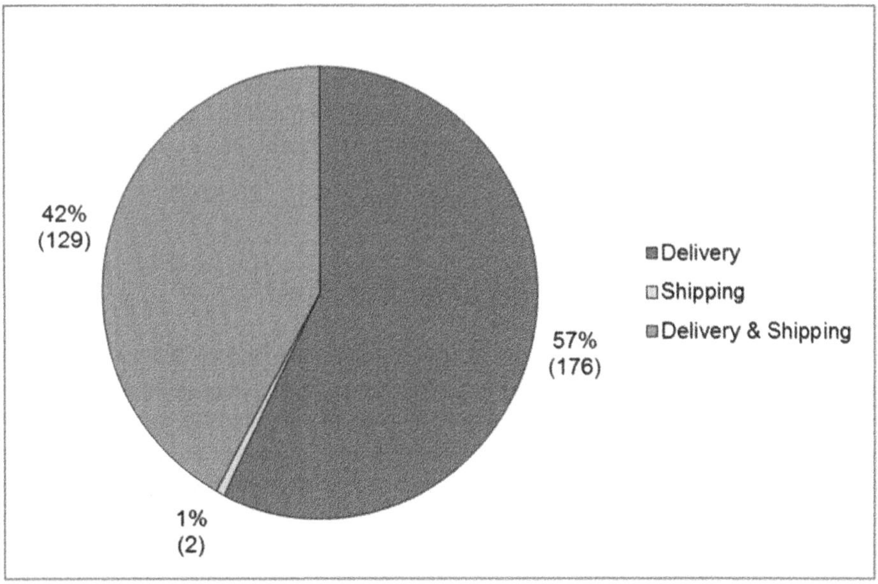

Fig. 1. Share of receivers and shippers of parcels (n = 307)

In Fig. 2 the share of receivers and shippers of parcels differentiated by sectors is shown. The figure shows that over 40% of those receiving exclusively are from the service sector. In addition, 12.4% of commercial clients from the gastronomy and 5% from the retail sector are also delivered exclusively. The proportion of commercial clients who are both delivered and shipped is 27.4% from the services sector, 7.4% from the retail sector and 5% from the gastronomy.

Fig. 2. Share of receivers and shippers of parcels differentiated by sectors (n = 307)

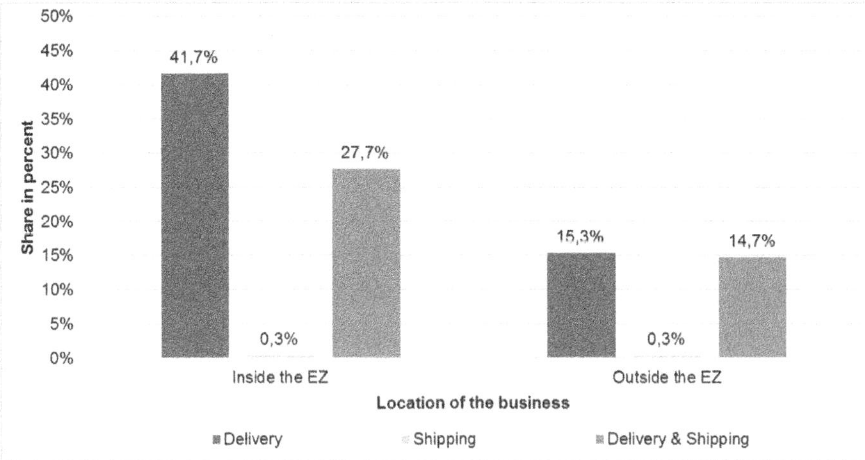

Fig. 3. Share of receivers and shippers of parcels differentiated by the location of the business (n = 307)

Now, the focus is on the location of commercial clients who receive or ship. The location in this context refers to whether the operating locations of the commercial clients are inside or outside the environmental zone (EZ) in Berlin. 41.7% of the clients who only receive parcels are located inside the EZ, 15.3% are outside the EZ. 27.7% who receive as well as ship have their operations within the zone, 14.7% outside of it. More than two thirds of commercial clients are therefore within the environmental zone (see Fig. 3).

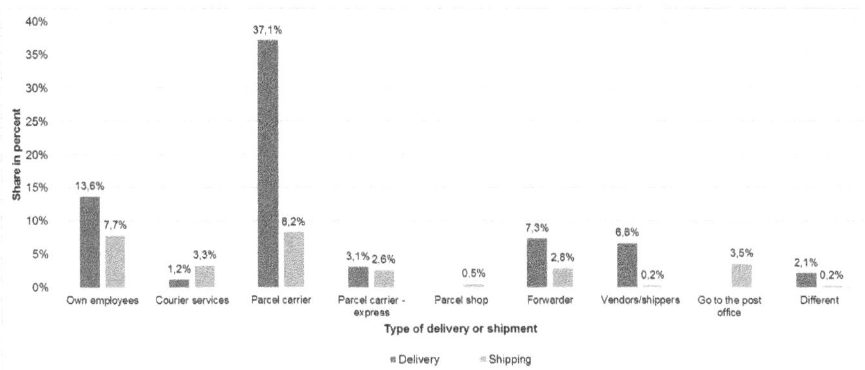

Fig. 4. Share of receivers and shippers differentiated by the type of delivery or shipment (n = 307)

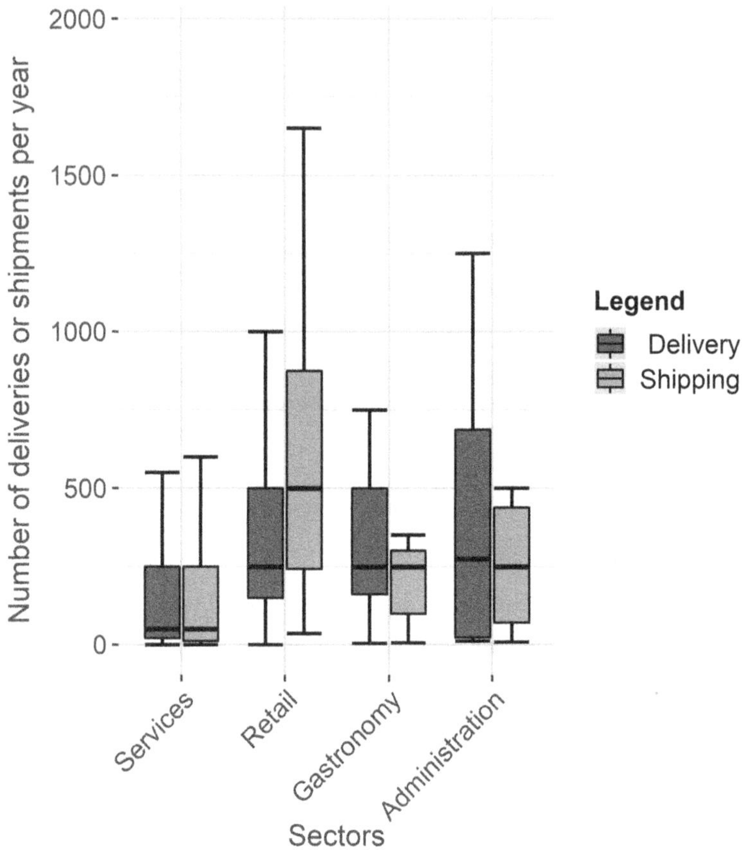

Fig. 5. Number of deliveries or shipments per year differentiated by sectors (n = 307)

The share of receivers and shippers differentiated by the type of delivery or shipment are discussed in the following. Over 37% of receivers are delivered by parcel carriers. Approximately 14% use their own employees for delivery. When shipping the type of shipments are divided evenly, but also in this case the parcel carriers with 8.2% and the own employees with 7.7% are the most frequently mentioned shipping types. The parcel shop is rarely used (in total only 0.5%) (see Fig. 4).

In Fig. 5 the number of deliveries and shipments per year differentiated by the four different sectors is presented. From these analyses, the outliers were excluded. The lowest delivery and shipping frequency, with a median of 50 deliveries per year, and the lowest level of dispersion, are in the service sector. As a result, high delivery or shipping transport volume is not expected in this sector. Highest median with 500 shipments per year and the highest dispersion is recorded in retail sectors. The delivery frequency in the retail sectors amounts to a median of 250 deliveries per year. In

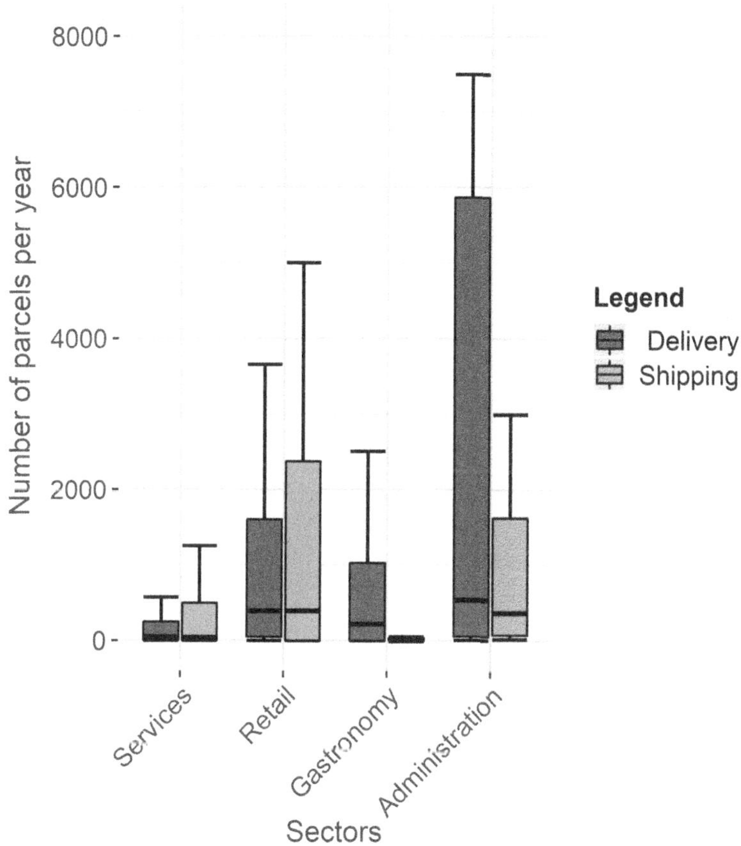

Fig. 6. Number of parcels per year differentiated by sectors (n = 307)

gastronomy, both 250 shipments per year (median) are delivered and shipped. However, there is less dispersion in shipping compared to delivery. Further detailed informations concerning the evaluation of this box-plot diagram are shown in Table 2.

Figure 6 illustrates the demanded number of parcels per year for the four different sectors. In the service sector, compared to the retail, gastronomy and administrative institutions, few parcels are delivered per year (median: 50), but also shipped (median: 40). In retail, 400 parcels per year (median) are delivered, as well as shipped. In the gastronomy, 225 parcels (median) are delivered per year, but no parcels are shipped. The administration receives most parcels per year (median: 542). The parcels shipped per year are 375 (median). In Table 3 the evaluation of this box-plot diagram is shown.

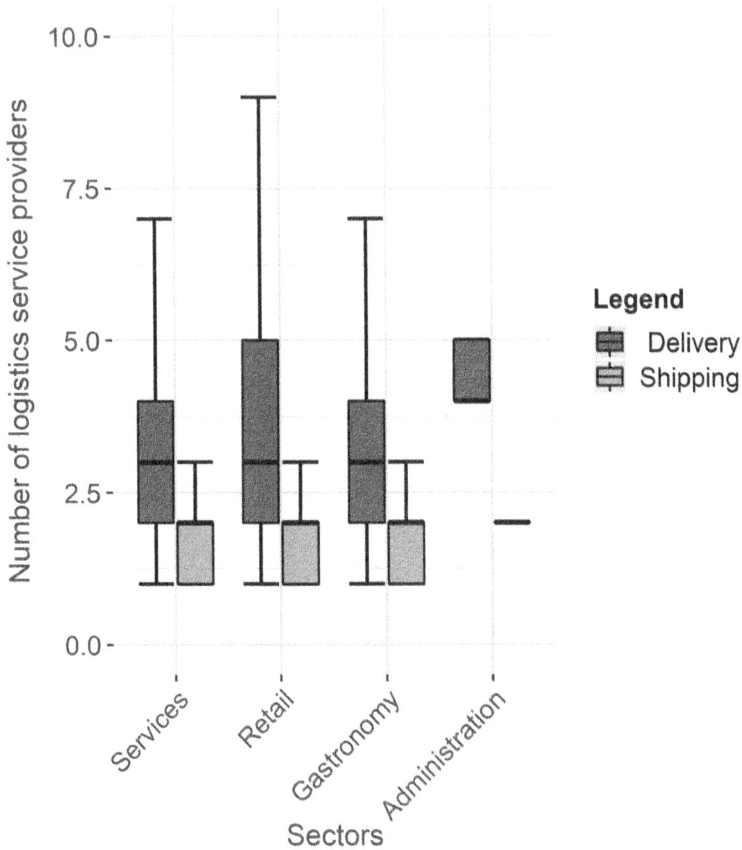

Fig. 7. Number of logistics service providers differentiated by sectors (n = 307)

Figure 7 shows the number of logistics service providers used by the four different sectors for delivery and shipment. It should be noted that clearly more logistics service providers across all four sectors are used for delivery than for shipment. Three logistics service providers (median) are used to deliver the service, retail and gastronomy. Administrative institutions use four logistics service providers for delivery (median). The median for all four sectors is two logistics service providers for shipment. There are no specific deviations between the sectors observed. As a result, the number of logistics service providers used does not depend on the individual sector. Table 4 shows the evaluation of this box-plot diagram.

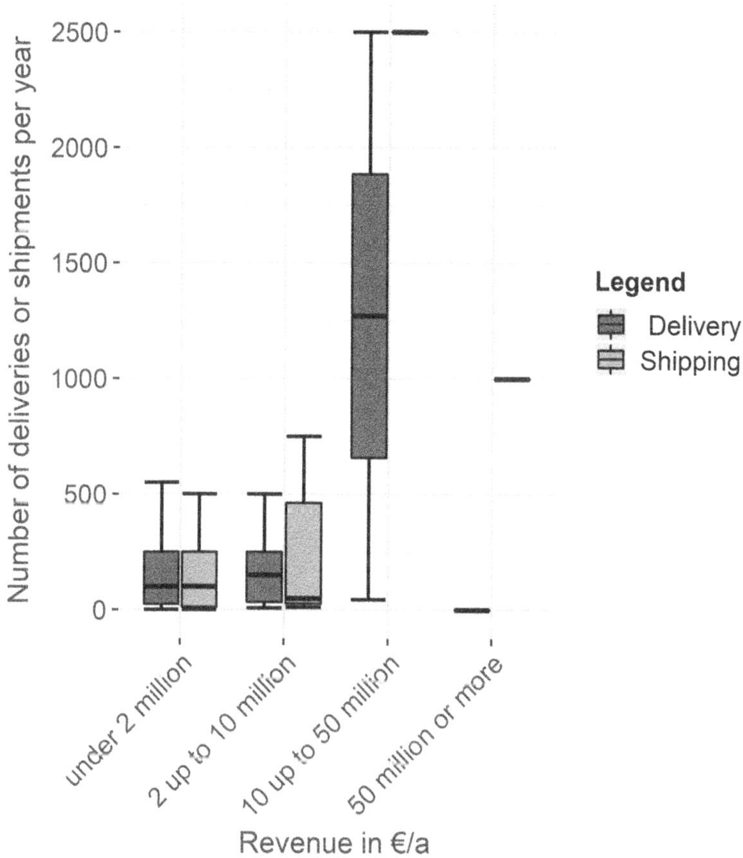

Fig. 8. Number of deliveries or shipments per year differentiated by revenue (n = 307)

Figure 8 shows how high the delivery or shipping frequency is depending on the generated re-venue of the company. When delivery, the frequency of deliveries per year is highest in the companies with an average revenue of 10 up to <50 million € per year (median: 1,272). The highest frequency of shipments in this group can also be observed (median: 2,500). Commercial clients with an annual revenue of less than 2 million € per year are delivered 100 times a year (median) and also shipped just as much. Companies with a revenue of between 2 up to <10 million € per year receive approximately 150 deliveries per year. On the other hand, they ship 50 times a year (median). The companies with a revenue of 50 million € per year and more stated that they do not receive any deliveries, instead making 1,000 shipments per year. A detailed evaluation of this box-plot diagram is shown in Table 5.

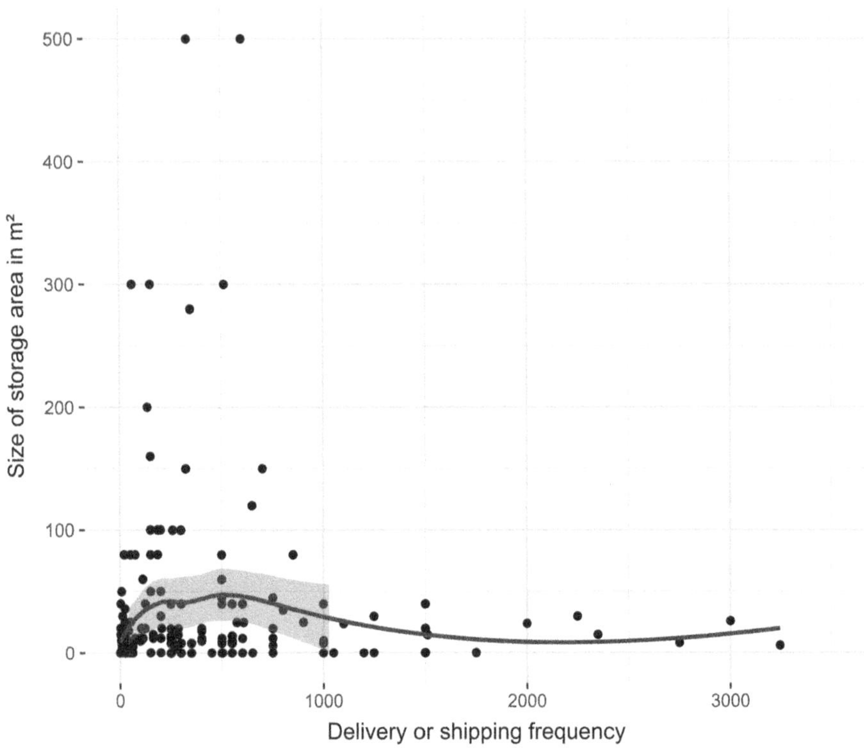

Fig. 9. Size of storage area in dependency to delivery or shipping frequency (n = 307)

Figure 9 shows how large the respective size of the storage area is in dependency to the delivery or shipping frequency. In this case, a slight descent of the size of storage area with increasing frequency of delivery can be seen. As a result, the smaller the storage area of the operating site, the more often the commercial clients must be delivered.

In conclusion, most of the commercial clients of the CEP service providers are only be delivered. These are primarily companies in the service sector. However, nearly two fifths of the surveyed companies also indicated that they both receive and ship deliveries. Also in this case, mainly companies of the service sector are represented. Only a small proportion of the sample ships exclusively. Inside the environmental zone are primarily commercial clients who either receive only deliveries or receive and ship deliveries. Most of the commercial clients are delivered by parcel carriers, followed by their own employees. When shipping, different types are used. The lowest number of deliveries and shipments per year, and even the smallest dispersion thereof, is in the service sector. By contrast, most shipments are recorded with the highest dispersion in the retail sector. In the gastronomy, the same number of deliveries per year are received as well as shipped. However, there is less dispersion in shipping compared to delivery. In the service sector, in comparison to the retail sector, gastronomy but also in the administration a few parcels per year are delivered, but also shipped. In retail, the same number of parcels are delivered per year, as well as shipped. In gastronomy, 225 parcels are delivered per year, but no parcels are shipped. The administration receives most parcels per year. The parcels shipped per year also take high values. With regard to the logistics service providers deployed, it should be noted that clearly more logistics service providers are being used of all four sectors for delivery rather than shipping. Three logistics service providers are used to supply service sector, retail and gastronomy. Two logistics service providers are used of all four sectors for shipping. There are no specific deviations between the sectors. As a result, the number of logistics service providers used does not depend on the individual sector. When delivery, the frequency of delivery per year is highest in the companies with an average revenue of 10 up to <50 million € per year. The highest frequency of shipments for this group is also observed when shipping. In addition, it can be stated that the size of the storage area is dependent on the delivery or shipping frequency. The smaller the size of the storage area of the operating location is, the more frequently the commercial client must be delivered.

6 Conclusions and Outlook

In conclusion, it can be stated that the presented empirical design for investigating the delivery and shipping behaviour of commercial clients is the first research approach, which focus on the commercial receivers and shippers in CEP market. By this approach, the research gap with regard to the less knowledge about the delivery and shipping behaviour of these stakeholders is closed. The description of the survey phase showed that there are a lot of issues to consider for carrying out a survey which focus on companies and administrations as subjects. In this paper, first descriptive analyses of the primary data collection focussing on the delivery and shipping behaviour of commercial clients of CEP service providers were presented.

In further steps, inductive analyses will be carried out based on the results of the descriptive analyses to identify correlations between the variables observed. Furthermore, the commercial clients will be classified with regard to their specific delivery and shipping behaviour. In this context, business specific characteristics of these clients will be used to group the specific behaviour. By cluster analyses, the hypothesis will be verified that homogeneous groups of commercial clients exist, who shows homogeneous delivery and shipping behaviour. Finally, CEP affine commercial clients will be identified.

Acknowledgement. We gratefully acknowledge the funding of the project "Behaviour based Multi-Agent Simulation of CEP markets at urban level to evaluate transport policy and logistics strategies and measures" enabled by the German Research Foundation (DFG) under contracts LI 1729/3-1 and CL 318/18-1.

Appendix

Table 2. Detailed evaluation of the box-plot diagram *Number of deliveries or shipments per year differentiated by sectors*

Sectors		Minimum	25% percentile	Median	75% percentile	Maximum
Services	Delivery	0	21.25	50	250	550
	Shipping	0	12	50	250	600
Retail	Delivery	0	150	250	500	1,000
	Shipping	36	242.5	500	875	1,650
Gastronomy	Delivery	4	162	250	500	750
	Shipping	6	100	250	300	350
Administration	Delivery	12	24	274	687.5	1,250
	Shipping	8	71.5	250	437.5	500

Table 3. Detailed evaluation of the box-plot diagram *Number of parcels per year differentiated by sectors*

Sectors		Minimum	25% percentile	Median	75% percentile	Maximum
Services	Delivery	0	6.5	50	250	572
	Shipping	0	0	40	500	1,250
Retail	Delivery	0	50	400	1,600	3,650
	Shipping	0	0	400	2,375	5,000
Gastronomy	Delivery	0	0	225	1,025	2,508
	Shipping	0	0	0	34	50
Administration	Delivery	8	57	542	5,875	7,500
	Shipping	24	80	375	1,625	3,000

Table 4. Detailed evaluation of the box-plot diagram *Number of logistics service providers differentiated by sectors*

Sectors		Minimum	25% percentile	Median	75% percentile	Maximum
Services	Delivery	1	2	3	4	7
	Shipping	1	1	2	2	3
Retail	Delivery	1	2	3	5	9
	Shipping	1	1	2	2	3
Gastronomy	Delivery	1	2	3	4	7
	Shipping	1	1	2	2	3
Administration	Delivery	4	4	4	5	5
	Shipping	2	2	2	2	2

Table 5. Detailed evaluation of the box-plot diagram *Number of deliveries or shipments per year differentiated by revenue*

Revenue per year		Minimum	25% percentile	Median	75% percentile	Maximum
<2 million €	Delivery	0	24	100	250	550
	Shipping	0	11.5	100	250	500
2 – <10 million €	Delivery	8	33	150	250	500
	Shipping	12	26.5	50	462.5	750
10 – <50 million €	Delivery	4	658	1,272	1,886	2,500
	Shipping	2,500	2,500	2,500	2,500	2,500
50 million € or more	Delivery	0	0	0	0	0
	Shipping	1,000	1,000	1,000	1,000	1,000

References

Bogdanski, R.: Nachhaltige Stadtlogistik durch KEP-Dienste. Studie über die Möglichkeiten und notwendigen Rahmenbedingungen am Beispiel der Städte Nürnberg und Frankfurt am Main. Hg. v. Bundesverbandes Paket und Expresslogistik e. V. (BIEK), Köln (2015). http://biek.de/index.php/studien.html?file=tl_files/biek/downloads/papiere/BIEK_Nachhaltigkeitsstudie_Innenstadtlogistik.pdf. Accessed 22 Sep 2016

Dabidian, P., Clausen, U., Denecke, E.: An investigation of behavioural and structural characteristics of CEP service providers and freight demand considering e-commerce in Germany. Transp. Res. Procedia, 2795–2804 (2016). https://doi.org/10.1016/j.trpro.2016.05.473

Esser, K., Kurte, J.: Kurier-, Express- und Paketdienste – Wachstumsmarkt und Beschäftigungsmotor. KEP-Studie 2016 – Analyse des Marktes in Deutschland. Eine Untersuchung im Auftrag des Bundesverbandes Paket und Expresslogistik e. V. (BIEK). Hg. v. Bundesverbandes Paket und Expresslogistik e. V. (BIEK), Köln (2016). http://biek.de/index.php/studien.html?file=tl_files/biek/downloads/papiere/BIEK_KEP-Studie_2016.pdf. Accessed 21 Sep 2016

Farag, S.: E-shopping and its interactions with in-store shopping. Ph.D. thesis, Faculty of Geosciences, Utrecht University, Utrecht (2006)

Farag, S., Schwanen, T., Dijst, M.: Empirical investigation of online searching and buying and their relationship to shopping trips. Transp. Res. Rec. J. Transp. Res. Board **1926**, 242–251 (2005). https://doi.org/10.3141/1926-28

Gould, J., Golob, T.F.: Shopping without travel or travel without shopping? An investigation of electronic home shopping. Transp. Rev. **17**(4), 355–376 (1997). https://doi.org/10.1080/01441649708716991

Handelsverband Deutschland (HDE): Frühjahrspressekonferenz - 12. April 2016. Hg. v. Handelsverband Deutschland (HDE), Düsseldorf (2016). http://einzelhandel.de/images/presse/konjunktur/Charts.pdf. Accessed 21 Sep 2016

Henschel, S.: Standortfaktoren im elektronischen Einzelhandel. Kurzfassung des Vortrags gehalten am 53. Deutschen Geographentag 2001. Hg. v. Wirtschaftsgeographie der Humbold-Universität zu Berlin (10) (2001). https://www.geographie.hu-berlin.de/de/abteilungen/wirtschaftsgeographie/AK-Einzelhandelsforschung/publikationen/Berichte%20des%20Arbeitskreises%20-%20Geographische%20Handelsforschung%20-%20Bericht%2010.pdf. Accessed 22 Sep 2016

Jaller, M., Wang, X., Holguin-Veras, J.: Large urban freight traffic generators. Opportunities for city logistics initiatives. J. Transp. Land Use **8**(1), 51–67 (2015). https://doi.org/10.5198/jtlu.2015.406

Krizek, K., Li, Y., Handy, S.: Spatial attributes and patterns of use in household-related information and communications technology activity. Transp. Res. Rec. J. Transp. Res. Board **1926**, 252–259 (2005). https://doi.org/10.3141/1926-29

Lenz, B., Kulke, E., Nerlich, M.R., Rauh, J., Vogt, W.: Produktion - Distribution - Konsum. Auswirkungen von Informations- und Kommunikationstechnologien (IKT) auf Wirtschafts- und Versorgungsverkehr. Springer, Heidelberg (2010)

Manner-Romberg, H.: Kompendium des Interaktiven Handels 2014/15. Unter Mitarbeit von Horst Manner-Romberg (MRU). Hg. v. Bundesverband E-Commerce und Versandhandel Deutschland e.V. (bevh) (2015)

Nerlich, M.R., Schiffner, F., Vogt, W.: Einfluss des Internets auf das Informations-, Einkaufs- und Verkehrsverhalten. In: Produktion - Distribution - Konsum, pp. 75–115 (2010)

Popp, M., Rauh, J.: Standortfragen im Zeitalter des E-Commerce. In: Ducar, D., Rauh, J. (Hrsg.) E-Commerce: Perspektiven für Forschung und Praxis. Geographische Handelsforschung, vol. 8, Passau (2003)

Ren, F., Kwan, M.-P.: The impact of geographic context on e-shopping behavior. Environ. Plann. B **36**(2), 262–278 (2009). https://doi.org/10.1068/b34014t

Sánchez-Díaz, I., Holguín-Veras, J., Wang, X.: An exploratory analysis of spatial effects on freight trip attraction. Transportation **43**(1), 177–196 (2016). https://doi.org/10.1007/s11116-014-9570-1

Schellenberg, J.: Endverbraucherbezogener e-commerce. Auswirkungen auf die Angebots- und Standortstruktur im Handel und Dienstleistungssektor. Geographische Handelsforschung, vol. 10. LIS-Verl., Passau (2005)

Schliephake, K., Schenk, T.A., Bozičnik, S. (Hrsg.): Verkehrsgeographie. Mobilität, Logistik und Infrastruktur im Spannungsfeld von Nachfrage und Angebot. Würzburger geographische Manuskripte, vol. 67. Geogr. Inst., Würzburg (2004)

Visser, E.J., Lanzendorf, M.: Mobility and accessibility effects of b2c e-commerce: a literature review. Tijdschr. Voor Econ. Soc. Geogr. **95**(2), 189–205 (2004)

Weltevreden, J., van Rietbergen, T.: E-Shopping versus city centre shopping. The role of perceived city centre attractiveness. Tijd. Voor Econ. Soc. Geog. **98**(1), 68–85 (2007). https://doi.org/10.1111/j.1467-9663.2007.00377.x

Zhou, Y., Wang, X.: Explore the relationship between online shopping and shopping trips. An analysis with the 2009 NHTS data. Transp. Res. Part A Policy Pract. **70**, 1–9 (2014). https://doi.org/10.1016/j.tra.2014.09.014

Urban Factories – Identification of Measures for Resource-Efficient Integration of Production Systems in Cities

Max Juraschek[1(✉)], Felix Kreuz[2], Michael Bucherer[3], Arnim Spengler[4], Sebastian Thiede[1], Christoph Herrmann[1], Alexander Schmidt[4], and Uwe Clausen[2]

[1] Chair of Sustainable Manufacturing and Life Cycle Engineering, Institute of Machine Tools and Production Technology (IWF), Technische Universität Braunschweig, Braunschweig, Germany
m.juraschek@tu-braunschweig.de

[2] Institute of Transport Logistics, TU Dortmund University, Dortmund, Germany

[3] Institute of Industrial Building and Construction Design (IIKE), Technische Universität Braunschweig, Braunschweig, Germany

[4] Institute of City Planning and Urban Design, University of Duisburg-Essen, Duisburg, Germany

Abstract. Cities are a hotspot for resource consumption. This is induced, among other activities, by housing, transportation, trade, the production of goods and the use of products and services. Industrial production and urban factories are an essential part of a city's economy although they are commonly associated with negative impacts, e.g. on the environment or traffic infrastructure. Through positive integration of existing production systems into the urban surrounding, the negative impacts of these urban factories can be lowered or eliminated, and even positive impacts can be achieved. In order to reach a higher degree of integration of different utilizations in urban areas, new methods and conceptual approaches are required to identify suitable measures and actions for urban located companies, city authorities and further stakeholders. For this purpose, an approach has been developed that is founded on research on urban areas and production systems.

Keywords: Urban factories · Urban logistics · Sustainable development

1 Introduction

Cities are the most vibrant and dynamic places in human civilization. Today, already the majority of the world's population is living in urban areas. This figure is prospected to reach more than 70% until the year 2050 [1]. There seems to be no single reason responsible for the specific attractiveness of cities, but rather the mixture of different functional elements and their diversity coming together in close proximity to each other. This density enables the generation of new products, new ways of value creation and fosters innovation [2, 3]. One important functionality of urban systems is the

U. Clausen et al. (Eds.): ICPLT 2019, LNLO, pp. 221–232, 2019.
https://doi.org/10.1007/978-3-030-13535-5_16

generation of products for local and global consumption. Within the spatial layers of a city, this production generally takes place in urban factories. Various interconnections can be found between factories located in cities and other functional elements of the urban space, although cities and urban factories currently coexist in most cases as independent isolated entities. The relationship of these systems is defined by a common spatial interface that can act as a barrier for interaction [4]. The reasons for this separation are subject to current research. While the spatial layers of a city can be analyzed with structural models from the field of urban planning (e.g. [5]) and factories with their spatial and functional structure are also well described in literature from an external and internal perspective (e.g. [6]), an integrated, holistic consideration of both systems is still seldom undertaken. However, in the emerging field of urban production it is recently increasingly recognized (e.g. [7–9]).

Factories as places of value creation are part of a local and the global economy and influence their surrounding urban system, while they are influenced by the city at the same time. During the operation of a factory the input flows of materials, energy and information are transformed into products, by-products, waste and emissions [10]. Existing urban factories can generally be distinguished in two main types depending on their background [11]. The first category are intended urban factories, which are purposefully located within an urban area in order to utilize one or more of the available functionalities of the surrounding city. Unintended urban factories fall into the second category, as these were originally not planned to operate in a densely populated urban environment and have been reached by a cities' growth. Regardless of their background, both types of factories are connected to their surroundings and share the same available resources. Empiric evidence shows that overcoming the spatial and functional separation can lead to efficiency gains and cost saving for urban located companies, for instance in the case of (urban) industrial symbiosis [12]. Interviews with companies producing in urban areas further suggest that there can be benefits for urban factories, yet also challenges arising from their locations, depending on the characteristics of the production system and the urban quarter surrounding them [13]. As a working hypothesis it can be stated that one of the main reasons for the lack of integration of factories within the urban systems are the large time differences in planning [14] and the lack of a holistic approach.

2 Methodology

A new systematic approach is required to enable the exchange of information, knowledge and material flows across existing physical and organizational barriers. The overall goal of the approach is to enhance the resource efficiency of the combined "Factory-City-System" by providing a systematic methodology for analysis of urban factories and subsequently for the identification of suitable measures for implementation. As a first step, the terminology needs to be established in the context of urban production:

- A **Resource** is a mean or enabler for the execution of a task or process. Resources can be either material or immaterial and are bound to time and financial capital.
- **Efficiency** as a goal means achieving a highest possible benefit with a lowest possible effort.

- **Resource efficiency** can be defined as the considerate utilization of resources in a way that this utilization has the greatest possible benefit.
- The utilization of a resource can be seen as **sustainable** if the rate of utilization is not exceeding the rate of regeneration for a resource and the impact of this process is staying within the carrying capacity of the resource providing system.

The systematic approach towards a holistic understanding of urban factories and the underlying methodology is based on eight key resources shared by cities and factories as shown in Fig. 1. All activities in the urban factory system influence the shared resources positively or negatively as explained in further detail in [15].

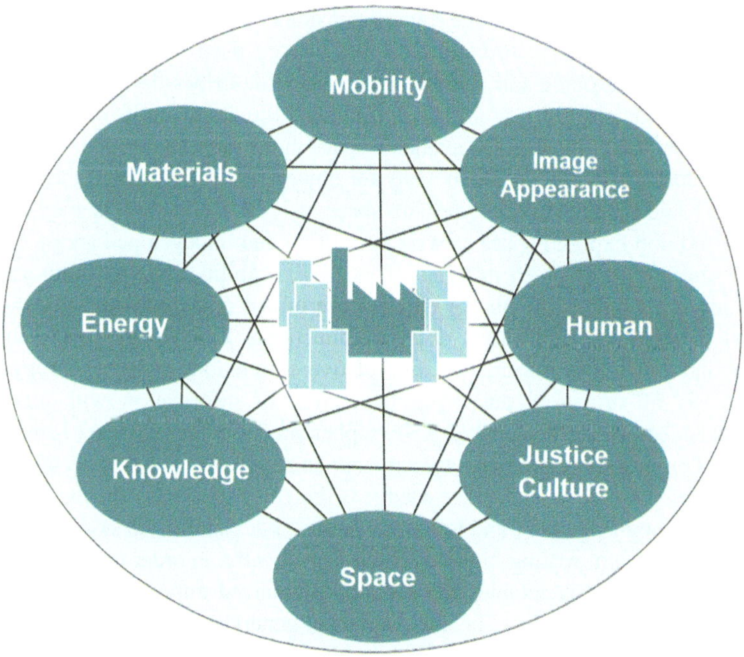

Fig. 1. The eight resources of the urban factory system based on [15].

2.1 Introduction of the "Resources of the Urban Factory"

Energy as a resource is required for the execution of all processes and actions. It can occur in different forms and can be transferred from one existing form of energy to another by conversion. i.e. from primary energy to useful energy, associated with conversion losses. The physical quantity of power establishes the relation of energy to the time dimension. The resource **Materials** comprises all physically existing matter. In particular, this includes all materials that can occur in the context of the Factory-City-System. The resource Materials stands in contrast to immaterial resources such as energy and information. Materials can be changed in their properties under the influence of energy and can be examined on different scales from the (sub-)atomic level to macroscopic material combinations.

All human actions that interact with the Factory-City-System are regarded as the resource **Human**. These include, in particular, the functions of labor, innovation generation and consumption that enable production in urban space. The goal of the efficient use of the resource Human is to improve the availability of workers and consumers in urban space and to use it as a creative element for the creation of the resource Knowledge. In addition, the resource Human in the sense of neighborhood is also an important factor for the ability of production to integrate into urban space. On the one hand, the human being as an asset in need of protection has to be considered during production operation, for example by limiting emissions. On the other hand, the integration capacity of production systems depends on the acceptance of the neighborhood, which is determined by various factors.

The resource **Space** is, in the two-dimensional view, represented as area and describes the part of the earth's surface that can be used for various purposes. In particular the sub-resource soil is considered, which describes the uppermost layer of the earth's surface and fulfils multiple environmental functions in its natural form. The resource Space also comprises the available space in the three dimensions height, width and depth. Space is indispensable in a city for objects and activities and there is often a conflict of objectives between types of space use. A distinction is made between enclosed and non-exposed space as well as built-up and undeveloped areas. The value of space and area is assessed in terms of location, usability and flexibility as well as quality and availability. The aim of efficient use of this resource is to fully exploit its potential in relation to the city's objective functions. **Knowledge** is an immaterial resource and consists of information and relative correlations of these. Information in turn consists of structured data. The interaction of information with energy and materials can be used to describe all known phenomena in the universe. Knowledge is regarded as an indispensable mean to carry out an action, to let a process take place or to solve problems. The available knowledge is thereby decisive for the available action alternatives. Knowledge is bound to human beings, but can also be stored by technical means, for instance in writing, figuratively or acoustically, in order to expand memory.

The resource **Justice** enables the socially guaranteed enforcement of claims and allows behavior and activities. The law consists of generally valid rules of conduct. The concrete enforceability of this rule is assumed here as a necessary condition to achieve justice. It is a part of the culture of a society. From the point of view of resources, **Culture** is the existence of (social) competences in the urban society. These competences can be used and unfold their effect on a material and immaterial level. **Mobility** as a resource enables the spatial transformation of objects through movement. The transport of people and goods is made possible by a mobility infrastructure. Intangible things, such as information, also require a material infrastructure for targeted spatial mobility (transmission). Various performance parameters can be considered to assess the quality of mobility. There is a high, spatially concentrated demand for mobility in urban areas.

The resource **Image** is understood as the effect of entities and objects on human consciousness and their associations. Image is an emotional, immaterial resource and indirectly influences the generation of actions and effects. The identity of a product, building, company or person results in connection with the **Appearance**. Identity is essentially determined by behavior, symbols, offered services and communication. It is essentially determined by the four elements of behavior, symbols, range of services and communication. The stakeholders (here companies and cities) transfer their identity to the environment both indirectly and directly through interaction and thus influence their image ("image/impression") in a variety of ways.

2.2 Introduction of a Three-Level Model

In order to enable a structured analysis and evaluation of the Factory-City-System and a subsequent operationalization through the deduction of improvement measures for resource efficiency, the eight key Resources of the Urban Factory need to be broken down into smaller elements. As shown in Fig. 2 a three-level model is introduced for this purpose evolving from each key resource (level I). On a more detailed layer, each resource consists of several sub-resources (level II). One step further, these sub-resources can be divided into characteristics (level III), which represent actual and potential exchanges of elements in the Factory-City-System.

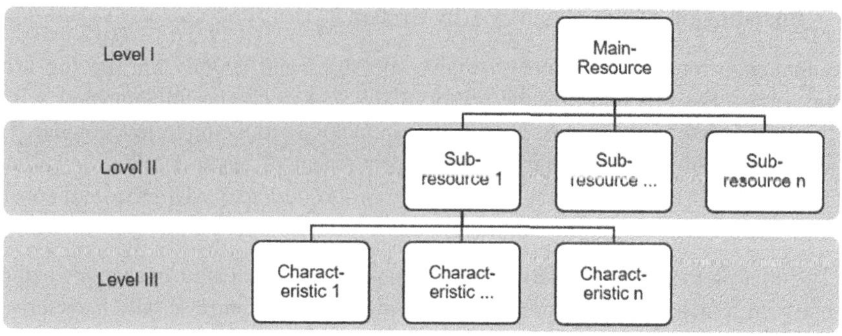

Fig. 2. Hierarchy of resources, sub-resources and characteristics for evaluation

Based on this approach, the implementation of measures for improving the resource-efficiency of the Factory-City-System can be related on the input and output characteristics of the system elements. To achieve more resource efficient states, it allows evaluating characteristics of production sites in urban areas by matching the interchange relationships, i.e. the supply and demand on both sides of the interface, the "factory fence". It has been established that at the key resource level an exchange between individual disciplines and stakeholders is possible and that if these are broken down far enough, the output of one stakeholder can be the direct input for the activities of another.

However, one major challenge is that characteristics cannot be operationalized on their own. They have to be set into the context of the specific Factory-City-System under investigation and cannot be effectively implemented through a purely generic approach. Examples are political persuasion, expert networks, culture, aesthetic sensation or knowledge – all these can be contextualized very differently depending on the system's state and history. The ability to utilize their potential relies on their specific manifestation, meaning that in order to operationalize the characteristics on the third level, different cases need to be considered.

In an ideal case, the characteristics of a specific resource can be operationalized and quantified. If a quantification is not possible but the characteristics can still be utilized, the finding of efficiency increasing actions and measures becomes more challenging. In this case, a rating of available alternatives cannot be based on quantifiable indicators, but in most cases a relative comparison is a suitable work-around. In the third case, neither an operationalization nor a quantification of characteristics are possible. This can be very challenging for the identification of suitable efficiency measures and in most cases only general guidelines can be offered.

Even once the operationalization of the characteristics on the third level has been determined for a specific case, the connection and the exchange of resources between the involved stakeholders is challenging since it also depends on the scientific disciplines examining it. Each discipline involved in the field of urban production has found its own way of looking at and dealing with the operationalization and quantification of its tasks. Establishing a common basis on an interdisciplinary level is therefore extremely important.

2.3 Introduction of the "Factory-City-System"

The challenge remains to identify suitable measures and actions for specific urban factory cases. For this purpose, the Factory-City-System can be implemented as supporting tool based on the resources of the urban factory and the three-level model. This allows support of a factory analysis and improvement of resource efficiency from a holistic point of view. As shown in Fig. 3, the proposed Factory-City-System consists of three main elements.

- The first element is the analysis of the existing and potential exchange relations between a factory and the urban surroundings. For this purpose, all characteristics are assessed for each resource regarding their potential connection with the other sub-system. These factors are evaluated based on their appearance in the specific case and can be classified on a simplified level as a (potential) offer or demand. For instance, if products are manufactured in a factory and offered to the local market, a demand for these products can be fulfilled through the connection of both systems. Similarly, if a factory offers waste heat this could fulfil the energy demand of residential buildings in close proximity. In both cases, the overall resource efficiency could be improved.
- The second element of the Factory-City-System is a collection of options for actions. It consists of a structured database of generic methods and measures that are described regarding their prerequisite, desired outcome and further information such as challenges, best-practice cases and involved stakeholders.
- As a third element of the system a method-based decision support functionality connects the other two elements allowing the matching of actual urban factories with the generic methods and measures. The steps executed for the decision support system are an assessment and evaluation leading to an analysis result for each specific case. In combination with a filtering and rating algorithm, a pre-structured collection of suitable methods and measures can be generated enabling a more resource-efficient operation of urban factories and their urban surroundings.

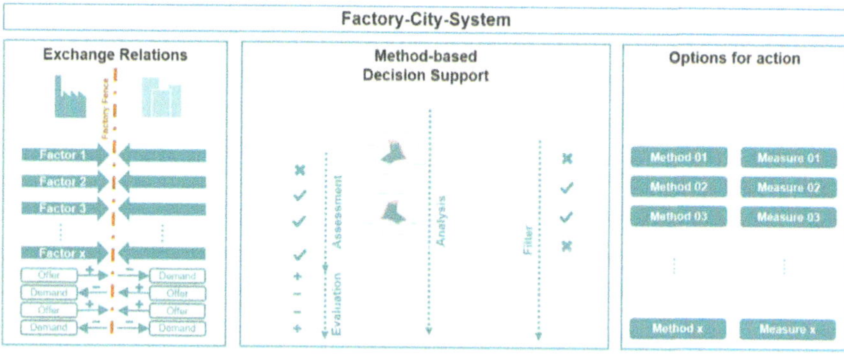

Fig. 3. The elements of the Factory-City-System

3 Implementation

The developed methodology has been implemented as a multi-platform software application and applied within a case study for evaluation. The generalized system architecture is shown in Fig. 4. The goal of the implementation is to make the methodology accessible for stakeholders involved in the urban factory system. All persons involved in the planning and operation of a (potentially) connected subsystem to the urban factory are considered as stakeholders. The focus lies on users from a company's perspective (i.e. from the urban factory), from an administrative perspective (i.e. from the local authorities) or a perspective from another involved discipline (i.e. logistics, architecture, energy design or city planning). The core challenge lies in the algorithm matching the specific urban factory case with suitable improvement measures from the data collection. This task of evaluating a specific case and subsequently identifying matching actions or measures regarding their suitability can be found in numerous applications with corresponding many different approaches. Examples for such decision support applications can be found for improving resource sharing in manufacturing networks [16], raising energy efficiency [17] or transparency [18] in production systems by providing appropriate actions and technologies, identifying matching partners in online dating [19] or in finding best matching partners for resource exchange in the aforementioned concept of Industrial Symbiosis [20]. In order to implement an interface to the Factory-City-System a matching approach of specific urban factory cases with based a suitable scoring algorithm is required. All efficiency measures from the data collection are assigned with parameters regarding their impact on specific resources, stakeholders and their spatial and temporal characteristics (e.g. short-term, local effect). The algorithm was developed as a deduction from formulated importance criteria derived in two workshops with experts from industry and research on urban production, noted in order of descending importance:

- Goal: Matching the user input requirements
- Reward for matching each by the user desired resources, stakeholders and spatial and temporal characteristics
- Penalty for non-fulfillment of a user input

- Penalty for over-fulfillment of a user input
- Over-fulfillment incurs lower penalties than non-fulfillment
- Resources are more important than stakeholders and characteristics

Additionally, a requirement was formulated that the scoring algorithm should ignore sections (i.e. resources, stakeholders or temporal/spatial characteristics), in which no user input was provided. The requirements and importance criteria were met during implementation with the following definitions:

$$so_m = t \cdot sr_m + u \cdot ss_m + v \cdot sc_m \tag{1}$$

with so_m : overall score for measure m; sr_m : resource matching score for measure m; ss_m : stakeholder matching score for measure m; sc_m : characteristics matching score for measure m; $m = 1, \ldots, n$ with n: number of measures in database; t, u, v: weighting parameters.

As an example for the calculation of the scoring elements, the resource matching score is represented with:

$$sr_m = w \cdot a_m + x \cdot b_m + y \cdot c_m + z \cdot d_m \tag{2}$$

with w, x, y, z as weighting parameters for representation of the importance criteria; a_m: scoring value for matching resources of measure m with the user input; b_m: scoring value for matching non-fulfilled resources of measure m with the user input; c_m: scoring value for over-fulfillment of resources of measure m regarding the user input; d_m: scoring value for under-fulfilling resources required from user input of measure m.

All scoring values are calculated by the counting of the meeting or failing the given requirements by comparing the user input with the assigned values of the measure under investigation. In the case of positive matching of resources (a_m) for a specific measure m, this can be written as:

$$a_m = \sum_{l=1}^{8} K_{m,l}, \text{ with } K_{m,l} = \begin{cases} 1 & \text{if } UR_{m,l} \wedge MR_{m,l} \\ 0 & \text{otherwise} \end{cases} \tag{3}$$

with $UR_{m.l}$: User input on resource l; $MR_{m,l}$: Associated impact with resource l of measure m.

The same procedure is conducted for all other scoring values. In several test runs good scoring results were achieved with the following values for the weighting parameters:

$$t = 1, u = 0.5, v = 0.5 \tag{4}$$

$$w = 1, x = 0, y = -0.5, z = -1 \tag{5}$$

The resulting scoring values are calculated for each measure on every user request and subsequently used as evaluation criteria for the identification of suitable measures for decision support.

The overall workflow of the exemplary application was designed based on the definition of user journeys according to an expedited design sprint [21]. All user journeys were framed by the overall goal of offering an analysis of a distinct Factory-City-System based on the user's input and subsequently recommending implementable measures to improve the resource efficiency of this system with regard to the eight key resources of the urban factory, as follows:

- As a first step, the user identifies his goals regarding the system's application, clarifying if specific conflicts should be solved between stakeholders, the efficiency of a distinct resource should be improved or if an overall, holistic analysis is required.
- Following the specification of the goal, the user provides case-specific data on the characteristics of the factory under investigation and the urban quarter.
- As a next step, this data is processed and connected to the pre-defined, generic measures and methods in the database. A rating of all available measures is conducted in the backend and the results are presented to the user in a structured way. With this background knowledge, actions can be identified to manage the flow of output and input resources which have not been utilized before.
- Based on this evaluation, recommended actions are offered to the user together with contextual information on potential benefits, challenges and reference cases as well as the associated stakeholders.

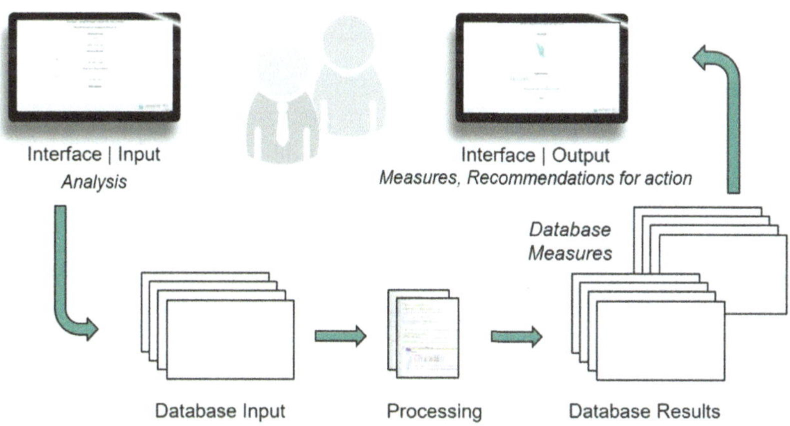

Fig. 4. Process flow for identification of suitable improvement measures within the Factory-City-System

For the implementation workflow, a multi-platform executable application was designed with an underlying database of efficiency measures gathered from literature, industry experts and best practice cases. It was conducted within a research project on urban factories involving partners from research and industry and focused on the applicability in real-world case studies. In this context, two different case studies were considered; (i) a large-scale steel making company located in a mixed industrial and residential area and (ii) a lab-scale factory for lithium-ion batteries connected to a research institution. Both case studies were located in Germany. The test procedure included data collection for the case studies and execution of the described workflow in order to assess the suitability of the efficiency improvement measures proposed by the system for each urban factory. Several different initial goals were given into the system as starting points and the results provided evaluated. The user interface for high-level entry of user input and an exemplary efficiency measure provided by the application are shown in Fig. 5. The developed application was found to support the identification process of potentially suitable efficiency improvement measures within the case studies as well as generally learning about possible measures and actions regarding urban factories.

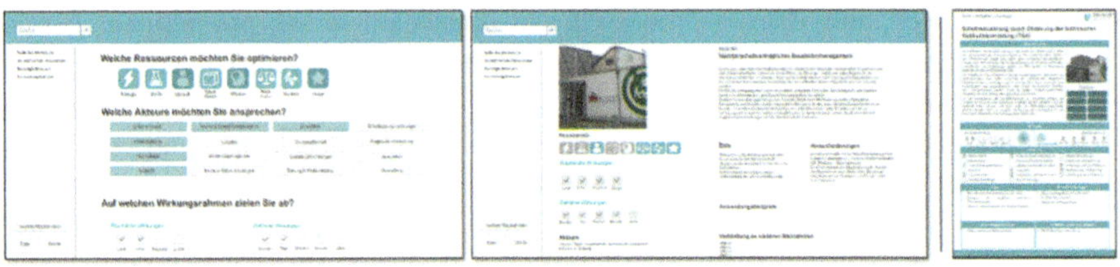

Fig. 5. Screenshots from implemented application showing one user interface (left) and an identified efficiency measure (middle). Additionally, a paper-based representation of an efficiency measure for urban factories is shown (right).

4 Summary and Outlook

Cities and factories are both complex systems. In the case of urban factories these systems are in direct proximity and relation to each other but are in most cases considered isolated from each other. This leaves a high potential for improving resource efficiency through the connection of the factories with their urban surroundings remaining untapped. For this reason, an approach is presented allowing a holistic, joint analysis of the Factory-City-System. At the core, eight key resources were identified and described. Based on these resources, a division in sub-elements was conducted and a demand and supply model designed. To operationalize this approach, a multi-platform software application was implemented and applied within two case studies for evaluation. Exchange relations within the eight key resources were analyzed based on the case studies' data regarding the opportunity to reduce negative effects and generate additional benefits. The implementation as an easy-to-use software application makes the developed approach accessible, allowing users with different backgrounds and intentions the identification of measures for the resource-efficient integration of production systems into their urban context. The major benefits of the developed system can be found in the structured availability of information for specific stakeholders and the possibility to analyze urban factories on an aggregated level while still being able to receive suitable recommendations. The quality of the provided results relies on the matching algorithm as well as on the database quality. Future research can improve the proposed approach by expanding and validating the database with efficiency measures in the Factory-City-System and improving the scoring algorithm and its weighting parameters.

The approach allows to examine individual elements of the Factory-City-System and to compare resource potentials (supply) and requirements (demand). As an outlook for future work, each element of the Factory-City-System needs to be investigated regarding the induced resource flows into their sub-system (consumption) and out of the sub-system (production). Further research on the transformation processes and structures on an element level will enhance the applicability of the identification of measures for the resource-efficient integration of production systems in cities.

Acknowledgement. The authors gratefully acknowledge the funding by the Federal Ministry for Economic Affairs and Energy (BMWi) within the 6th Federal Government's Energy Research Program (EnEff:Stadt, Grant 03ET1311A-D).

References

1. United Nations Department of Economic and Social Affairs: World Urbanization Prospects. World Urban. Prospect. 2014 Revis. Highlights, p. 32 (2014)
2. Jacobs, J.: The Economy of Cities. Random House, New York (1969)
3. Carlino, G.A., Chatterjee, S., Hunt, R.M.: Urban density and the rate of invention. J. Urban Econ. **61**(3), 389–419 (2007)
4. Juraschek, M., Vossen, B., Hoffschröer, H., Reicher, C., Herrmann, C.: Urban factories: ecotones as analogy for sustainable value creation in cities. In: 1. interdisziplinäre Konferenz zur Zukunft der Wertschöpfung, pp. 135–145 (2016)
5. Reicher, C.: Städtebauliches Entwerfen. Springer Fachmedien Wiesbaden, Wiesbaden (2014)
6. Wiendahl, H.-P., Reichardt, J., Nyhuis, P.: Handbuch Fabrikplanung (2014)
7. Verein Deutscher Ingenieure e.V.: Handlungsfelder VDI-Initiative Stadt:Denken Bausteine für die Stadt der Zukunft (2015)
8. Läpple, D.: Produktion zurück in die Stadt? In: Kronauer, M., Siebel, W. (eds.) Polarisierte Städte. Soziale Ungleichheit als Herausforderung für die Stadtpolitik, pp. 129–150 (2013)
9. van Capelleveen, G., Amrit, C., Yazan, D.M.: A literature survey of information systems facilitating the identification of industrial symbiosis. In: Otjacques, B., Hitzelberger, P., Naumann, S., Wohlgemuth, V. (eds.) From Science to Society. Progress in IS. Springer, Cham (2018)
10. Westkämper, E., Löffler, C.: Strategien der Produktion- Technologien, Konzepte und Wege in die Praxis (2016)
11. Juraschek, M., Thiede, S., Herrmann, C.: Urbane Produktion Potenziale und Herausforderungen der Produktion in Städten. In: Corsten, H., Gössinger, R., Spengler, T.S. (eds.) Handbuch Produktions- und Logistikmanagement in Wertschöpfungsnetzwerken, pp. 1113–1133. De Gruyter, Berlin, Boston (2018)
12. Van Berkel, R., Fujita, T., Hashimoto, S., Fujii, M.: Quantitative assessment of urban and industrial symbiosis in Kawasaki, Japan. Environ. Sci. Technol. **43**(5), 1271–1281 (2009)
13. Juraschek, M., Becht, E.J., Büth, L., Thiede, S., Kara, S., Herrmann, C.: Life cycle oriented industrial value creation in cities. Procedia CIRP **69**, 94–99 (2018)

14. Sonntag, R., Voigt, A.: Planungsleitfaden Zukunft Industriebau. Ganzheitliche Integration und Optimierung des Planungs- und Realisierungsprozesses für zukunftsweisende und nachhaltige Industriegebäude. Abschlussbericht. Teil D: Planungssystematik. Fraunhofer IRB Verlag (2011)
15. Juraschek, M., et al.: Die Ressourcen der urbanen Fabrik: Definitionen und Erläuterungen aus dem Forschungsprojekt Urban Factory (2018)
16. Kádár, B., Egri, P., Pedone, G., Chida, T.: Smart, simulation-based resource sharing in federated production networks. CIRP Ann. **67**(1), 503–506 (2018)
17. Krones, M., Müller, E.: An approach for reducing energy consumption in factories by providing suitable energy efficiency measures. Procedia CIRP **17**, 505–510 (2014)
18. Posselt, G.: Towards Energy Transparent Factories. Springer International Publishing, Cham (2016)
19. Rudder, C.: Inside OKCupid: The math of online dating. TEDEd Presentation (2013). http://ed.ted.com/lessons/inside-okcupid-the-math-of-online-dating-christian-rudder. Accessed 1 Sept 2016
20. Van Capelleveen, G., Amrit, C., Yazan, D.M.: A literature survey of information systems facilitating the identification of industrial symbiosis (2017)
21. Knapp, J., Zeratsky, J., Kowitz, B.: Sprint: How to Solve Big Problems and Test New Ideas in Just Five Days. Simon and Schuster, New York (2016)

Implementation of Public Supply Chain Management: A Case Study of the Nelson Mandela Bay Municipality in South Africa

Cynthia Ngxesha[1], Ozias Ncube[1], Chengedzai Mafini[2](✉), and Shephard Mugwenhi[2]

[1] Graduate School of Business Leadership, University of South Africa, Pretoria, South Africa

[2] Faculty of Management Sciences, Vaal University of Technology, Vanderbijlpark, South Africa

chengedzaim@vut.ac.za

Abstract. This paper investigated the implementation of supply chain management by the Nelson Mandela Bay Municipality in the Eastern Cape Province of South Africa. Specifically, the study analysed relationships between public supply chain management, supply chain performance and service delivery. The study employed a quantitative research design in which a survey questionnaire was distributed to 68 supply chain role players drawn from the Nelson Mandela Bay Municipality. Data were analysed using a combination of descriptive and inferential statistics. The results of the study showed that the application of public supply chain management within the municipality was in its advanced stages. Positive correlations were observed between the various public supply chain management strategies, performance and service delivery. Regression analyses indicated that only three public supply chain management strategies, namely people, processes and risk management positively influenced supply chain performance. However, the implementation of various supply chain management strategies exerted no influence on service delivery.

Keywords: Supply chain management · Service delivery · Supply chain performance · Nelson Mandela Bay Municipality

1 Introduction

Soon after the establishment of democracy in 1994, South Africa as a country identified a need to revisit its socio-economic and political order. The new government faced a myriad of public sector challenges that included the imperative to transform the local government. This led to the introduction of new regulations and policies, in line with the constitution. One such result was the formulation and promulgation of Chapter 13 of the constitution, which states that the National Treasury is responsible for ensuring transparency, accountability and sound financial controls in the management of public sector funds. The chapter also mandated the National Treasury to facilitate the Division of Revenue Act and to monitor the implementation of provincial budgets (The Constitution of the Republic of South Africa 1996). The National Treasury was further

U. Clausen et al. (Eds.): ICPLT 2019, LNLO, pp. 233–245, 2019.
https://doi.org/10.1007/978-3-030-13535-5_17

mandated by the parliament to support the optimal allocation and utilisation of resources in all spheres of government to eradicate poverty. In 2003 the National Treasury introduced the Municipal Finance Management Act No. 56 of 2003 (MFMA) with the aim of modernising the budget, accounting and financial management practices by placing local government finances on a sustainable footing in order to maximise the capacity of municipalities to deliver services to their respective communities (Municipal Finance Management Act 2003).

The MFMA Circular No. 1 of (2004:2), linked the Integrated Development Plan to the budget process, introduced new accounting standards, established audit committees and strengthened internal controls. The act also provided for improved procurement and supply chain management (SCM), performance reporting, staff competency levels and new mechanisms to resolve financial problems and misconduct. The provisions of the MFMA took effect on 1 July 1994, except for specific provisions that were officially delayed (Government Gazette 26510, 2004). Before the adoption of SCM, procurement was characterised by a lack of interpretation, accountability and implementation of the preferential procurement policy adopted after the first democratic elections (Ambe 2014).

This study aimed to investigate the implementation of SCM by the Nelson Mandela Bay Municipality (NMBM). The NMBM is the local government overseeing the city of Port Elizabeth in the Eastern Cape Province of South Africa. In its quest to evaluate the impact of the supply chain strategy adopted by the municipality, the study explored the perceptions of NMBM staff regarding the implementation status as well as relationships between public SCM strategies, supply chain performance and service delivery. Supply chain performance (SCP) refers to the extended supply chain's activities in meeting end-customer requirements, including product availability, on-time delivery, and all the necessary inventory and capacity in the supply chain to deliver that performance responsively (Hausman 2004). Service delivery is concerned with the provision of a product or service, by a government or government body to a community that it was promised to, or which is expected by that community (Riekert 2001). Over the years, SCM in South African municipalities has been characterised by fraud and corruption, despite the efforts of the National Treasury to ensure optimum implementation of the strategy (Ambe 2016). To a greater extent, the intended outcomes for the adoption of SCM (e.g. SCP and service delivery) have not fully been achieved. The National Treasury (2015) reports that despite the adoption of SCM policies and strategies in municipalities, its implementation remains a challenge. Further research is necessary then to ensure that currently existing information gaps in this area are addressed, which facilitates the development of possible solutions.

2 Literature Review on Public Supply Chain Management in South Africa

Supply chain management (SCM) has been defined as the coordination of production, inventory, location and transportation among the role-players in a supply chain to realise the best mix of responsiveness and efficiency for end users (Chandrasekaran 2012). When executed within the various arms of government, SCM is generally referred to as public SCM. According to Quinot and Arrowsmith (2013), in South Africa, there historically existed a lack of coherence in the scope of application of the primary procurement legislation, which led to the development and emergence of SCM. Ngcamphala and Ambe (2016) noted that SCM was initiated to support procurement reforms and to bring consistency in the implementation of procurement policy as well as ensuring the accountability of all parties. This initiative is guided by the SCM framework which has six key linked elements that are based on flows in the supply chains from demand management, acquisition management, logistics management, disposal management risk management and supply chain performance (Mhlongo 2014). Migro and Ambe (2008) further assert that the goals of public SCM in South Africa are to add value to all the stages of the process: from the demand of goods and services to their acquisition, managing the logistics process and finally after use and disposal. In doing so, deficiencies in public SCM practices especially concerning activities such as procurement contract management, inventory and assert control as well as obsolescence planning were addressed (Migro and Ambe 2008).

Within the context of South Africa, several legislative pieces have been developed to regulate public SCM. The leading laws include the Public Finance Management Act 1 of 1999, Preferential Procurement Policy Framework Act 5 of 2000, Broad-Based Empowerment Act 53 of 2003 and the Prevention and Combating of Corrupt Activities Act 12 of 2004 (Dlamini and Ambe 2012). However, public SCM within municipalities in the country faces numerous problems. Some of these include the conflict between professional municipal workers and political appointees, poor planning, ineffective internal processes, ineffective leadership, demoralised employees, corruption and fraud, amongst others (KPMG South Africa 2016). Due to these problems, most municipalities are unable to fulfil their mandate, which creates financial leakages running into billions of South African rands, internal conflicts and service delivery protests by the public who are demanding a change of circumstances (Ambe and Badenhorst-Weiss 2012; Dlamini 2016). The current study is intended to address these challenges by generating information that may be applied to overcome the SCM-related problems facing the South African public sector.

In this study, six public SCM strategies are considered, which are policies, people, processes, technology, governance and risk management. Supply chain policies are concerned with the guidelines and rules formulated or adopted by an organisation to facilitate the efficient and effective implementation of SCM (Wible et al. 2014). People refer to human resources such as management, employees and other stakeholders that develop and implement SCM (Gómez-Cedeño et al. 2015). Supply chain processes are the internal procedures within the area of SCM employed in the day-to-day running of the organisation (Croxton et al. 2001). Supply chain technology refers to use in SCM of any systems, methods and any apparatus developed through scientific means (Cegielski et al. 2012). Supply chain governance relates to the system through which policies are implemented with the intent to create greater efficiency in SCM (Rickey et al. 2010). Supply chain risk management involves the application of actions to manage threats in the supply chain, with the aim to reduce exposures and to ensure the continuity of the organisation (Wieland and Wallenburg 2012).

3 Conceptual Model and Research Propositions

To achieve the aim of the research, the conceptual model presented in Fig. 1 is put forward. The model presents SCM as the predictor variable which has six sub-dimensions, namely policies, people, processes, technology, governance and risk management. In turn, public SCM leads to supply chain performance (SCP) and service delivery. Also, the model proposes that SCP leads to service delivery.

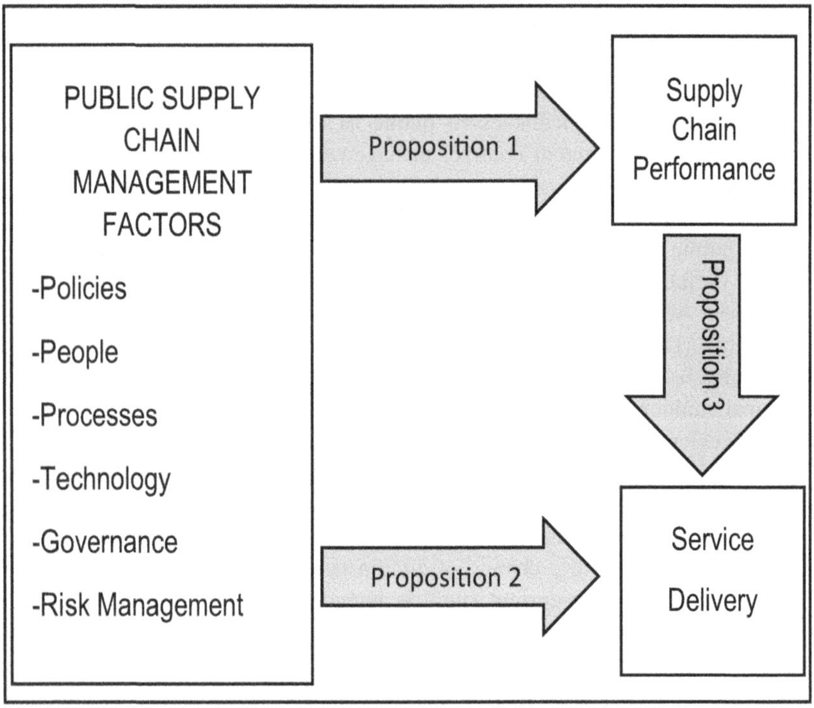

Fig. 1. Conceptual model for the study

As presented in the conceptual model, the study tested the following three propositions;

Proposition 1: There is a relationship between the implementation of public SCM and SCP at the NMBM

Proposition 2: There is a relationship between the implementation of public SCM and service delivery at the NMBM

Proposition 3: There is a relationship between SCP and service delivery at the NMBM.

4 Methodology

The study followed a quantitative approach using the cross-sectional survey design. The final sample was composed of 68 SCM role-players drawn from the NMBM. The NMBM classifies its SCM role-players as Senior Managers, Bid Committee members, NMBM employees with Purchasing Authority and SCM practitioners. Random sampling was used to select relevant respondents.

Data were collected using a self-administered questionnaire distributed using the drop and collect method. The questionnaire was composed of closed-ended questions which were used to measure variables and the demographical details of respondents such as education and training, employment experience and business unit details. All variables under consideration were measured using a five-point Likert type scale anchored by (1) strongly disagree (2) disagree (3) Neutral (4) agree (5) strongly agree.

The data collected were analysed with the aid of the Statistical Packages for the Social Sciences (SPSS version 24.0). The demographic profile of respondents was analysed using descriptive statistics, while Pearson correlations and regression analyses were performed to test for relationships between public SCM, SCP and service delivery. The measurement scales used in the study were tested for validity and reliability. According to Diamantopoulos and Schlegelmilch (2000), the validity of a measure is the extent to which a research instrument is free of systematic and random errors. This study focused on content and construct validity since the instrument selected created the basis of these two validities. Content validity was established through a review of the questionnaire by a panel of experts in the field of SCM. Construct validity was tested through Pearson correlations. The results (Table 2) showed positive correlations between the constructs, which confirmed that construct validity was satisfactory in this study. To test for reliability, the Cronbach's Alpha coefficient was used. All measurement scales used in this study attained Cronbach alpha values greater than the recommended minimum threshold of 0.7 (Hassan et al. 2013), which confirms that instrument reliability was satisfactory in this study.

5 Results of the Study

5.1 Biographic Information

Of the total of 80 questionnaires that were distributed to potential respondents within the NMBM, 68 responded, which produced a response rate of 85%. Nearly 35% of the respondents were in possession of a diploma, while degree and post-graduate degree holders constituted almost 13% and members with professional affiliations constituted approximately 15% of the total respondents. The analysis in terms of the distribution of respondents by directorate shows that a majority 53% were from the Budget and Treasury Directorate while the other 47% came from other directorates ranging from Corporate Services, Economic Development, Tourism and Agriculture, Electricity and Energy, Safety and Health to mention a few. Also, the study considered the distribution of respondents by service in years working within SCM. The study revealed that nearly 40% of the respondents stated that they had been working within the area of SCM in the NMBM for periods ranging between five and ten years. Approximately 29% of the respondents declared that they had been working within SCM in the municipality for less than five years while nearly 15% indicated they had been working in SCM for more than five years.

5.2 Perceptions of Respondents Regarding the Implementation of Public Supply Chain Management in the Municipality

The perceptions of respondents on the implementation of public SCM in the municipality are presented in Table 1.

Table 1. Perceptions of respondents toward the implementation of public supply chain management

SCM attribute	Mean score	Std deviation	Comment
Adoption of SCM systems	4.0	0.961	Good
Familiarity with SCM goals and mission	3.6	1.096	Good
Clarity of SCM strategy	3.2	1.004	Moderate
Communication of SCM objectives	3.1	1.077	Moderate
Provision of demand management	3.1	1.392	Moderate
Provision of acquisition management	4.0	0.836	Good
Provision of logistics management	3.8	0.874	Good
Provision of disposal management	3.3	1.250	Moderate
Provision of SCM performance management	3.2	1.288	Moderate
Provision of SC risk management	2.9	1.331	Moderate

An analysis of Table 1 shows that the adoption of SCM systems, familiarity with SCM goals and mission, provision of acquisition management and provision of logistics management scored reasonably high ratings. All SCM elements recorded mean scores above 3.5 and compared favourably to the maximum possible score of 5. These results depict that respondents perceived that the implementation of public SCM in the municipality had progressed positively and was in its advanced stages.

5.3 Correlation Analysis

Associations between SCM strategies, SCP and service delivery were established using the Pearson Correlation coefficient. The results are presented in Table 2.

The results presented in Table 2, show moderate ($r = 0.3 - 0.49$) positive correlations between service delivery and supply chain performance, service delivery process, service delivery and people, service delivery and technology and service delivery and risk management. All of the above moderate positive correlations were significant at the 0.01 significance level. Strong positive correlations ($r => 0.5$) emerged between, people, processes, technology, SCM governance and risk management. All of these correlations were significant at 0.01 level. However, the association between SCP and

Table 2. Pearson correlations

Variable	Correlation	[1]	[2]	[3]	[4]	[5]	[6]	[7]	[8]	[9]
SCM policy [2]	Correlation	.329	1							
	Sig.	.006								
SCM service delivery [3]	Correlation	.262	.161	1						
	Sig.	.032	.194							
People [4]	Correlation	.499	.295	.429	1					
	Sig.	.000	.017	.000						
Processes [5]	Correlation	.559	.361	.477	.553	1				
	Sig.	.000	.003	.000	.000					
Technology [6]	Correlation	0.404	0.256	0.374	0.463	0.710	1			
	Sig.	0.001	0.041	0.002	0.000	0.000				
Governance [7]	Correlation	0.391	0.325	0.186	0.380	0.432	0.395	1		
	Sig.	0.001	0.007	0.131	0.002	0.000	0.001			
Performance [8]	Correlation	0.596	0.281	0.492	0.680	0.728	0.629	0.518	1	
	Sig.	0.000	0.020	0.000	0.000	0.000	0.000	0.000		
Risk management [9]	Correlation	0.556	0.199	0.392	0.527	0.548	0.621	0.643	0.807	1
	Sig.	0.000	0.103	0.001	0.000	0.000	0.000	0.000	0.000	0.00

SCM policy is positive but weak with a correlation of +0.28 which is significant at 0.05 significance level.

The above results testify that construct validity was satisfactory in this study (Westen and Rosentha 2003). Furthermore, the results suggest the existence of linear associations between public SCM strategies, SCP and service delivery. Hence, when one of the strategies increases, the intensity of the other individual strategies will also increase.

5.4 Regression Analysis

To test the propositions put forward in this study, multiple regression analysis was applied. Regression analysis is a statistical procedure for estimating relationships between a set of independent (predictor) variables and outcome (dependent) variables (Lindley 1987). In this study, three regression models were computed, each representing the three propositions put forward for testing. The enter method, a procedure in which all independent and dependent variables are entered in a single step, was followed.

5.4.1 The Relationship Between Public Supply Chain Management and Supply Chain Performance

Multiple linear regression analysis was applied to test proposition 1, which suggested that there is a relationship between public SCM and SCP at the NMBM. The six strategies of public SCM were entered in the regression model as the independent variables, while SCP was entered as the dependent variable. The results are reported in Table 3.

Table 3. Regression model 2: Public supply chain management strategies and supply chain performance

Independent variable: Public SCM	Dependent variable: Supply chain performance		
	Standardised coefficient	T	Sig.
	Beta		
SCM policy	0.013	0.188	0.851
People	0.227	2.793	0.007
Processes	0.365	3.822	0.000
Technology	−0.055	−0.592	0.556
Governance	−0.064	−0.775	0.442
Risk management	0.534	5.248	0.000
$R = 0.510$: Adjusted $R^2 = 0.453$: $F = 14.44$			

An analysis of Table 3 shows that the public SCM factor ($R^2 = 0.453$) explained approximately 45% of the variance in SCP. The remaining 55% of the variance is explained by other factors that were not considered in this study.

Proposition 1 was partially supported since the results of the regression analysis show that three public SCM strategies predicted SCP. The first one was the people dimension ($\beta = 0.227$; $t = 2.793$; $p = 0.007$), which suggests that having more qualified and motivated employees in the municipality leads to improved SCP. The second strategy that predicted SCP is the processes dimension ($\beta = 0.365$; $t = 3.822$; $p = 0.000$). This result depicts that municipal SCP is likely to be greater when the processes in place in the organisation are effective and efficient. The third strategy that predicted SCP is risk management. The results of the regression analysis showed a strong relationship ($\beta = 0.534$; $t = 5.248$; $p = 0.000$) between risk management and SCP. This result illustrates that an emphasis on risk management leads to greater SCP within NMBM.

A notable result is that since it emerged with the highest beta value, risk management emerged as the most critical predictor of SCP. This implies that amongst the public SCM strategies considered in this study, it would be logical for greater attention to be directed towards improving risk management, as this leads to higher SCP. Overall, the result of this study demonstrates that SCP at NMBM is significantly (at 5% level) influenced by people, processes and risk management.

The remaining three strategies, namely supply chain policy ($\beta = 0.013$; $t = 0.188$; $p = 0.851$), technology ($\beta = -0.055$; $t = -0.592$; $p = 0.556$) and SCM governance ($\beta = -0.064$; $t = -0.775$; $p = 0.442$) were statistically insignificant since their p values were greater than 0.05. This result suggests that despite the association between them (Table 2), these three strategies do not influence the performance of the supply chain in the municipality.

5.4.2 The Relationship Between Public Supply Chain Management and Service Delivery

Multiple linear regression analysis was also applied to test proposition 2, which suggested that there is a relationship between public SCM and service delivery at the NMBM. Service delivery was entered into the regression model as the dependent variable while the six public SCM strategies were entered as independent variables. The results are shown in Table 4.

Table 4. Regression model 2: Public supply chain management strategies and service delivery

Independent variable: Public SCM	Dependent variable: Service delivery		
	Standardised coefficient	T	Sig (p)
	Beta		
SCM policy	0.018	0.142	0.089
People	0.171	1.057	0.295
Processes	0.274	1.365	0.178
Technology	−0.025	−0.144	0.886
Governance	−0.140	−0.897	0.137
Risk management	0.172	0.740	0.462
R = 0.552 Adjusted R^2 = 0.313 F = 11.21			

An analysis of Table 4 shows public SCM (Adjusted $R^2 = 0.313$) explained approximately 31% of the variance in service delivery. The remaining 69% of the variance is explained by other factors that were not considered in this research.

Proposition 2 was not supported in this study since all of the relationships tested in the regression model were statistically insignificant at the 0.05 significance level. These results demonstrate that the implementation of the public SCM strategies does not influence service delivery by the municipality. This result is surprising, given the expectation that the public SCM strategies included in this study would typically be related to service delivery, as shown in several previous studies (Livhuwani 2012; Kaluki 2015; Mwangi 2017). Although correlation analysis (Table 2) revealed positive associations between them, none of the public SCM strategies predicted service delivery. Service delivery then would perhaps be connected to other factors that were not considered in this study.

5.5 The Relationship Between Supply Chain Performance and Service Delivery

Multiple linear regression analysis was further applied to test proposition 3, which suggested that there is a relationship between SCP and service delivery at the NMBM. Service delivery was entered into the regression model as the dependent variable, while SCP was entered as the independent variable. The results are reported in Table 5.

Table 5. Regression model 3: Supply chain performance and service delivery

Independent variable: Supply chain performance	Dependent variable: Service delivery		
	Standardised coefficient	T	Sig.
	Beta		
(Constant)		1.134	0.021
Supply chain performance	0.513	3.654	0.000
R = 0.281: Adjusted R2 = 0.174: F = 4.112			

The results of the regression analysis in Model 3 in Table 5 reveal that the single predictor variable, SCP (adjusted R^2 = 0.174) explained approximately 17% of the variance in the service delivery scale. The remaining 83% of service delivery is explained by other factors that were not considered in this study.

Proposition 3, which suggested that there is a relationship between SCP and service delivery was supported. The regression analysis showed a strong positive relationship (B = 0.513; t = 1.134; p = 000) between SCP and service delivery. This result implies that higher SCP within NMBM is expected to result in improved service delivery. Thus, in order to improve service delivery, municipal authorities should concentrate on enhancing the performance of public SCM since the study has produced a connection between gains in this area and service delivery.

6 Conclusions and Managerial Implications

This investigated the implementation of public SCM by the NMBM and to test the relationships between public SCM, SCP and service delivery in that municipality. Descriptive statistics revealed positive and advanced implementation of public SCM management in areas such as the adoption of SCM systems, the familiarity of role players with SCM mission and goals, clarity of SCM strategy, communication of SCM objectives, provision of demand management, provision of acquisition management, provision of logistics management, provision of disposal management, provision of SCM performance management and provision of supply chain risk management. Application of the Pearson correlation analysis showed moderate to strong correlations between the six SCM strategies (supply chain policy, people, processes, risk management, technology and SCM governance), SCP and service delivery. Regression analyses indicated that three public SCM strategies, namely people, processes and risk management influenced the performance of the municipal supply chain. However, the six public SCM strategies employed by the municipality did not affect service delivery.

The study has managerial implications. The municipality should design a structure aligned to the public SCM strategies considered in this study. Financial resources should be made available to support the implementation of this SCM structure. The structure should be supported with suitably qualified SCM practitioners, and leadership should support SCM staffing and skills development programs. The NMBM should further make use of independent and professional recruitment agencies in order to ensure that qualified and experienced SCM practitioners are recruited. A SCM communications strategy reflecting clear SCM objectives, and encourage the participation of all stakeholders should be developed and implemented. The municipality should also update its systems and ensure that the systems in use support the SCM strategy.

7 Limitations and Suggestions for Further Research

The study is limited in that the sample size (n = 68) is small and that the only one municipality (NMBM) was involved. Future studies could be conducted using amplified sample sizes and involve several municipalities. Since the study revealed that there is no relationship between public SCM and service delivery, it is also perhaps necessary to investigate this result further by conducting specific studies on the interplay each of the public SCM strategies and service delivery. A mixed method approach could be more suitable to ensure that both qualitative and quantitative insights are gathered.

References

Ambe, I.M.: Differentiating supply chain strategies: the case of light vehicle manufacturers in South Africa. Probl. Perspect. Manag. **12**(4), 415–426 (2014)

Ambe, I.M.: Insight into supply chain management in a municipal context. Public Munic. Financ. **5**(2), 20–29 (2016)

Ambe, I.M., Badenhorst-Weiss, J.A.: Supply chain management challenges in South African public sector. Afr. J. Bus. Manage. **6**(44), 11003–11014 (2012)

Cegielski, C.G., Jones-Farmer, L.A., Wu, Y., Hazen, B.T.: Adoption of cloud computing technologies in supply chains: an organizational information processing theory approach. Int. J. Logist. Manag. **23**(2), 184–211 (2012)

Chandrasekaran, N.: Supply Chain Management: Process, System and Practice, 2nd edn. Oxford University Press, New Delhi (2012)

Croxton, K.L., García-Dastugue, S.J., Lambert, D.M., Rogers, D.L.: The supply chain management processes. Int. J. Logist. Manag. **12**(2), 13–36 (2001)

Defee, C.C., Stank, T.P.: Applying the strategy-structure-performance paradigm to the supply chain environment. Int. J. Logist. Manag. **16**(1), 28–50 (2005)

Diamantopoulos, A., Schlegelmilch, B.B.: Taking the fear out of Data Analysis. Cengage Learning EMEA, Canada (2000)

Dlamini, W.: Determining procurement best practices in South African comprehensive universities. Masters dissertation, University of South Africa, Pretoria (2016)

Dlamini, W., Ambe, I.: The influence of public procurement policies on the implementation of procurement best practices in South African universities. UNISA, Pretoria (2012)

Gómez-Cedeño, M., Castán-Farrero, J.M., Guitart-Tarrés, L., Matute-Vallejo, J.: Impact of human resources on supply chain management and performance. Ind. Manag. Data Syst. **115** (1), 129–157 (2015)

Hassan, M., ElBeheiry, M.M., Hussein, K.N.: Drivers and Barriers facing adoption of Green Supply Chain Management in Egyptian Food and Beverage Industry, International Transport and Logistics (2013). http://gccbs2013.aast.edu/newgcc/images/pdf/driversandbarriersfacingadoptionofgreensupplychainmanagementinegyptianfoodandbeverageindustry.pdf. Accessed 21 Nov 2018

Hausman, W.H.: Supply chain performance metrics. In: Harrison, T.P., Lee, H.L., Neale, J.J. (eds.) The Practice of Supply Chain Management Where Theory and Application Converge, vol. 62, pp. 61–73. Springer, Boston (2004). https://doi.org/10.1007/0-387-27275-5_4

Kaluki, B.: Supply chain management practices and service delivery in humanitarian organisations in Kenya. Masters, Dissertation, School of Business. University of Nairobi, Kenya (2015). http://erepository.uonbi.ac.ke/bitstream/handle/11295/92988/Kaluki,%20Beth_Supply%20chain%20management%20practices%20and%20service%20delivery%20in%20humanitarian%20organisations%20in%20Kenya.pdf?sequence=3. Accessed 24 Dec 2018

KPMG South Africa: Making Higher Education Procurement Work (2016). https://assets.kpmg.com/content/dam/kpmg/za/pdf/2016/11/Making-Higher-Education-Procurement-Work.pdf. Accessed 30 Apr 2018

Lambert, D.M.: Supply chain management: processes, partnerships, performance, chapter 1 supply chain management. Supply Chain. Manag. J. **3**, 1–22 (2015)

Lindley, D.V.: Regression and correlation analysis. In: New Palgrave: A Dictionary of Economics, vol. 4, pp. 120–23 (1987)

Livhuwani, T.V.: Impact of the supply chain management on service delivery: the case study of Provincial Department of Economic, Development, Environment and Tourism in the Limpopo Province. Masters dissertation, School of Management and law. University of Limpopo (2012). http://ulspace.ul.ac.za/bitstream/handle/10386/704/tshamaano_vl_2012.pdf?sequence=1&isAllowed=y. Accessed 24 Dec 2018

Malhan, S.: Effects of email – supplier relationship management (e-SRM) on improving supply chain management plant of alloy steel Isfahan (2015). http//www.visijurnal.com. Accessed 22 Nov 2018

Mhlongo, N.F.: Transparency in supply chain management: a South African local government case analysis. Masters dissertation. Stellenbosch University (2014). http://scholar.sun.ac.za/handle/10019.1/86626. Accessed 12 Nov 2018

Mwangi, J.N.: Quality, supply chain management and service delivery in the electricity supply firms in Kenya. Masters dissertation, School of business, University of Kenya (2017). http://erepository.uonbi.ac.ke/bitstream/handle/11295/103264/Mwangi_Quality%2C%20Supply%20Chain%20Management%20and%20Service%20Delivery%20in%20the%20Electricity%20Supply%20Firms%20in%20Kenya.pdf?sequence=1&isAllowed=y. Accessed 24 Dec 2018

Naidoo, R.: Corporate Governance, 2nd edn. Lexis Nexis Cape Town, Durban (2009)

National Treasury: Public Supply Chain Management Review. The National Treasury Republic of South Africa (2015). http://www.treasury.gov.za/publications/other/SCMR%20REPORT%202015.pdf. Accessed 15 Nov 2018

Ngcamphalala, T.K.T., Ambe, I.M.: Policies and regulations guiding procurement practices in the South African commuter bus sector. J. Contemp. Manag. **13**, 1204–1224 (2016)

Pauw, J.C., Woods, G., van der Linde, G.J.A., Fourie, D., Visser, C.B.: Managing Public Money Systems from the South, 2nd edn. Heinemann Publishers, Sandton (2009)

Quinot, G., Arrowsmith, S.: Public Procurement Regulations in Africa. Cambridge University Press, New York (2013)

Republic of South Africa: The Constitution of the Republic of South Africa, Act 108 of 1996. Government Printers, Pretoria (1996)

Republic of South Africa: Municipal Finance Management Act, (Act 56 of 2003). Government Printers, Pretoria (2003)

Republic of South Africa: National Treasury Supply Chain Regulations. Government Gazette. 27636. Pretoria: Government Printers (2005)

Rickey, G., Roath, A., Whipple, J., Fawcett, S.: Exploring a governance theory of supply chain management: barriers and facilitators to integration. J. Bus. Logist. **31**(1), 237–256 (2010)

Riekert, D.: Batho Pele/customer care. course in effective service delivery: enhancing the capacity of existing and emerging public service managers for effective service delivery. Unpublished course notes (2001)

South Africa: National Treasury 2003. Policy Strategy to Guide Uniformity in Procurement Reform Processes in Government. http://www.treasury.gov.za/legislation/pfma/supplychain/Policy%20to%20Guide%20Uniformity%20in%20Procurement%20Reform%20Processes%20in%20Government.pdf. Accessed 13 Nov 2018

Westen, D., Rosentha, R.: Quantifying construct validity: two simple measures. J. Pers. Soc. Psychol. **84**(3), 608–618 (2003)

Wible, B., Mervis, J., Wigginton, N.: Rethinking the global supply chain. Science **344**(6188), 1100–1103 (2014)

Wieland, A., Wallenburg, C.M.: Dealing with supply-chain risks: linking risk management practices and strategies to performance. Int. J. Phys. Distrib. Logist. Manag. **42**(10), 887–905 (2012)

Complexity-Oriented Evaluation of Production Systems for Online-Switching of Autonomous Control Methods

Daniel Mueller(✉), Julian Graefenstein(✉), David Scholz(✉), and Michael Henke

Chair of Enterprise Logistics, TU Dortmund University, Leonhard-Euler-Straße 5, 44227 Dortmund, Germany
{mueller, graefenstein, scholz, henke}@lfo.tu-dortmund.de

Abstract. Production control and its characteristics can be seen as a key factor for handling enhancing dynamic and complex circumstance of production systems. Especially autonomous control methods (ACMs) are considered as an expedient solution for dealing with increasing dynamic and complexity of production and logistic processes. The logistic efficiency of the different ACMs varies greatly depending on the conditions of the respective production scenario and its current complexity and dynamic. This fact motivates to switch the recently used ACM in a given production system during runtime. The central key requirement for this prospective procedure is the appropriate evaluation of the complexity state of production systems. In this paper, current methods for complexity oriented evaluation of production systems and its focused complexity drivers are analyzed. After this, the relevance and interdependencies of possible complexity drivers in the field of autonomous control is focused and represented by an influencing matrix. In addition, simulation studies regarding applicable complexity drivers and used production scenarios in the field of autonomous control are analyzed. As as result, it could be derived that especially dynamic and external complexity drivers should be addressed during online-switching of ACMs.

Keywords: Dynamics · Complexity · Evaluation of complexity · Complexity drivers · Autonomous production systems · Autonomous control · Flexibility · Adaptivity

1 Introduction

The dynamics and complexity of the business environment are increasingly rising as a result of more volatile market conditions, shorter product life cycles and continuous product adaption [1–3]. In order to meet customer, the capability to adapt immediately and flexibly to changing conditions becomes an essential competitive factor [1, 4]. In this context, production control as an interface between the vertical operational processes and the horizontal processes along the value chain is a key element [5].

Autonomous control methods (ACMs) are considered as a promising concept for dealing with increasing dynamic and complexity of production logistical processes by a

U. Clausen et al. (Eds.): ICPLT 2019, LNLO, pp. 246–264, 2019.
https://doi.org/10.1007/978-3-030-13535-5_18

positive emergence of the overall system through a distributed and flexible handling [6, 7]. Windt and Hülsmann define autonomous control generically as a process of decentralized and autonomous decision-making of interacting elements in heterarchical structures and non-deterministic systems [8]. In the context of production logistics, autonomous control is characterized by the capability of logistic objects to interact with each other and to process information to make and execute decisions on their own based on the current system state [9]. The main focus of ACMs is on the situational and autonomous machine assignment (dispatching) of production orders on the shopfloor [10, 11]. Autonomous control is a core component of the vision of an intelligent factory (Smart Factory) and the current development towards the internet of things (IoT).

Numerous ACMs have established in literature [11, 12]. Their logistic efficiency varies greatly depending on the conditions of their application, i.e. the respective production scenario and the current production situation [13]. This finding motivates to switch the recently used ACM in a given production system during runtime depending on the current production situation to ensure maximum logistical efficiency. In contrast to classical control methods, in which physical dependencies represent an invincible obstacle for changing the control method, short-term and almost effortless online-switching of ACM is made practicable due to their exclusive data dependency. In order to capture the current production conditions for online-switching of ACMs, an uniform description form consisting of parameters, that converts a production and logistics system into a manageable and aggregated form of system description, is necessary.

Following decisive works [14–16], the determining factors in the research field of production management to describe production systems are complexity and dynamics. According to de Beer, there is almost no feature of a production system that cannot be described by complexity or dynamics [13]. Further works also showed that the concept of complexity-oriented production system description and evaluation is suitable for mapping the production conditions that determine the performance of ACM [17–19]. They also emphasize that the relation between the logistical target achievement of ACMs is highly influenced by different factors of complexity and dynamics. The challenge now is to map complexity and dynamics in such a way that a production system is described precisely enough despite the principle-related information aggregation of the substitutive approach. Thus, sufficiently meaningful conclusions for the change of ACMs should be able to be derived at runtime. However, the number of single factors of dynamics and complexity must be limited to keep the parameter scope to be investigated manageable, e.g. for simulation studies.

For this reason, this paper aims to identify the single internal and external complexity and dynamics factors of production systems that decisively determine the logistical performance of ACMs. Based on an introduction to the fundamentals of dynamics and complexity in production and logistics, this work presents and classifies existing description and evaluation models and examines their suitability for the formulated problem. A fundamental distinction into a top and bottom view is made according to the basic structure and the pursued objective of models. It is shown that only bottom view approaches are useful for the complexity-based switch of ACMs.

In this context, the work of Grundstein et al. has been identified as a particularly suitable approach [20]. It provides an extensive collection of complexity factors, but their weighting, i.e. the relevance for the logistic performance of ACMs, does not consider any interdependencies between the single factors. Therefore, this work extends the weighting of Grundstein by determining the interdependencies between the factors using an influencing matrix to enable a more differentiated consideration. These are then compared with factor evaluations done by existing simulation studies. In this way, dynamic and external complexity drivers could be identified as the most significant that have to be regarded prioritized when examining online-switching of ACMs.

2 Complexity and Dynamics in Production and Logistics

In the context of production and logistics, complexity is of great importance. The term can be characterized as the multilayeredness of objects and emphasizes the interaction of diversity and connectedness [21]. Further, complexity is defined as the ability of a system to assume many states or behaviors and proportionally to the amount of information required to describe the system and resolve uncertainty associated with the system [22]. Following on from this, cybernetics expresses complexity by the measure variety, which describes the number of possible states of a system [23]. The concept of complexity is generally very widely used and is often equated with complexity in general usage. However, complexity means more than that problems or structures are complicated in their setup. Generally systems are perceived as complex when they can no longer be described simply. In this case, Scherer refers to subjective complexity [24]. According to Schuh, complexity is characterized by the fact, that it can be analyzed, shaped and controlled only to a limited extent, whereas complicatedness only refers to facts and relationships that are rather difficult to understand [25].

Westphal describes complexity only to the static structure of a system, which is made up of the variety and diversity of the elements of a system. Variety refers to the number and relations of the elements and diversity to their difference and heterogeneity. Complexity additionally includes the components dynamics and uncertainty. A system according to Westphal is therefore only described as complex if it has one or more characteristics of complexity, dynamics and uncertainty [26]. Complex systems thus differ from complicated systems by their time behavior, i.e. their dynamics.

Thus dynamics is an inherent property of complexity and is determined by the temporal variability of external influencing variables and system immanent variables [32]. According to system theory, a system is described by elements, properties and relations [13]. The system-theoretical approaches to describe complexity have in common that they separate complexity into a structural and a dynamic complexity component (see Table 1). Structural complexity is usually described by the number of system elements and their interrelationships, while dynamic complexity is usually represented by the variability of individual system variables or the system state over time.

Table 1. Dimensions of system theoretical approaches for the characterization of complexity

Author	Structural complexity	Dynamic complexity
De Beer [13]	• number and properties of elemts • element relations	• change in quantities over time (Simplified referred to as dynamics)
Scherer [24]	• number of elements • networking elements	• feedback Loops dynamic and non-linear behaviour
Patzak [27]	• number of elements • number of relations	
Ulrich und Probst [28]	• number of elements • networking elements	• number of different system statuses (variablity over time)
Scherer und Dobberstein [29]	• number and properties of elements (element complexity) • braiding of element relationships (connection complexity)	• time-dependent interdependencies between system states, environmental facts and system-internal influencing variables (state of structural change)
Schuh [30]	• number and diversity of elements • networking elements	• changeability over time (variety of possible behaviours; changeability of the effects between elements)

Based on this fundamental work, other authors also classify static and dynamic parts of complexity in the application context. Thus Blunck and Windt also distinguish the static from the dynamic complexity in the field of research of this contribution, the autonomous control of production logistic processes [31]. Engelhardt understands static complexity as the long-term design of a production environment (e.g. resource configuration), while dynamic complexity results from short-term changes in production structures (e.g. machine failure, customer demand) [33]. Both forms of complexity influence the effectiveness and efficiency of production planning and control [31, 34, 35].

Scholz-Reiter and Tervo, who have established a cycle of complexity and dynamics, also obtain an equivalent result: dynamics emerges from complexity and complexity emerges from dynamics [36]. Thus there is not only a causal relationship between dynamics and complexity, but dynamics is an integral part of complexity. This work is based on this concept of the relationship between dynamics and complexity. Complexity is seen as the totality of static and dynamic parts of complexity, so that dynamics always represents a property inherent in complexity. However, it should be noted that in single works a separation of dynamics and complexity as properties of a system is made [13, 27]. Accordingly, complexity is a merely structural characteristic and dynamics represents an independent characteristic of a system.

3 Methods for Complexity-Oriented Evaluation of Production and Logistic Systems

A wide range of approaches and methods for assessing complexity can be found in literature. Additionally, there is a great diversity of approaches for defining and characterizing complexity, which is outlined by a high degree of heterogeneity in terms of both complexity and dynamics and the underlying theories. They show numerous methodological approaches for assessing complexity in various fields of application.

A method for the quantification of complexity that is suitable for the performance evaluation of ACMs and providing clues for complexity-oriented changes of ACMs must meet various requirements. An important characteristic is based on a holistic approach and assessment of complexity. There are numerous manifestations of complexity of production systems, most of which can be assigned to the superordinate categories product, process and resource [37]. The complexity drivers to be assigned to these categories are very diverse and include, for example 29 drivers according to Grundstein, which have to be differentiated and to be weighted individually [20]. Since the complexity drivers and performance-determining parameters for self-control procedures are very diverse and interdependent, no specific complexity drivers can be excluded in the addressed field of application. Especially the interactions between the individual complexity drivers have to be considered for a comprehensive complexity assessment [38]. The necessity of a view of complexity that is as holistic as it can be is explicitly emphasized and is also pursued in the context of this paper [39].

In addition to the scope of consideration of underlying static complexity drivers, the consideration of dynamic complexity components is very important in the context of a holistic approach (see Sect. 2). Especially in the field of application of the complexity-related performance evaluation of ACMs, dynamic influencing variables, such as fluctuating order quantities or processing times, can be regarded as directly affecting the performance. The methods to be found in the literature differ besides depending upon the area of application in the way of the preparation of complexity characteristic values and regarding their interpretation. Several approaches are characterized by the effort to combine complexity parameters into aggregated complexity indicators [e.g. 20, 40]. These procedures can be defined as target-oriented for the identification of complexity states for production systems. However, the problem of an increased aggregation of complexity parameters lies in the ambiguous ability to interpret them. This results from the fact that different system states can lead to the same aggregated complexity key figure.

In the field of the addressed application of the evaluation and the selection of suitable ACMs, aggregated indicators cannot provide a sufficient basis for evaluation, since no clear conclusions can be drawn about the underlying system configuration [13, 40]. The approaches examined and the resulting key indicators of complexity are characterized by different levels of detail with regard to the identifiability and interpretability of complexity. This is of great importance for decision support in the selection of a suitable ACM.

Author	Field of investigation					Method					Application				Field of application			
	Product	Process	Resource	Aggregation	Dynamic	Entropy	vector based	Graph theory	Simulation	Mathematics	Identification	Designing	Planning&Operation	Monitoring	ACM	Supply Chain	Process optimization	Cost accounting
Lasch & Gießmann 2009 [41]		●	◐	◐						●	●		●					
Turner & Williams 2005 [42]	●	◐		◐	◐				●		◐	●	◐			●		
Lechner 2012 [43]	●	◐	◐	◐	◐					●	●	◐	●				●	●
Matsson 2016 [44]	◐	◐	●	◐	◐					●	●	◐					●	
Grundstein et al. 2015 [20]	●	●	●	●	●					●	●		●		●			
Wang et al. 2011 [45]		◐	◐	●	◐	●					●						●	
Frizelle et al. 1995 [46]		●	◐	◐		●					●						●	
Ashby 1956 [23]		●	◐	◐			●				●						◐	◐
Kersten et al. 2005 [47]	◐	◐	◐	◐	◐		●				●	●		◐			●	
Gerschberger et al. 2012 [48]	●	●	◐	◐	●		●				●	◐	●			●		
Scholz-Reiter et al. 2006 [49]	◐	●	●	◐	●		●				●	◐			◐		◐	
Meepetchdee & Shah 2007 [50]	◐	●	◐	●	◐			●			●	●				●		
Lübke 2007 [51]	●	◐								●	●	◐					◐	●
Kota et al. 2000 [52]	●			●						●	●	◐					◐	●
Wildemann 2011[40]	●	●	●	●	●					●	●	●	●	●			●	
Hanenkamp 2004 [53]	●	●	●	◐			●				●	●	●	●			●	
Schuh 2005 [30]	◐	●	●	●	◐					●	●	◐						●
Kaplan & Anderson 2009 [54]	◐	●		◐						●	●	◐					◐	●
Brecht 2003 [55]	●	◐		◐						●	●	◐						●
Horvath & Meyer 1998 [56]	●	●		◐						●	●	◐						●
Jeschke 1997 [57]	●	●		◐						●	●							●
Puhl 1999 [58]	◐	●	◐	◐	◐					●	●	◐		●			◐	●
Braithwaite & Samakh 1998 [59]	●	◐		◐						●	●	◐						●
Schaffner 2010 [60]	●			◐						●	●							●
Bräutigam 2004 [61]	●	●		◐						●	●							●
Heina 1999 [62]	●	●		◐						●	●							●
Bartuschat 1995 [63]	●	◐		◐	◐					●	●							●
Bohne 1998 [21]	◐	◐	◐	◐						●	●	◐						●
Dalhöfer 2009 [64]	◐	◐	◐		◐					●	●	◐						●
Feldhütter 2018 [65]	●	●	●	◐	●		●				●		◐	◐		●	◐	
Tarride 2013 [66]		◐		◐				●			●					●		
Becker et al. 2014 [67]		●	◐	◐				●			●					◐	◐	
Marti 2007 [68]	●	◐		◐						●	●							●
Bliss [2000]	●	◐		◐	◐					●	●	◐	◐			◐	●	◐
Kirchhof 2003 [70]	◐	◐	◐	◐	◐					●	●	◐	◐	◐		●	◐	
Meyer [71]	◐	◐	◐	◐						●	●	◐		◐			●	◐

● is considered ◐ is partly considered

Fig. 1. Systematization of approaches to complexity assessment

Figure 1 classifies the approaches of complexity assessment found in the relevant literature, according to the characteristics of the aforementioned holistic orientation and their underlying methodology. These are supplemented by the consideration of dynamics as an individual area of evaluation to correspond to the described understanding of complexity. In addition, a classification and separation into the phases of complexity management (identification, designing, planning & operation, monitoring) as well as the corresponding field of application of the investigated approach were made.

Methods for complexity quantification from different fields of application were taken into account for the investigation, i.e. also from areas which cannot be directly assigned to the field of application of ACMs. The reason for that is that only few approaches in the addressed field of application exist and that even approaches that are not directly used in the field of ACM are to be examined for their adoptability.

The investigated approaches differ at first sight with regard to the considered complexity areas product, process, resource as well as the consideration of dynamic complexity aspects. The degree of a holistic focus of individual criteria differs between the considered approaches. The approaches, which are characterized by a high degree of holistic consideration of individual criteria, follow a kind of "top view" orientation. This orientation is characterized by the fact that it is based on a rather superordinate complexity management on the basis of an aggregated representation of quantified complexity drivers. The aggregation of the different drivers within the product, process and resource areas can be done by different methods (entropy, vector-based, etc.).

In contrast to this rather holistic view or "top view" orientation in the complexity-oriented evaluation of production systems, other approaches focus on the effects of individual respectively fewer complexity drivers on the performance or cost structures of the systems. The associated approaches, which follow the opposite type of top-view orientation in their consideration of complexity, are characterized by a much more detailed procedure regarding the identification and quantification of complexity factors. They are called "bottom-view" oriented approaches and consider further phases in complexity management, for example, the complexity-oriented derivation of design measures. The investigation of a single complexity driver and its effects can therefore be carried out in much more detail than with a holistic respectively top-view of several different drivers. Aggregation individual drivers to a single complexity key figure always leads, depending on the calculation method, to a certain loss of information and quality content of the statement of this aggregated key figure.

The holistic nature of the approaches differs greatly depending on the field of application (ACM, supply chain, process optimization, cost accounting), but also in the use in the respective phases of complexity management. If approaches for the determination of complexity costs are applied mainly in the field of cost accounting, they focus mainly on the investigation fields product and process (e.g. Horvath and Meyer 2003). Other approaches focus on an aggregated complexity indicator to optimize the existing system. However, only individual fields are examined more closely instead of covering all three fields in their entirety. An optimization on the basis of the investigated complexity takes place either in the field of design, i.e. the restructuring of the given status quo (e.g. Turner and Williams [42]) or in the area of planning and control. A direct use of the system-specific complexity for the application field of ACMs addressed in this paper, however, is only considered in the work of Grundstein [20].

This approach pursues a comprehensive consideration of a multitude of complexity drivers in the three fields product, process and resource, which are relevant for the determination of complexity in combination with the consideration of a change of ACMs. In addition, the necessary consideration of dynamic complexity components is also part of this approach. The use of the methodology in the addressed research field of the evaluation of ACMs on the basis of the respective complexity state of the system is however only partly suitable. While the scope of the considered complexity drivers provides a good starting point for the complexity assessment in the application field, the weighting and prioritization of the individual complexity parameters and in particular the aggregation to compressed parameters must be critically questioned. Especially the interactions of the individual drivers were not investigated. However, these interactions are an essential component for a significant complexity indicator as a basis for decision-making for changing ACMs [38].

Overall, it can be stated that only one approach could be identified in the direct field of the investigated application. In contrast to the other methods found, this one is characterized by an appropriate scope of complexity drivers, in particular also dynamic components were considered. In the following, the results and the identified complexity drivers will be used as a basis, but will be considered in a much more differentiated way. Among other things, this includes a necessary evaluation of the interactions of the individual complexity drivers.

4 Evaluation of Complexity Drivers in the Field of Autonomous Control

In the context of a differentiated consideration of individual complexity drivers, Grundstein et al. defines an extensive literature-based and systematized compilation of complexity drivers, that serves as a morphologic pattern for the complexity-oriented description and evaluation of production systems [20]. Based on this comprehensive work, a weighting of the individual complexity drivers is required, as they have different effects on a system. This evaluation also serves as a preliminary stage for filtering the complexity drivers, that are estimated to have a more significant influence on the systems' behavior. This is also a very crucial measure to ensure a manageable parameter scope for simulative investigations. In the work of Grundstein et al., weighting of each criterion is done on a scale from 1 to 3 according to its presumed significance for the system. A weighting of 3 represents criteria, that are necessary to classify a basic production system and occurring disruptions.

The weighting is given for information and criteria, that are crucial to characterize scheduling approaches. A weighting of 1 contains additional constraints, which are neither compulsory for the characterization of a production system or a scheduling approach [20]. Weighting is therefore determined by the systemic level of complexity drivers. However, this evaluation procedure completely neglects interdependencies of complexity drivers, although it can be assumed, that these have a great importance for the systems' behavior and the decision for a suitable ACM. Their high significance has already been shown and pointed out in several studies [11, 72, 73]. In summary, it is important to additionally consider interdependencies during weighting.

Table 2. Influencing matrix of complexity drivers

				(1)	(2)	(3)	(4)	(5)	(6)	(7)	(8)	(9)	(10)	(11)	ΣA
production system		(1)	number of production stages	-	0	1	0	1	0	0	0	0	1	1	**4**
		(2)	max. number of stations per stage	0	-	0	1	2	0	0	0	0	1	2	**6**
		(3)	due dates characteristics	0	2	-	2	1	3	0	0	0	3	0	**11**
		(4)	order availability	0	0	0	-	0	1	0	0	0	0	0	**1**
		(5)	material flow - complexity	0	0	1	2	-	0	0	0	0	0	1	**4**
		(6)	material flow - homogeneity	0	1	0	2	2	-	0	0	0	1	1	**7**
dynamic influences	external	(7)	deviation from expected due dates	0	1	3	2	1	2	-	1	1	3	3	**17**
		(8)	deviation from expected demanded quantity	0	3	3	2	1	3	3	-	1	0	3	**19**
		(9)	deviation from expected demanded product types	2	1	3	1	3	3	1	1	-	2	3	**20**
	internal	(10)	mean availability of correct information	0	0	2	2	1	0	0	0	0	-	1	**6**
		(11)	mean resource availability	0	3	1	3	2	1	0	0	0	0	-	**10**
		ΣP		**2**	**11**	**14**	**17**	**14**	**13**	**4**	**2**	**2**	**11**	**15**	

Legend

Value	Intensity	Description
0	No impact	No impact identifiable
1	Weak impact	Has impact, can induce complexity
2	Medium impact	Contributes to an increase in complexity
3	Strong impact	Influences significantly, increase complexity significantly, strong correlation

Therefore, the weighting according to Grundstein et al. [20] is extended in the following by additional consideration of interdependencies between the complexity drivers. This is limited to drivers whose weighting was rated with 2 or 3 in the first evaluation stage. The combined consideration of both evaluation stages is intended to enable a more differentiated and thus more precise weighting of complexity drivers. When examining interdependencies of system components, a distinction has to be made between direct and indirect interdependencies [74]. While a direct interdependency is defined as the effect between a systems component pairing, an indirect interdependency describes how a component affects a component via other components. For the evaluation of direct interdependencies, selected complexity drivers are mapped in a influencing matrix and evaluated using the evaluation scheme according to [75] (cf. Table 2).

The due date characteristics (3) have the greatest influence of the non-dynamic influencing variables, in particular on the material flow - homogeneity (6) as well as on the mean availability of correct information (10). This is based on the fact that a very

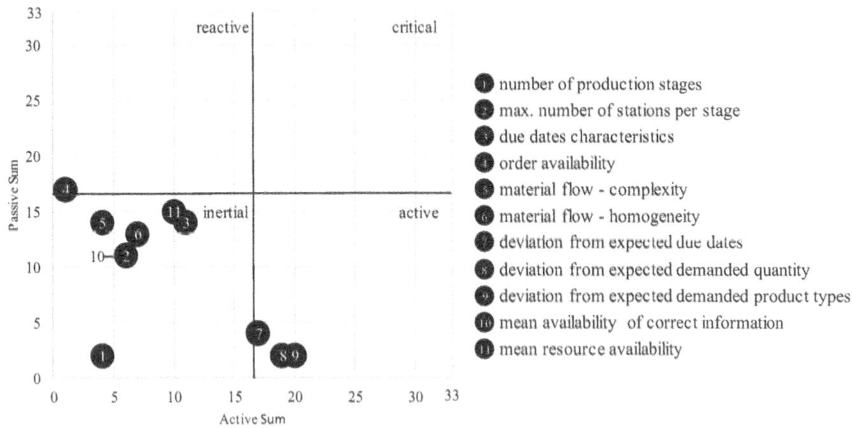

Fig. 2. Classification of interdependencies of considered complexity drivers

flexible due date (3) can cause the corresponding lot formation to vary considerably and thus also has a significant influence on complexity. On the other hand there is the order availability (4) which has little or no active influence on the other drivers. However, the order availability (4) is affected by almost all other drivers, which explains the high value of the passive sum. If the material flow (5, 6) or also the deviations of the expected delivery dates (7) are varied, this also influences the order availability (4) and in this context also favours the increase in complexity. This means that the order availability (4), as an influencing factor, has a minor impact on the other drivers. Apart from the consideration of non-dynamic drivers, it becomes obvious that dynamic complexity drivers have a strong direct internal and external influence on the other drivers. An important reason for this is the external dynamic complexity drivers influence almost every other driver. These dynamic drivers have reached the highest values and additionally illustrate the relevance of dynamics in terms of complexity. The figures also show that these drivers cannot all be equally weighted and that a consideration of the interactions between each other is necessary in order to obtain the best possible statement about the complexity of the system.

In order to provide better information about the individual values of the drivers, the results of the drivers are clustered in a coordinate system, taking into account the active (ΣA) and passive (ΣP) sum.

This enables to determine, how influencing or sensitive an element is in total. The active sum indicates the influence that an element has on all other elements and thus on the overall system. The passive sum is a measure of the degree to which an element is influenced by the overall system. It should be noted that the interdependencies between elements are local, but the effects are usually global. As shown in Fig. 2, complexity drivers can be classified as inertial, reactive, active and critical based on the calculation of active and passive sums. In this way, indirect interdependencies are considered at least in a very aggregated way. Resulting accuracy can be regarded as sufficient with regard to the purpose of this paper. This methodical approach is based on the works of [41, 71].

The evaluation of the active and passive sums of the respective complexity drivers shows a clear result. The dynamic external influencing variables (7, 8, 9) can be identified as active drivers. They have a strong influence on all other drivers and should, therefore, be the focus of further investigations. In addition to that only one driver (1) is in a far distance of the other drivers and is a highly recognizable inertial driver. The evaluation shows that the inertial driver (1) has almost no influence on any other driver and in the same course is not influenced by others, so this driver could be ignored in future investigations. The rest of the drivers are located very close to the upper left quadrant of the reactive drivers but are still in the inertial quadrant. These drivers interact with a large number of the examined drivers but are influenced in the same way by the active drivers (7, 8, 9). The dispersion of the individual drivers within the quadrant is striking. The scattering of the individual drivers within the quadrant is conspicuous, which suggests that they are influenced to different degrees. This dispersion also reinforces the statement that the drivers of Grundstein cannot all be examined with the same weighting, since they all vary greatly, especially in their interactions with each other.

Finally, it is important to point out that the quadrant of the critical drivers is empty. No drivers have both strong influence on others and are influenced at the same time by other drivers. On the one hand, this is advantageous, since the handling of the drivers and their influences is significantly simplified for further investigations. On the other hand, it should be questioned whether there are drivers that may belong to these quadrants, but which have not been detected in the present research.

In the following section, previous simulation studies of complexity drivers in the field of autonomous control are presented and compared with antecedent results. This survey is based on Table 3, in which the studies were itemized according to the applied production scenario and the varied complexity drivers. It has been pointed out that most studies were done in the context of flexible flowshops. This production form describes a structure similar to the workshop principle, but without returns. The system contains several production stages, each with multiple workstations per stage. The system can be scaled arbitrarily in both dimensions. In this scenario, an autonomous control has to decide for one workstation per stage. In single works, there are also simulative studies of autonomous control using industrial real manufacturing systems and data. For example, Scholz-Reiter et al. analyses the effects of ACMs based on a loading terminal [76].

It is noticeable in most studies that the system load, in the form of the (intermediate) arrival rate, is assumed to be predominantly sinusoidal, i.e. periodically fluctuating. A performance characterization within the evaluation of simulation execution is usually done based on KPI's of production logistics like the work-in-process, machine utilization, throughput time of orders and due date reliability. In most cases, the average throughput time of orders is used here. [77] also examines the dynamics of production systems using the power density spectra of the mean throughput time.

It becomes apparent that the selection of complexity drivers to be examined has been done systematically to some degree in existing studies, but above all to a certain extent arbitrarily and thus subjectively determined. None of the studies listed in Table 3 carried out a preliminary investigation about the selection of complexity drivers to be examined depending on their estimated influence for the systems' logistical performance. It can be stated that usually different ACMs are usually compared with each

Table 3. Simulation studies of complexity drivers and used production scenarios in the field of autonomous control

author	production system							production control				demand fluctuation		
	systems' design			operating times						machine assignement				
	FFS	real system	Number of workstations	station dependent	variant dependent	machine reliability	flexibility level	systems degree of autonomy	order release method	CP-SH	ACM	number of different product types	(inter) arrival rate (level of workload)	rush orders
Scholz-Reiter et al. 2009 [2]		●				●		●		●	●			
Grundstein et al. 2015 [10]	●					●			●		●		●	
Grundstein et al. [11]		●		●	●	●				●	●	●		●
Windt et al. 2010 [12]	●		●			●					●			
De Beer 2008 [13]	●	●	●								●		●	
Scholz-Reiter et al. 2006b [49]	●		●								●	●		
Grundstein et al. 2015 [72]	●					●				●	●		●	●
Fernandes et al. 2018 [73]	●			●		●					●		●	
Scholz-Reiter et al. 2007 [76]	●	●				●				●	●			
Hülsmann et al. 2006 [78]	●		●										●	
Schukraft et al. 2015 [79]	●					●				●	●		●	●
Scholz-Reiter et al. 2006 [80]	●		●								●	●		
Windt et al. 2010 [81]		●					●				●			
Scholz-Reiter et al. 2005 [82]	●												●	
Scholz-Reiter et al. 2005b [83]	●												●	
Scholz-Reiter et al. 2010 [84]	●		●							●	●		●	
Boyaci 2013 [85]	●	●	●								●		●	
Philipp 2014 [86]		●	●							●	●			
Scholz-Reiter et al. 2007 [87]	●										●		●	

● is considered

other. A comparison of ACMs with conventional production planning based on scheduling heuristic is only done by a few works. The so-called Queue Length Estimator (QLE) is the most frequently used ACM and can be seen as the benchmark method. This can be explained by the high efficiency of the ACM with at the same time trivial functionality so that a low implementation expenditure and high traceability is ensured.

An alternative benchmark are random decisions, which do not follow specific criteria. As shown in Table 3, the most considered complexity drivers for mapping the static and dynamic complexity of production conditions in simulations studies are machine availability, the number of workstations and parameters of the system load. In some works, rush orders and the heterogeneity of the production program, i.e. the number of product variants, are also applied parameters.

In most cases, the simulation approach used is ceteris paribus, i.e. only one parameter value is varied under constant conditions of the other parameters. Only a few studies use so-called complexity profiles. These are combinations of different values of the single complexity drivers in such a way that that no value combination exists more than once. However, the simulative parameter studies do not consider more than three complexity drivers. This limitation of the parameter scope probably aims at limiting the experimental effort to a manageable level, which in turn motivates the objective of this work. Even if the simulation results in the field of autonomous control are nowadays quite comprehensive, it is hardly possible to derive quantitative conclusions about the relevance of single complexity drivers due to the limited comparability of the studies. This is due to the fact that the conditions under which the simulation studies were carried out differ too much, especially on a more detailed level (e.g. product types, processing and set-up times, strategy of queue processing). However, it is possible to derive some qualitative findings for comparison with the interdependencies already assessed.

While the evaluation of the interdependencies has shown that especially external complexity drivers are to be considered as prioritized, in simulative studies often internal complexity drivers as machine availability and the number of workstations are also varied. However, simulation results show in common that the influence of internal complexity drivers on the logistic system performance is rather low compared to external influences. In particular, the impact of the number of machines as a representation of static complexity is almost negligible. Also according to the simulation results, dynamic complexity drivers have a significantly higher influence than static complexity drivers and should therefore be examined with priority. Both the evaluation of interdependencies and the simulation studies reveal that complexity drivers, whose dynamics are externally caused, have the greatest influence.

According to Table 3, the drivers of demand fluctuation are thus critical for the use of ACM. The external and dynamic drivers deviation of expected due dates, deviation of demanded quantity and deviation of product types prioritized according to the evaluation of interdependencies have not yet been fully addressed in the simulation studies. In most cases, the studies vary the mean value and the amplitude of the sinusoidal arrival rate and thus only deal with the deviation of demanded quantity. According to Table 3, the deviation of expected due dates can be represented by rush orders, but is only treated very rarely. The deviation of product types is considered also only very occasionally. Combined and under consideration of the interdependencies these three drivers have not yet been studied. Thus it is important to map these in a complexity profile in future studies and to determine the logistical performance of different ACMs as a function of these market-oriented complexity drivers. This enables to determine the range of the performance-suitable areas of the single ACMs, which in turn can be used to characterize the sensitivity of the control methods against value changes of the complexity drivers. This sensitivity represents an additional potential decision criterion for a suitable situational change of the ACM in addition to the logistical performance.

5 Conclusion and Outlook

In summary, ACMs have become increasingly important in a dynamic and complex production environment. There are numerous approaches in the literature which deal in particular with the simulative performance evaluation of ACMs. The parameters respectively complexity drivers that are analyzed for the particular investigations are not very comprehensive, do not necessarily follow the relevance of the individual parameters in the fields of application and do not consider the interactions of the individual parameters. Nevertheless, the complexity-oriented observation and evaluation of production systems are target-oriented in terms of the performance evaluation ACMs. The approaches in the field of complexity assessment provide good evidence for the application of complexity drivers. However, there are significant drawbacks regarding the relevance of individual complexity drivers as well as in the aggregation and interpretation of complexity parameters.

By the shown approach the relevance of complexity drivers can be derived systematically regarding the performance evaluation of ACMs. At the same time, there is a need for further research regarding the influence of driver constellations on the logistical performance of ACMs. The knowledge about this is necessary in order to derive the overarching research objective of developing an adaptive ACM. In this first approach, we derived, that the external and dynamic complexity drivers are the most significant and thus must be regarded prioritized, when online-switching of ACMs is analyzed.

Acknowledgment. The authors gratefully acknowledge the financial support of the Deutsche Forschungsgemeinschaft (DFG, German Research Foundation) within the Research Training Group GRK-2193 "Adaption Intelligence of Factories in a Dynamic and Complex Environment".

References

1. Nyhuis, P., Wiendahl, H.-P.: Fundamentals of Production Logistics. Theory, Tools and Applications. Springer, Berlin (2009)
2. Scholz-Reiter, B., Görges, M., Philipp, T.: Autonomously controlled production systems—Influence of autonomous control level on logistic performance. CIRP Ann. **58**(1), 395–398 (2009). https://doi.org/10.1016/j.cirp.2009.03.011
3. Kim, J.-H., Duffie, N.A.: Backlog control design for a closed loop PPC system. CIRP Ann. **53**(1), 357–360 (2004). https://doi.org/10.1016/S0007-8506(07)60716-1
4. Windt, K., Böse, F., Philipp, T.: Autonomy in production logistics. Identification, characterisation and application. Robot. Comput.-Integr. Manuf. **24**(4), 572–578 (2008). https://doi.org/10.1016/j.rcim.2007.07.008
5. Lödding, H.: Handbook of Manufacturing Control. Fundamentals, Description, Configuration. Springer, Heidelberg (2013)
6. Scholz-Reiter, B., Lappe, D., Toonen, C., Ruthenbeck, C.: Development of a hybrid control approach for automotive logistics based on close to real-time information management. Int. J. Syst. Appl. Eng. Dev. **5**(4), 545–552 (2011)
7. Vec, M. (Hrsg) Selbstorganisation: ein Denksystem für Natur und Gesellschaft. Böhlau, Köln (2006)
8. Windt, K., Hülsmann, M.: Changing paradigms in logistics- understanding the shift from conventional control to autonomaus cooperation and control. In: Windt, K., Hülsmann, M. (Hrsg) Understanding Autonomous Cooperation and Control in Logistics. The Impact of Autonomy on Management, Information, Communication and Material Flow, pp. 1–16. Springer, Berlin (2007)
9. Philipp, T., de Beer, C., Windt, K., Scholz-Reiter, B.: Evaluation of autonomous logistic processes - analysis of the influence of structural complexity. In: Windt, K., Hülsmann, M. (Hrsg) Understanding Autonomous Cooperation and Control in Logistics. The Impact of Autonomy on Management, Information, Communication and Material Flow, pp. 357–360. Springer, Berlin (2007)

10. Grundstein, S., Schukraft, S., Scholz-Reiter, B., Freitag, M.: Coupling order release methods with autonomous control methods – an assessment of potentials by literature review and discrete event simulation. Int. J. Prod. Manag. Eng. **3**(1), 43 (2015). https://doi.org/10.4995/ijpme.2015.3199
11. Grundstein, S., Freitag, M., Scholz-Reiter, B.: A new method for autonomous control of complex job shops – integrating order release, sequencing and capacity control to meet due dates. J. Manuf. Syst. **42**, 11–28 (2017). https://doi.org/10.1016/j.jmsy.2016.10.006
12. Windt, K., Becker, T., Jeken, O., Gelessus, A.: A classification pattern for autonomous control methods in logistics. Logist. Res. **2**(2), 109–120 (2010). https://doi.org/10.1007/s12159-010-0030-9
13. de Beer, C.: Untersuchung der Dynamik von selbststeuernden Prozessen in produktionslogistischen Systemen anhand ereignisdiskreter Simulationsmodelle. Zugl.: Bremen, Univ., Diss., 2008. Schriftenreihe, Bd 9. GITO-Verlag, Berlin (2008)
14. Wiendahl, H.-H.: Situative Konfiguration des Auftragsmanagements im turbulenten Umfeld. Zugl.: Stuttgart, Univ., Diss., 2002. IPA-IAO-Forschung und -Praxis, Bd 358. Jost-Jetter, Heimsheim (2002)
15. Ueda, K., Lengyel, A., Hatono, I.: Emergent synthesis approaches to control and planning in make to order manufacturing environments. CIRP Ann. **53**(1), 385–388 (2004). https://doi.org/10.1016/S0007-8506(07)60722-7
16. Helbing, D.: Modeling and optimization of production process. Lessons from traffic dynamics. In: Radons, G., Neugebauer, R. (Hrsg) Nonlinear Dynamics of Production Systems, pp. 85–105. Wiley-VCH, Weinheim, Cambridge (2004)
17. Philipp, T., Windt, K.: Evaluation of autonomously controlled logistic processes. In: Proceedings of 5th CIRP International Seminar on Intelligent Computation in Manufacturing Engineering, Ischia, 18–21 July 2006
18. Philipp, T., Böse, F., Windt, K.: Autonomously controlled processes - characterisation of complex production systems. In: Cunha, P., Maropoulos, P. (Hrsg) Proceedings of 3rd CIRP Conference in Digital Enterprise Technology, Setubal, Portugal (2006)
19. Windt, K.: Ermittlung des angemessenen Selbststeuerungsgrades in der Logistik – Grenzen der Selbststeuerung. In: Nyhuis, P. (Hrsg) Beiträge zu einer Theorie der Logistik, pp. 349–372. Springer, Heidelberg (2008)
20. Grundstein, S., Schukraft, S., Scholz-Reiter, B., Freitag, M.: Evaluation system for autonomous control methods in coupled planning and control systems. Procedia CIRP **33**, 121–126 (2015). https://doi.org/10.1016/j.procir.2015.06.023
21. Bohne, F.: Komplexitätskostenmanagement in der Automobilindustrie. Identifizierung und Gestaltung vielfaltsinduzierter Kosten. Deutscher Universitätsverlag, Wiesbaden, s.l. (1998)
22. Klir, G.J.: Facets of systems science, 2. Aufl. IFSR international series on systems science and engineering, Bd 15. Kluwer Acad./Plenum Publ., New York (2001)
23. Ashby, W.R.: An Introduction to Cybernetics. Chapman & Hall Ltd., New York (1957)
24. Scherer, E.: The reality of shop floor control—approaches to systems innovation. In: Scherer, E. (Hrsg) Shop Floor Control—A Systems Perspective, pp. 3–26. Springer, Heidelberg (1998)
25. Schuh, G., Krumm, S., Amann, W.: Chefsache Komplexität. Navigation für Führungskräfte. Springer, Wiesbaden (2013)
26. Westphal, J.R.: Komplexitätsmanagement in der Produktionslogistik. Ein Ansatz zur flussorientierten Gestaltung und Lenkung heterogener Produktionssysteme. Gabler Edition Wissenschaft. Deutscher Universitätsverlag, Wiesbaden, s.l. (2001)
27. Patzak, G.: Systemtechnik - Planung Komplexer Innovativer Systeme. Grundlagen, Methoden, Techniken. Springer, Berlin (1982)

28. Ulrich, H., Probst, G.J.B.: Anleitung zum ganzheitlichen Denken und Handeln. Ein Brevier für Führungskräfte, 4. Aufl. Haupt, Bern (1995)
29. Scherer, E., Dobberstein, M.: Komplexität in der Produktion beherrschen? io Manag. Z. **65** (3), 60–64 (1996)
30. Schuh, G.: Produktkomplexität managen. Strategien; Methoden; Tools, 2. Aufl. Carl Hanser Verlag GmbH & Co. KG, s.l. (2005)
31. Blunck, H., Windt, K.: Komplexität schafft Spielraum für Selbststeuerung. Werkstatttechnik Online **103**(2), 109–113 (2013)
32. Kieser, A.: Organisationsstruktur. Schmalenbachs Zeitschrift für betriebswirtschaftliche Forschung, no. 25, pp. 569–590 (1974)
33. Engelhardt, P.R.: System für die RFID-gestützte situationsbasierte Produktionssteuerung in der auftragsbezogenen Fertigung und Montage. Dissertation, TU München (2015)
34. Windt, K., Philipp, T., Böse, F.: Complexity cube for the characterization of complex production systems. Int. J. Comput. Integr. Manuf. **21**(2), 195–200 (2008). https://doi.org/10.1080/09511920701607725
35. ElMaraghy, W., ElMaraghy, H., Tomiyama, T., Monostori, L.: Complexity in engineering design and manufacturing. CIRP Ann. **61**(2), 793–814 (2012). https://doi.org/10.1016/j.cirp.2012.05.001
36. Scholz-Reiter, B., Tervo, J.T.: Kausaler Zusammenhang von Komplexität und Dynamik in der Produktion. In: Lucas, K. (Hrsg) Kausalität in der Technik. Vorträge im Rahmen der wissenschaftlichen Sitzungen der Technikwissenschaftlichen Klasse am 24. Februar, 5. Mai und 18. Oktober 2006, pp. 1–3. Berlin-Brandenburgische Akad. der Wiss, Berlin (2007)
37. van Brussel, H., Wyns, J., Valckenaers, P., Bongaerts, L., Peeters, P.: Reference architecture for holonic manufacturing systems: PROSA. Comput. Ind. **37**(3), 255–274 (1998). https://doi.org/10.1016/S0166-3615(98)00102-X
38. Gießmann, M., Lasch, R.: Der Einfluss der Beschaffungskomplexität auf den Logistikerfolg. In: Bogaschewsky, R., Eßig, M., Lasch, R., Stölzle, W. (Hrsg) Supply Management Research, pp. 149–196. Gabler, Wiesbaden (2010)
39. Schoeneberg, K.-P.: Komplexität – Einführung in die Komplexitätsforschung und Herausforderungen für die Praxis. In: Schoeneberg, K.-P. (Hrsg) Komplexitätsmanagement in Unternehmen, pp. 13–27. Springer, Wiesbaden (2014)
40. Wildemann, H., Voigt, K.-I.: Komplexitätsindex-Tool. Entscheidungsgrundlagen für die Produktprogrammgestaltung bei KMU. TCW, München (2011)
41. Lasch, R., Gießmann, M.: Ganzheitliche Ansätze zum Komplexitätsmanagement – eine kritische Würdigung aus Sicht der Beschaffungslogistik. In: Bogaschewsky, R., Eßig, M., Lasch, R., Stölzle, W. (Hrsg) Supply Management Research, pp. 195–231. Gabler, Wiesbaden (2009)
42. Turner, K., Williams, G.: Modelling complexity in the automotive industry supply chain. J. Manuf. Technol. Manag. **16**(4), 447–458 (2005). https://doi.org/10.1108/17410380510594525
43. Lechner, A.: Modellbasierter Ansatz zur Bewertung vielfaltsinduzierter Logistikkomplexität in der variantenreichen Serienfertigung der Automobilindustrie. Zugl.: Dortmund, Techn. Univ., Diss., 2011. Unternehmenslogistik. Verl. Praxiswissen, Dortmund (2012)
44. Mattsson, S., Tarrar, M., Fast-Berglund, Å.: Perceived production complexity – understanding more than parts of a system. Int. J. Prod. Res. **54**(20), 6008–6016 (2016). https://doi.org/10.1080/00207543.2016.1154210
45. Wang, H., Zhu, X., Wang, H., Hu, S.J., Lin, Z., Chen, G.: Multi-objective optimization of product variety and manufacturing complexity in mixed-model assembly systems. J. Manuf. Syst. **30**(1), 16–27 (2011). https://doi.org/10.1016/j.jmsy.2011.03.002

46. Frizelle, G., Woodcock, E.: Measuring complexity as an aid to developing operational strategy. Int. J. Oper. Prod. Manag. **15**(5), 26–39 (1995). https://doi.org/10.1108/01443579510083640
47. Blecker, T., Kersten, W., Meyer, C.: Development of an approach for analyzing supply chain complexity. In: Blecker, T., Friedrich, G. (eds.) Mass Customization. Concepts – Tools – Realization. Proceedings of the International Mass Customization Meeting 2005, pp. 47–59 (2005)
48. Dekkers, R., Gerschberger, M., Engelhardt-Nowitzki, C., Kummer, S., Staberhofer, F.: A model to determine complexity in supply networks. J. Manuf. Technol. Manag. **23**(8), 1015–1037 (2012). https://doi.org/10.1108/17410381211276853
49. Scholz-Reiter, B., Philipp, T., de Beer, C., Windt, K., Freitag, M.: Einfluss der strukturellen Komplexität auf den Einsatz von selbststeuernden logistischen Prozessen. In: Pfohl, H.-C., Wimmer, T. (Hrsg) Steuerung von Logistiksystemen - auf dem Weg zur Selbststeuerung, Konferenzband zum 3. BVL-Wissenschaftssymposium Logistik, pp. 11–25. Deutscher Verkehrs-Verlag, Hamburg (2006)
50. Meepetchdee, Y., Shah, N.: Logistical network design with robustness and complexity considerations. Int. J. Phys. Distrib. Logist. Manag. **37**(3), 201–222 (2007). https://doi.org/10.1108/09600030710742425
51. Lübke, E.: Lebenszyklusorientiertes Produktstrukturmanagement. Eine theoretische und empirische Untersuchung. Zugl.: München, Techn. Univ., Diss., 2006, 1. Aufl. TCW Wissenschaft und Praxis, Bd 43. TCW Transfer-Centrum, München (2007)
52. Kota, S., Sethuraman, K., Miller, R.: A metric for evaluating design commonality in product families. J. Mech. Des. **122**(4), 403 (2000). https://doi.org/10.1115/1.1320820
53. Hanenkamp, N.: Entwicklung des Geschäftsprozesses Komplexitätsmanagement in der kundenindividuellen Serienfertigung. Ein Beitrag zum Informationsmanagement in mehrdimensional modellierten Produktionssystemen. Zugl.: Bochum, Univ., Diss., 2004. Schriftenreihe des Lehrstuhls für Produktionssysteme/Institut für Automatisierungstechnik, Lehrstuhl für Produktionssysteme, Bd 2004, 5. Shaker, Aachen (2004)
54. Kaplan, R.S., Anderson, S.R.: Time-driven activity-based costing. A simpler and more powerful path to higher profits, 3. Aufl. Harvard Business School Press, Boston (2009)
55. Brecht, U.: Die prozeßkostenrechnung. In: Vollmuth, H.J., Pepels, W. (Hrsg) Kosten senken und Leistungen steigern durch marktorientiertes Kostenmanagement. expert-Verl., Renningen (2003)
56. Horvath, P., Mayer, R.: Prozesskostenrechnung. Der neue Weg zu mehr Kostentransparenz und wirkungsvolleren Unternehmensstrategien. Controlling **4**, 214–219 (1998)
57. Jeschke, A.: Beitrag zur wirtschaftlichen Bewertung von Standardisierungsmaßnahmen in der Einzel- und Kleinserienfertigung durch die Konstruktion. Inst. für Konstruktionslehre, Maschinen- u. Feinwerkelemente, Braunschweig (1997)
58. Puhl, H.: Komplexitätsmanagement. Univ., Lehrstuhl für Fertigungstechnik und Betriebsorganisation, Kaiserslautern (1999)
59. Braithwaite, A., Samakh, E.: The cost-to-serve method. Int. J. Logist. Manag. **9**(1), 69–84 (1998). https://doi.org/10.1108/09574099810805753
60. Schaffer, J.: Entwicklung und Optimierung eines treiberbasierten Modells zur Bewertung varianteninduzierter Komplexitätskosten in industriellen Produktionsprozessen. Zugl.: Lüneburg, Univ., Diss., 2010. Leuphana-Universität Lüneburg, Institut für Produkt- und Prozessinnovation, Bd 1. Sierke, Göttingen (2010)
61. Bräutigam, L.-P.: Kostenverhalten bei Variantenproduktion. Schriften zum Produktionsmanagement. Deutscher Universitätsverlag, Wiesbaden (2004)

62. Heina, J.: Variantenmanagement. Kosten-Nutzen-Bewertung zur Optimierung der Variantenvielfalt. Gabler Edition Wissenschaft. Deutscher Universitätsverlag, Wiesbaden, s.l. (1999)
63. Bartuschat, M.: Beitrag zur Beherrschung der Variantenvielfalt in der Serienfertigung. Zugl.: Braunschweig, Techn. Univ., Diss., 1994. Schriftenreihe des IWF. Vulkan-Verl., Essen (1995)
64. Dalhöfer, J., Rall, K.: Komplexitätsbewertung indirekter Geschäftsprozesse. Zugl.: Hamburg-Harburg, der Techn. Univ., Institut für Werkzeugmaschinen, Roboter und Montageanlagen, Diss., 2009, 1. Aufl. Schriftenreihe des Instituts für Werkzeugmaschinen, Roboter und Montageanlagen der Technischen Universität Hamburg-Harburg, Bd 19. Shaker, Aachen (2009)
65. Feldhütter, V.: Beitrag zur modellbasierten Bewertung der Komplexität in der Montagelogistik der Automobilindustrie. Technische Universität Dortmund (2018)
66. Iván Tarride, M.: The complexity of measuring complexity. Kybernetes **42**(2), 174–184 (2013). https://doi.org/10.1108/03684921311310558
67. Becker, T., Meyer, M., Windt, K.: A manufacturing systems network model for the evaluation of complex manufacturing systems. Int. J. Prod. Perform. Manag. **63**(3), 324–340 (2014). https://doi.org/10.1108/IJPPM-03-2013-0047
68. Marti, M.: Complexity Management, 1. Aufl. DUV Deutscher Universitäts-Verlag, s.l. (2007)
69. Bliss, C.: Management von Komplexität. Ein integrierter, systemtheoretischer Ansatz zur Komplexitätsreduktion. Unternehmensführung und Marketing, Bd 35. Gabler Verlag, Wiesbaden, s.l. (2000)
70. Kirchhof, R.: Ganzheitliches Komplexitätsmanagement. Grundlagen und Methodik des Umgangs mit Komplexität im Unternehmen. Beiträge zur Produktionswirtschaft. Deutscher Universitätsverlag, Wiesbaden, s.l. (2003)
71. Meyer, C.M., Kersten, W., Nedeß, C.: Integration des Komplexitätsmanagements in den strategischen Führungsprozess der Logistik. Zugl.: Hamburg-Harburg, Techn. Univ., Institut für Logistik und Unternehmensführung, Diss., 2006 u.d.T.: Meyer, Christian Martin: Komplexitätsmanagement in der Logistik, 1. Aufl. Schriftenreihe Logistik der Kühne-Stiftung, Bd 12. Haupt, Bern (2007)
72. Grundstein, S., Schuhkraft, S., Freitag, M., Scholz-Reiter, B.: Planorientierte autonome Fertigungssteuerung. Simulationsbasierte Untersuchung der Planeinhaltung und der logistischen Zielerreichung. Werkstatttechnik Online **105**(4), 220–224 (2015)
73. Fernandes, N.O., Martins, T., Carmo-Silva, S.: Improving materials flow through autonomous production control. J. Ind. Prod. Eng. **35**(5), 319–327 (2018). https://doi.org/10.1080/21681015.2018.1479895
74. Gäng, J.: Berücksichtigung von Wechselwirkungen bei Zuverlässigkeitsanalysen. Zugl.: Stuttgart, Univ., Diss., 2011. Berichte aus dem Institut für Maschinenelemente, Antriebstechnik, CAD, Dichtungen, Zuverlässigkeit, Bd 136. Univ. Inst. für Maschinenelemente, Stuttgart (2012)
75. Velte, C.J., Wilfahrt, A., Müller, R., Steinhilper, R.: Complexity in a life cycle perspective. Procedia CIRP **61**, 104–109 (2017). https://doi.org/10.1016/j.procir.2016.11.253
76. Scholz-Reiter, B., de Beer, C., Böse, F., Windt, K.: Evolution in der Logistik - Selbststeuerung logistischer Prozesse. In: 16. Deutscher Materialfluss-Kongress -Intralogistik bewegt- mehr Effizienz, mehr Produktivität, pp. 179–190. VDI Verlag, Düsseldorf (2007)
77. Scholz-Reiter, B., Freitag, M., de Beer, C., Jagalski, T.: Analysing the dynamics caused by autonomously controlled logistic objects. In: Proceedings of the 2nd International Conference on Changeable, Agile, Reconfigurable and Virtual Production (2007)

78. Hülsmann, M., Scholz-Reiter, B., Freitag, M., Wycisk, C., de Beer, C.: Autonomous cooperation as a method to cope with complexity and dynamics? – a simulation based analyses and measurement concept approach. In: Bar-Yam, Y. (ed.) Proceedings of the International Conference on Complex Systems (ICCS 2006), Boston, MA, USA (2006)
79. Schukraft, S., Grundstein, S., Freitag, M., Scholz-Reiter, B.: Strategies for the coupling of autonomous control and central planning: evaluation of strategies using logistic objectives achievement and planning adherence, pp. 121–126 (2015). https://doi.org/10.1109/smc.2015.34
80. Scholz-Reiter, B., Freitag, M., de Beer, C., Jagalski, T.: The influence of production networks' complexity on the performance of autonomous control methods. In: Teti, R. (ed.) Intelligent Computation in Manufacturing Engineering 5. Proceedings of the 5th CIRP International Seminar on Computation in Manufacturing Engineering (CIRP ICME 2006), pp. 317–320. University of Naples, Naples (2006)
81. Windt, K., Jeken, O., Becker, T.: Selbststeuerung in der Produktion. In: ZWF, vol. 105, no. 5, pp. 439–443 (2010). https://doi.org/10.3139/104.110319
82. Scholz-Reiter, B., Freitag, M., de Beer, C., Jagalski, T.: Modelling and analysis of autonomous shop floor control. In: 38th CIRP International Seminar on Manufacturing Systems – CIRP ISMS 2005 (2005)
83. Scholz-Reiter, B., Freitag, M., de Beer, C., Jagalski, T.: Modelling dynamics of autonomous logistic processes. Discrete-event versus continuous approaches. CIRP Ann. **54**(1), 413–416 (2005). https://doi.org/10.1016/s0007-8506(07)60134-6
84. Scholz-Reiter, B., Rekersbrink, H., Görges, M.: Dynamic flexible flow shop problems—scheduling heuristics vs. autonomous control. CIRP Ann. **59**(1), 465–468 (2010). https://doi.org/10.1016/j.cirp.2010.03.030
85. Boyaci, P.: Gestaltungsrahmen zur Implementierung von Selbststeuerung in der variantenreichen Großserienfertigung der Automobilzulieferindustrie. Dissertation. Universität Kassel, Kassel (2013)
86. Philipp, T.: Messung und Bewertung selbststeuernder Produktionssysteme. Dissertation. Universität Bremen, Bremen. Fachbereich Produktionstechnik (2014)
87. Scholz-Reiter, B., Freitag, M., de Beer, C., Jagalski, T.: Analysing the dynamics caused by autonomously controlled logistic objects. In: Proceedings of the 2nd International Conference Changeable, Agile Reconfigurable and Virtual Production (2007)

Using the Digital Shadow for a Prescriptive Optimization of Maintenance and Operation

The Locomotive in the Context of the Cyber-Physical System

Julian Franzen[1(✉)], Jannis Stecken[1], Raphael Pfaff[2], and Bernd Kuhlenkötter[1]

[1] Chair of Production Systems, Ruhr-Universität Bochum, Universitätsstr. 150, 44801 Bochum, Germany
franzen@lps.rub.de

[2] University of Applied Sciences FH Aachen, Goethestr. 1, 52064 Aachen, Germany

Abstract. In competition with other modes of transport, rail freight transport is looking for solutions to become more attractive. Short-term success can be achieved through the data-driven optimization of operations and maintenance as well as the application of novel strategies such as prescriptive maintenance. After introducing the concept of prescriptive maintenance, this paper aims to prove that vehicle-focused applications of this approach indeed have the potential to increase attractiveness. However, even greater advantages can be activated if data from the horizontal network of the vehicle is available. Drawing on the state of the art in research and technology in the field of cyber-physical systems (CPS) as well as digital twins and shadows, our work serves to design a system of systems for the horizontal interconnection of a rail vehicle and to conceptualize a draft for a digital twin of a locomotive.

1 Introduction

The transport of freight by rail makes up around 20 percent of the absolute transport volume in Germany [1]. Although there is an increase in the total transport volume, the share of rail freight transport has been steadily decreasing since 2015, whereas the share of road transport is constantly increasing. This development is alarming in view of congested roads and the advantages of the railway transport system. Especially in context of the discussion on sustainable and safe means of transport, it is obvious and essential to put rail transport into the center of scientific attention (Table 1).

U. Clausen et al. (Eds.): ICPLT 2019, LNLO, pp. 265–276, 2019.
https://doi.org/10.1007/978-3-030-13535-5_19

Table 1. Share of the individual modes of transport in the total transport volume [1]

Goods transport	Unit	2015	2016	2017
Rail transport	1000 t	367.314	363.512	348.559
Road transport	1000 t	3.035.333	3.111.819	3.161.781

2 Motivation

The reduced attractiveness of rail transport is mainly due to the high costs of operation and maintenance. While the introduction of innovative train protection systems like ETCS and massive infrastructure developments may be a partial solution to maintain competitiveness, these actions are very capital intensive and must be considered in the long run. Therefore, no short-term improvement is to be expected.

However, significantly faster benefits in the form of reduced life-cycle costs can be activated by optimizing operation and maintenance. While preventive maintenance strategies still dominate the railway vehicle sector nowadays, predictive approaches are located in the field of research. Although there are also commercially available solutions, they usually consider individual components or subsystems of a rail vehicle without taking the system as a whole into consideration. A reduction of the life cycle costs on system level through the use of innovative forms of maintenance does not take place.

2.1 Prognostics and Health Management

The procedure for the implementation of innovative maintenance strategies such as predictive maintenance consists of the data acquisition, data processing, analysis of the data and a recommendation of a specific action (see Fig. 1). The research discipline that deals with intelligent data and model-driven maintenance based on this approach is called prognostics and health management [2].

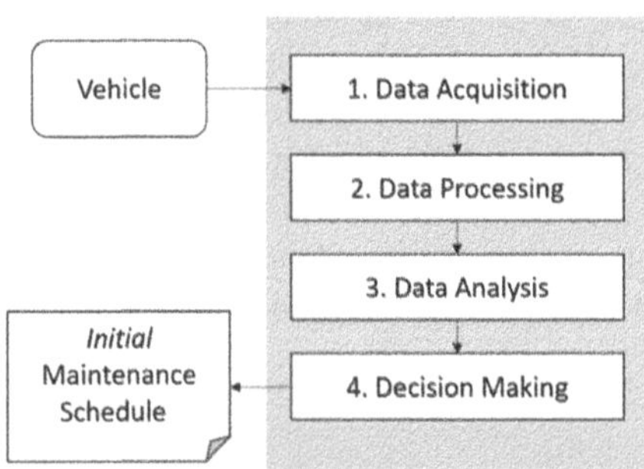

Fig. 1. General process of prognostics and health management [2]

2.2 Forms of Data Analysis

The most important difference between forms of maintenance is found in the step of data analysis. Davenport [3] illustrates on the basis of numerous cross-disciplinary examples that a huge competitive advantage can be achieved by integrating intelligent analysis methods into one's own business processes. For this purpose, a breakdown into four basic analysis forms is made according to Fig. 2. Most of these kinds of analytics can be found in the context of maintenance.

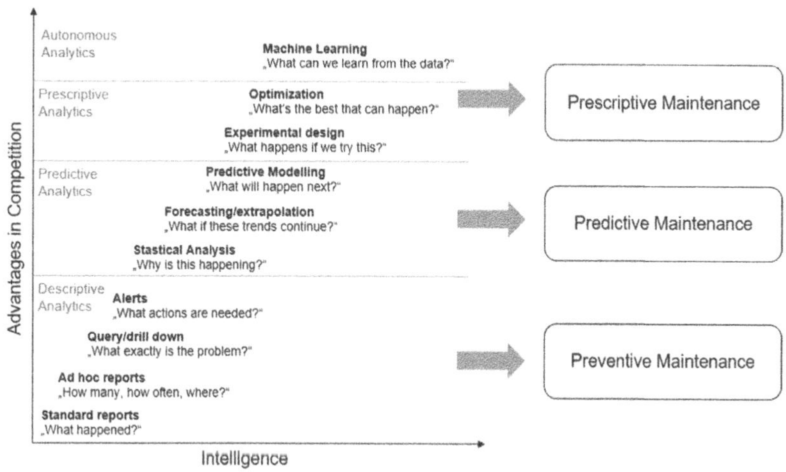

Fig. 2. Competitive advantage through the use of intelligent data analytics [3, 4]

In simplified terms, preventive maintenance uses a wide variety of descriptive and observation-based methods to determine the expected failure time of a target component or system from historical data. Preventive maintenance is then scheduled for a trusted time prior to failure.

Predictive analysis as the basis of the maintenance variant is gaining in intelligence and, using statistical methods, describes trends in the data collected. Thus, it is possible to investigate the question of what will happen next if an identified trend continues. Finally, the failure time of the unit under consideration is predicted by specifying a confidence interval to derive a corresponding maintenance action.

Even more complex forms of analysis, specifically prescriptive and autonomous analysis, are currently not considered. Simplified, prescriptive analysis includes a problem-specific method for automated decision-making regarding a target achievement criterion. The number of possible actions to achieve this goal is typically of such an enormous size that the human being, due to the limitation of his cognitive abilities and the inclination to intuition, is not able to select the optimal solution [5]. While this form of analysis may be considered prior art in the area of routing (calculating the fastest or shortest route), it has not yet taken maintenance planning into account.

2.3 Prescriptive Maintenance

Prescriptive analysis seeks answers to the question: what is the best that can happen? The best maintenance strategy for a railway operator includes the following features:

- optimal costs (lowest lifecycle costs)
- maximum availability
- safety.

According to Franzen [4], the implementation of prescriptive maintenance first of all means the scheduling of an initial optimal maintenance plan on the basis of assumptions about the operation of the target vehicle. These assumptions can be experiences of the operator, but also historical data, for example from condition monitoring systems.

An essential difference to previously known strategies is then based on the interpretation of the vehicle in operation. While only the condition of individual components is investigated in predictive approaches, the prescriptive approach according to Franzen et al. [6] interprets the vehicle as a proactive maintenance control element. If the objectives of the initial maintenance planning are not met, the system autonomously initiates corrective actions. Corrective actions may be to reduce the average speed or to harmonize acceleration and braking operations, and technically to manipulate the drive train (see Fig. 3).

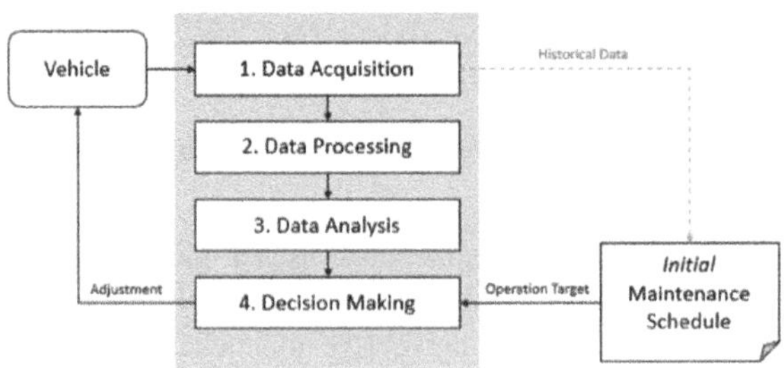

Fig. 3. Framework for the application of prescriptive maintenance

The choice of an adjustment action can be described by a complex optimization problem as shown in the configuration set of the optimization problem is made of the available manipulation activities; the objective function is defined by the mechanical and control properties of the target system.

However, additional boundary conditions arise from outside the vehicle. In addition to static legal requirements, e.g. from occupational or functional safety requirements of the railway system, the operational context of the vehicle forms the decisive dynamic limitation of the solution space. If, for example, it has been found that the operation needs to be carried out with less load, because assumptions of the initial planning are not fulfilled, the operation can only be adjusted as far as the operational context permits, i.e. the vehicle remains efficient. This information then comes from outside the

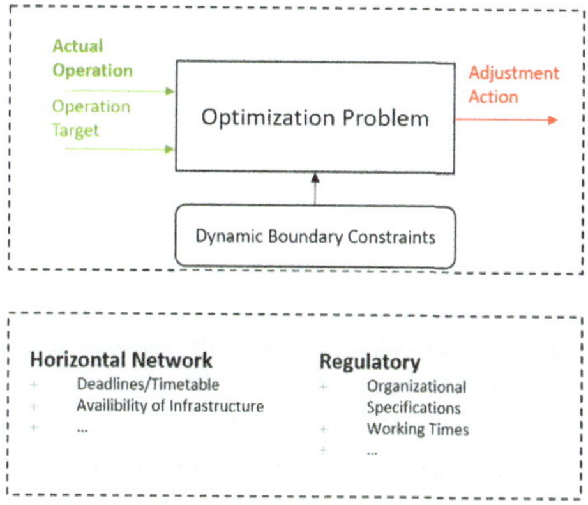

Fig. 4. Optimization problem for determining an adjustment action (conceptual)

vehicle from the horizontal integration with other systems such as other vehicles, the infrastructure and the customers of the transport system such as production systems.

In order to carry out optimizations within the system on a higher level, it is necessary to provide a global availability of data in a system of systems. While such approaches are still lacking for rail transport but are mandatory for the application of prescriptive maintenance strategies, similar ideas are much more advanced and mature in industry. The next chapter therefore provides an overview of the current state of research in this area in order to finally assess which needs of the transport system can already be met by existing systems through adaptation.

3 State of Scientific Research

In recent years, no term has shaped the industry as much as Industry 4.0. In many areas, the exact definition and effects of Industry 4.0 are discussed. Barthelmäs et al. [7] summarize the previously researched catchwords in connection with Industry 4.0 into the three clusters: flexibility, automation and networking. These three clusters can also be applied in the railway industry. Maintenance and operation must become more flexible by softening rigid maintenance intervals and adapting operation individually to the various travel profiles.

The aim must be for the locomotive to automatically determine the necessary maintenance and adaptation of its operation on the basis of data. Networking is the basis for automation. An automatically optimized operation must have all data available, which is provided by a vertical and horizontal data continuity.

3.1 CPS and CPPS

Purely physical products are no longer sufficient to handle the increasing demands on different fields such as production, energy or mobility. The concept of cyber-physical systems (CPS) therefore combines the physical world with the digital world by being able to receive, process and transmit data from the physical product via the Internet.

Since the 1970s, the concept of extending physical products with the ability to process information has already existed. CPS extend this concept by providing this information over the Internet and thus creating new possibilities of use [8]. For example, sensor data from machine components can be made available to the user via platforms. The data obtained in this way can be used to identify and solve problems in an early stage or to optimize operations.

If you take the concept of CPS and transfer it to production systems, you get so-called cyber-physical production systems (CPPS). By linking the different components of a production system with each other and the permanent availability of data, completely new production possibilities are created. The vision is to establish production systems that can organize or optimize themselves. [9] Therefore, the classical automation pyramid, as shown in Fig. 5, is broken down and a decentralized automation network is created.

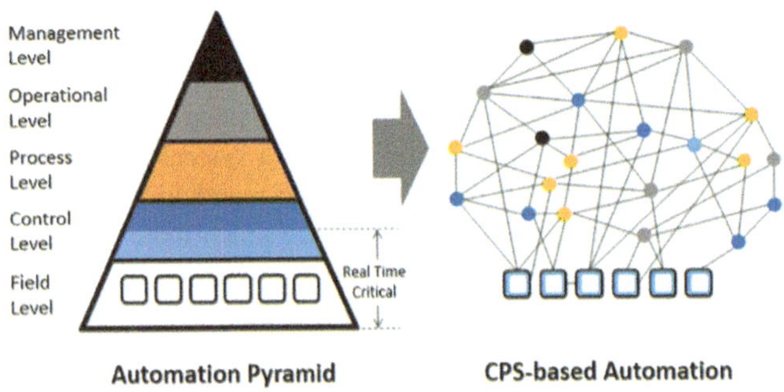

Fig. 5. Transformation from automation pyramid to decentralized automation network [8]

However, what stays the same is the typical field level with real-time critical components such as the PLC. At higher levels, however, the decentralized distribution is typical of CPPS. In the future, however, it is conceivable that real-time-critical applications could also be made available in a decentralized manner via platforms or an automation cloud. [8] This significantly increases the scalability of a production system and its flexibility.

3.2 Digital Twin and Digital Shadow

The term digital twin was first defined by the NASA as "an integrated multi-physics, multi-scale, probabilistic simulation of a vehicle or system that uses the best available physical models, sensor updates, fleet history, etc. to mirror the life of its flying twin. It is ultra-realistic and may consider one or more important and interdependent vehicle systems" [10]. In the course of time, the term from the aerospace industry has also been transferred to other areas. In addition to the term "digital twin", the term "digital shadow" is also often used referring to existing literature. The basic idea of both terms is to create a virtual image of its real counterpart. This can be used to provide information about the current state, to analyze past states or to forecast future states with the help of the data.

Bauernhansl et al. [11] define the digital shadow as "a sufficiently accurate" representation of the processes "in production, development and adjacent areas with the purpose of creating a real-time analysis basis for all relevant data".

Negri et al. [12] give a detailed analysis of the different definitions of the digital twin used in the literature between 2012 and 2016. The analysis shows that the term digital twin is used in many different areas and in completely different ways.

The problem with the concept of the digital twin or digital shadow is that research in this area is still in its infancy and there are therefore few implementations of this concept. Kritzinger et al. [13] give a categorical analysis of the use of the terms "digital twin" and "digital shadow" in the literature between 2014 and 2018. The results show that 55% of publications present only concepts with only small case studies.

It has been shown that the terms digital twin and digital shadow are often used contrary to each other and that there is no uniform definition similar to that of "Industry 4.0". Therefore, a definition of the two terms that apply and are used for this paper is given here. The digital twin describes the virtual image of a component, a system or a process which has no connection to reality. The digital twin can emerge both from the digital shadow, in which models are generated from data obtained, or also during the engineering process. The digital twin can be used, for example, to perform offline simulations. The digital twin consists of different models, which can be seen as different views on the real counterpart. For example, the digital twin of a machine can consist of different models for geometry, electrical interconnection or energy consumption. The digital twin describes the collection of these models, which can be used in the respective context as required.

Due to the missing link with reality, however, change in the physical world is not automatically transferred into the digital world. This leads to the fact that after some time the digital twin is incongruent with its real part.

The digital shadow, however, has the link between the digital image and the real world. It can be used to collect data, evaluate it and provide information about past, present and future states. Furthermore, it can be used to create models from the collected data, which can be combined to a digital twin. The data generated by the interconnections within CPPS can be used to create a digital shadow.

4 Conception of a Digital Twin of a Locomotive

As described above, a CPPS represents a combination of several cyber-physical systems in the context of production systems in order to carry out higher-level optimizations. Similarly, participants in a transport system or their cyber-physical representations can be combined to form a cyber-physical transport system (CPTS) to apply prescriptive maintenance. The basic prerequisite for the use of a CPTS is the presence of CPS of the individual participants. The following chapters present the development of a draft for the CPS of a locomotive. In a first step, a corresponding concept for the collection of data and the proactive operational adaptation is presented. The following sections provide a proposal for the structure of the digital shadow according to EN 15380-2 [14] and show the extension to a digital twin for enabling prescriptive maintenance.

4.1 Data Acquisition and Operation Adjustment

The majority of locomotives used for freight transport has an age of up to thirty years. Since these locomotives are often only rudimentarily equipped with sensors, the solution for data acquisition should be retrofittable.

Therefore, the system should be able to collect data from already existing sensors as well as retrofitted solutions. This principle corresponds to classical condition monitoring solutions. As special requirements, however, the system must also be able to implement adjustment actions. This means that, in addition to unidirectional communication with sensors, bidirectional communication with active components, which act as control element for the implementation of prescriptive maintenance, must also be operated.

Figure 6 shows a suitable architecture for a system for bidirectional communication on the vehicle which has been developed in the research project a^3-Lok.[1]

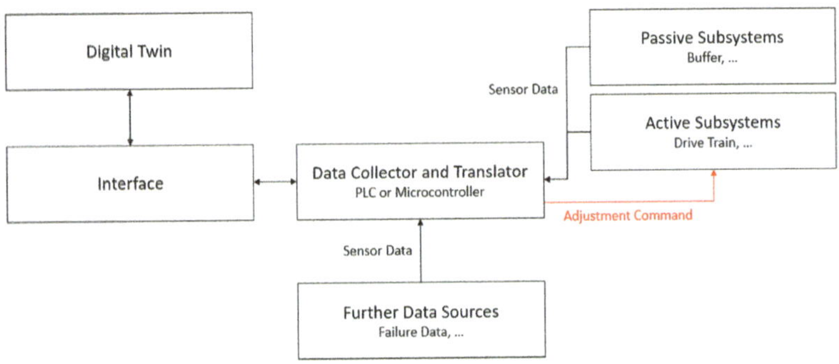

Fig. 6. Architecture for data acquisition and operation adjustment

4.2 Digital Shadow

The digital shadow fulfils the function of storing data from the operation and making it available to other participants in the CTPS. Without wanting to shed any more light on the legal aspects, the prompt obligation to adhere to the EU regulatory "Entity in Charge of Maintenance" [15] results in further requirements, in particular with regard to which data must be stored. Table 2 shows the requirements for data storage using digital shadows.

The vehicle's configuration is at the centre of the analysis. It is therefore obvious to structure and represent the vehicle at the component level. In order to create data consistency, it is important that all vehicles in the overall system under consideration have a uniform description in the form of a standard. It makes sense to use the contents

[1] a^3-Lok is a research project from Ruhr-Universität Bochum as well as the Westfälische Lokomotiv Fabrik GmbH & Co. KG (Hattingen, Germany) and the Ikado GmbH (Aachen, Germany). It is funded by the German Federal Ministry of Energy and Economic Affairs within the funding line "Zentrale Innovation Mittelstand", funding reference: ZF4060716SS7.

Table 2. Requirements for the digital shadow of a locomotive

No.	Requirement
1	Knowledge of the configuration
2	Storage and provision of operation data
3	Storage and provision of fault and maintenance data
4	Storage and provision of the operating status
5	Knowledge of vehicle and component condition

of EN 15380-2 to structure the vehicle. Due to the status of a standard, vehicle manufacturers and operators are already familiar with it. The structure is generally applicable, so it can also be used for descriptions, e.g. of wagons, track-laying machines and other vehicles in the system (see Fig. 7).

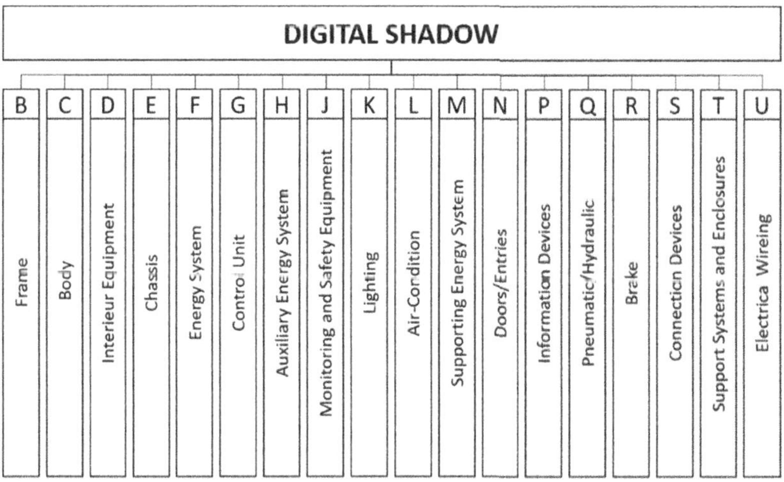

Fig. 7. Draft of the structure of a digital shadow/twin for a railway vehicle

The structure of the vehicle at the top level initially contains substructure systems B - U, to which the components assigned to them are then attached. This is sketched as an example in Fig. 8. The individual fields result from the above requirements.

4.3 Digital Twin

The digital shadow defines the data structure and thus forms an essential basis for the implementation of prescriptive maintenance. However, the CPS only becomes a useful tool for prescriptive maintenance if a function layer is added to it. This extension turns the digital shadow into a digital twin.

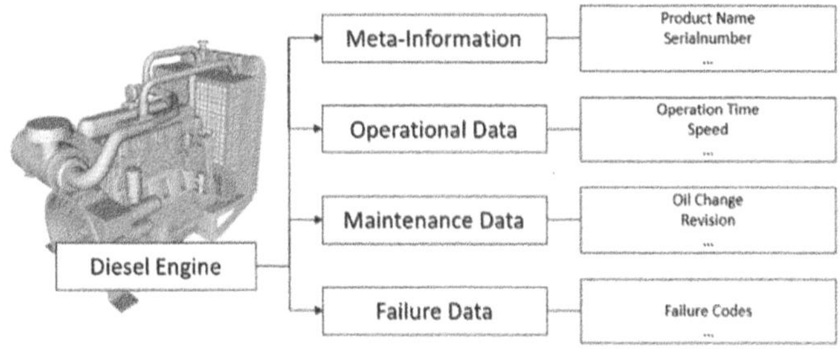

Fig. 8. Exemplarily description of a diesel engine on component level

This extension is carried out using the example of the optimization problem in Fig. 4. As part of prescriptive maintenance as defined in Chapter 2, an initial maintenance plan is defined on the basis of assumptions about operation. If a deviation from these assumptions is noticed, the digital twin is to be used to find out an optimal adjustment of the operation so that the vehicle remains efficient in its horizontal network, but nevertheless achieves the originally set goals in the best possible way. Thus, the digital twin is now used for experiments that cannot be carried out on the real vehicle. For example, the parameters of active participants in the architecture in Fig. 6 (e.g. the drive train) can be changed virtually to evaluate the result. The optimization then selects the solution with the best results.

To carry out these simulations, however, models are necessary which describe, for example, the change in the remaining useful lifetime or the performance of the vehicle as a function of the adjustment action.

5 Conclusion and Outlook

This paper has sketched a system of systems (CPTS) using the horizontal integration of a rail vehicle. Moreover, it has unfolded the interpretation of a locomotive as CPS. Based on this, a first draft for the digital twin of a locomotive has been developed for the application of prescriptive maintenance strategies on the basis of EN 15380-2. Numerous research efforts in this area prove the relevance of increasing attractiveness through innovative maintenance and operating strategies. While this paper considers prescriptive maintenance on a conceptual level, the research project a^3-Lok of Ruhr-Universität Bochum investigates the concrete technical implementation with the focus on the vehicle [4]. A further example is the research project Freight Wagon 4.0 of the FH Aachen [16, 17] with the aim to enable the freight wagon to operate autonomously as part of the CPTS and to communicate with other participants such as CPPS. Even if the research efforts relate to different vehicle categories, in both contexts the digital twin of the vehicle can be interpreted as the basis for the optimisation of operation and maintenance.

Before the digital twin, as described in this paper, can fully unfold its full advantages outside a prototype status, the sustainable development of a holistic CPTS in real operation is necessary. To this end, further participants must be identified and, as described in this paper for the locomotive, equipped with a suitable standard for a uniform description.

The way in which communication takes place within the network, how the decisions of autonomous network participants are logged and whether the network uses distributed, decentralized or central intelligence have not yet been examined. What is of further interest is the question of the extent to which autonomous driving, i.e. the reduction of human influence on operation, favours the application of prescriptive maintenance.

Finally, it should be noted that, on the one hand, the continuous availability of the data of the CTPS participants is a basic prerequisite for the complete application of prescriptive maintenance and, on the other hand, many platform-based business models (e.g. mobility as a service, automated maintenance relaying, etc.) can emerge outside this field, which are able to bring about enormous increases in the efficiency of the overall system.

References

1. Statista. Modal Split im Güterverkehr der EU im Jahr 2016 nach Verkehrsträgern (Anteil an der Transportleistung)
2. Atamuradov, V., Medjaher, K., Dersin, P., Lamoureux, B., et al.: Prognostics and health management for maintenance practitioners - review, implementation and tools evaluation. Int. J. Prognostics Health Manage. **8**, 1–31 (2017)
3. Davenport, T.H., Harris, J.G.: Competing on Analytics: The New Science of Winning. Harvard Business Review Press, Boston (2017)
4. Franzen, J., Kuhlenkötter, B.: Präskriptive Analyse im Kontext der Instandhaltung von Schienenfahrzeugen. In: Eisenbahntechnische Rundschau
5. Milchman, A., Fang, N.: Prescriptive analytics: a short introduction to counterintuitive intelligence (2017)
6. Franzen, J., Pinders, U., Lingen, M., Kuhlenkötter, B.: Entwicklung eines Systems zur Senkung der Lebenszykluskosten durch Einsatz proaktiver Wartungs-strategien. In: Tagungsband der 16. Internationalen Schienenfahrzeugtagung
7. Barthelmäs, N., Flad, D., Haußmann, T., Kupke, T., et al.: Industrie 4.0 – eine industrielle Revolution? In: Industrie 4.0, p. 33. Springer Fachmedien Wiesbaden, Wiesbaden (2017)
8. Bettenhausen, K.D., Kowalewski, S.: Cyber-Physical Systems: Chancen und Nutzen aus Sicher der Automation (2013)
9. Monostori, L.: Cyber-physical production systems: roots. Expect. R&D Challenges **17**, 9 (2014)
10. Shafto, M., Conroy, M., Doyle, R., Glaessgen, E., Kemp, C., LeMoigne, J., Wang, L.: DRAFT Modeling, Simulation, Information Technology & Processing Roadmap: Technology Area 11 (2010)
11. Bauernhansl, T., Krüger, J., Reinhart, G., Schuh, G.: WGP-Standpunkt Industrie 4.0, Darmstadt (2016)
12. Negri, E., Fumagalli, L., Macchi, M.: A review of the roles of digital twin in CPS-based production systems. Proc. Manuf. **11**, 939 (2017)
13. Kritzinger, W., Karner, M., Traar, G., Henjes, J., et al.: Digital Twin in manufacturing: a categorical literature review and classification. IFAC PapersOnLine **51**, 1016 (2018)

14. Deutsches Institut für Normung. Railway applications - Designation system for railway vehicles - Part 2: Product groups; German version EN 15380-2, 2006(15380-2)
15. Europäische Union: Verordnung (EU) Nr. 445/2011 der Kommission vom 10. Mai 2011 über ein System zur Zertifizierung von für die Instandhaltung von Güterwagen zuständigen Stellen: 445/2011 (2011)
16. Pfaff, R., Shahidi, P., Enning, M.: Connected freight rail rolling stock: a modular approach integrating sensors, actors and cyber physical systems for operational advantages and condition based maintenance. In: Proceedings of the Asia Pacific Conference of the Prognostics and Health Management Society (2017)
17. Schmidt, B., Enning, M., Pfaff, R.: Güterwagen 4.0 – Der Güterwagen für das Internet der Dinge. In: Eisenbahntechnische Rundschau

Reliable Power Supply for Smart Telematics in Freight Vehicles

Anadi Shankar Jha[1(✉)], Irmgard Buder[2], Amir Duri[2], and Dirk Bruckmann[2]

[1] Delhi Technological University, New Delhi, India
anadi217@gmail.com
[2] Hochschule Rhine-Waal, Kamp Lintfort, Germany

Abstract. In modern rail freight there is a growing need for automation. Smart telematics form an important part of automation. With the increasing trend to include sensors and data communication technologies in freight wagons, there is a need to have a reliable power supply for components of automation. The useful lifespan of the power supply solution needed here is nearly 8 years while keeping in mind the boundary conditions for maintenance time in freight wagons. This paper deals with telematics use cases and applications and describes feasible solutions for an energy supply.

1 Introduction

According to Jain et al. [1] currently the European single wagon load (SWL) in rail freight is at the risk of disappearance as is does not fulfil the shippers changing requirements. Galonske et al. [2] found, as a result of the ViWaS project, the main reasons for the decreasing demand in SWL. They mention the insufficient service quality of SWL, additional costs for wagon handling in the sidings and poor utilization rates of resources, like trains and wagons. Thus, new approaches to improve the current weaknesses of SWL are required. Galonske et al. [2] tested and proofed advanced technologies in the field of SWL to streamline the last mile operation, to improve the efficiency of asset usage, to increase transport quality and to address fully new markets. According to the results of the ViWaS project has approved that the modularity of wagons as well as the comprehensive use of telematics may increase the competitiveness of SWL in this fields.

New telematics components for SWL freight wagons were developed to measure movements of wagons, positions, status, intensities of shocks, humidity etc. In the ViWaS approach the energy supply for all sensors was transferred by using batteries. The restricted storage capacity of batteries resulted in either a regular need to replace/recharge the batteries or the energy demand restricts the data transfer rates. This paper assesses the installation of an independent and rechargeable energy supply for permanent supervision of the freight wagons as a solution to the above-mentioned problem.

Sensors are required for the process of monitoring the goods from a remote location. These sensors help measure parameters like humidity, temperature, weight on

U. Clausen et al. (Eds.): ICPLT 2019, LNLO, pp. 277–287, 2019.
https://doi.org/10.1007/978-3-030-13535-5_20

wagon etc. and their data can be processed further at a control center. For the data from the sensors to be processed, it must be transferred to a control center.

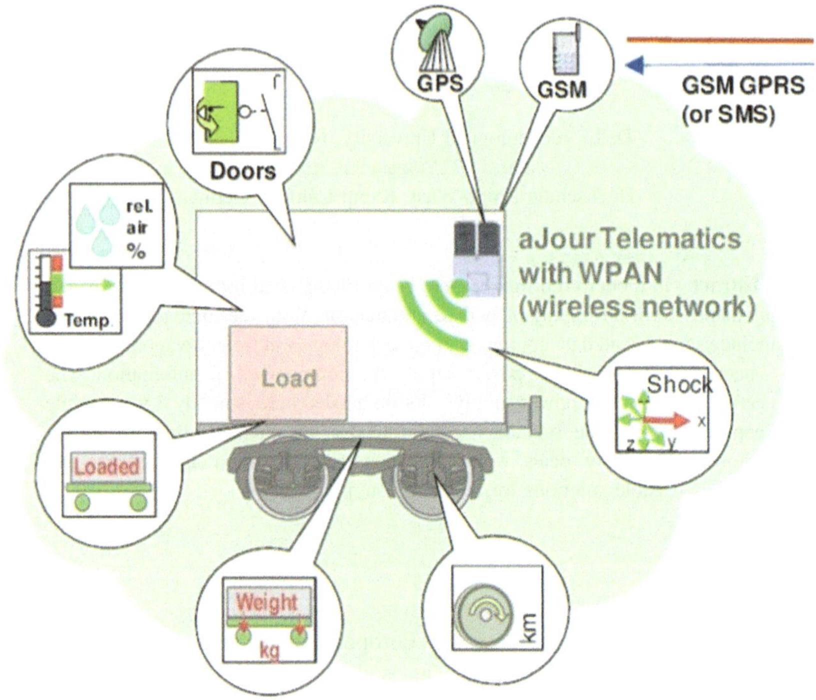

Fig. 1. Sensors used in freight vehicles [3]

2 Requirements for Power Supply

A reliable power supply is required to ensure the functioning of electronic devices. As freight wagons usually require maintenance once every 6 to 8 years, the regular maintenance processes cannot be used to recharge or replace batteries in shorter intervals. Between the maintenance works the wagons float free through the entire European rail freight network. Especially in foreign countries it is rather difficult to locate and dispatch the freight wagons [4]. A shortening of the maintenance cycles will cause additional empty runs for the transfer to the workshop of the wagon keeper. Furthermore, the effort to dispatch the wagons increases. Thus, a shorter service interval will increase the operational costs of the freight wagons for the wagon keeper. The wagon keepers want to avoid any additional maintenance work for the power supply systems.

Furthermore, it is difficult to draw power from the grid for such applications. The current state of the art of power supply in freight trains is either from overhead power lines or from generators in diesel locomotives. As there is yet no power supply between the freight wagons in a train, it is also impossible to use this approach for power supply. Thus, the boundary conditions in this case are very unfavorable.

Nevertheless, it is the need of the hour to develop a reliable power supply for sensors. This paper outlines the use of an axle generator for charging a battery backup taking the boundary conditions of running time, standing time, loading and unloading time as inputs of design.

3 Sensors

A number of sensors are used in freight vehicles. Sensors are most commonly used to measure pressure, temperature, shock, differential speed, distance and accelerometers. Figure 1 shows the types and placement of sensors. It also shows an aJour Compact Telematic unit [15] which takes in sensor inputs and transfers data via WPAN wireless channel.

The pressure and temperature sensors can be used to check the humidity inside the wagon and to see whether the temperatures are within safe levels. The shock sensor is an event driven sensor which should inform of the shocks and jerks on the transported product. Differential speed sensors are used while braking. The distance sensor is used to present the mileage of the vehicle. Accelerometers are used to check the speed of the freight wagon. All this information is essential to monitor and control the quantity and quality of goods transported.

The data shown in Table 1 is obtained from the sensor datasheets. The data shown here are only maximum values and hence highest power consumption is taken into consideration.

The output power is calculated from the datasheet readings as well as taking the operating voltage as 5 V from the battery. The datasheets mention the supply voltage and current when there is no load on the sensor i.e. the load end is an open circuit. Therefore, any other additional loading conditions result in additional power consumption. The operating voltage range of sensors can be accordingly set by using on-board voltage regulators which can adjust for a 6 volts (V) battery supply.

The operating temperature refers to the temperature of the chip itself. The storage temperature, on the other hand, is the temperature at which the device can be safely stored.

The working temperature should be within the safe limits. Reliability for the provision of energy from a battery is also affected by the operational temperature of the devices which also needs to be taken into consideration.

Table 1. Sensor datasheet [5–10]

Type of sensor	Vmax	I max	P max	Temperature
Pressure (MPXV7002)	4.75 V–5.25 V	10 mA	52.5 mW	10C to 60C (Operating) (−30C) to 100C (Storage)
Temperature (LM35A)	5 V	0.056 mA	0.28 mW	−55 °C to 150 °C
Accelerometer (ADXL335)	1.8 V–3.6 V	0.350 mA	1.75 mW	−65 °C to 150 °C
Differential Speed Sensor (TLE4941plusC)	4.5 V–20 V	Low Range 8.4 mA High Range 16.8 mA	42 mW 84 mW	−40C to 125C
SHOCK SENSOR (PKGX-14-4010)	2.5 V–5 V	1.5 mA	7.75 mW	−40C to 85C (operation)
DISTANCE SENSOR (GP2Y0A21YK0F)	4.5 V–5.5 V	30 mA	150 mW	(−10C) to 60C (operation) (−40C) to 70C (Storage)

4 Wireless Communication Networks

There are two kinds of wireless technologies being studied here namely GSM and Zigbee for the purpose of short range and long-range communication. Their power consumptions are simulated for different parameters of data packets, delay time etc. according to the freight wagon requirements.

GSM is a popular mode of wireless communication used in mobile phones for long range communication. It operates at either a 900 MHz or 1800 MHz frequency band.

GSM data consumption is studied on GL865 module which is a quad band GSM|GPRS module based on the latest release from Intel, with a 2G cellular chipset. The method of taking the measurements is by inserting a 0.1-Ω resistance in the GSM modules. The sensor communicates around 1–3 times a day. The data packet size discussed here has a size of 112 bytes. The GL865 works at a voltage level of 3.3 V for this experimentation.

The obtained graph in Fig. 2 [3] shows that during the process of registration of a network and updating location high pulses in current, the consumption is observed. From the graph, the current consumption of the module is nearly 20 mA for the ideal state. From a total of 26 location updates a consumption of 2.71 mAh is required, while the network registration amounts to 0.232 mAh and the module consumption is 456 mAh for a day.

Zigbee was used to analyze short range communication [10]. It is the IEEE standard 802.15.4 for communication protocol for short-range, low-energy Wireless Personal Area Networks (WPANs).

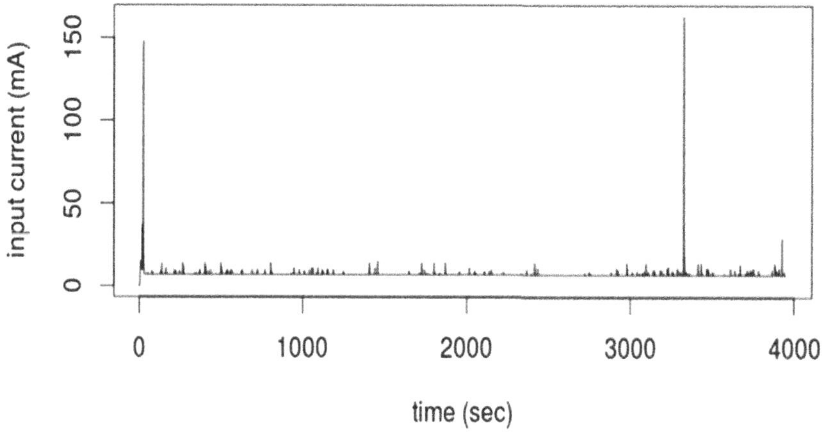

Fig. 2. Current consumption of GL865 module [3]

The Zigbee protocol is implemented on modules based on low power Atmel Atmega1281 8bit microcontroller and Atmel AT86RF230 transceiver. The radio part is designed for the radio range 2.4 GHz ISM. The method of measuring the power consumption values is obtained by connecting a 1-Ω resistance. The test resistance is used to measure the drain current.

As seen from Fig. 3, the ideal current consumption from the graph is nearly 2 mA. It is seen that the major power consumption is taken up by the processor and the RF circuits. Here the module is powered at 3 V. Figure 2 shows the development of input current while sending and receiving regular data packets of bytes.

5 Sources of Energy

There can be various sources of energy for meeting the above energy requirements. Some of them are solar power, regenerative braking, axle generators etc.

Solar standalone system can be used to supply power for certain times. However, taking the case of Germany, the sunlight hours are quite few in winters for eliciting appropriate power out from the solar panels. Adding to this are the problems of the freight wagon being parked in shade, and top less structures of the wagons. Hence Solar PV systems might be used as an auxiliary system to meet the power consumption demands of smart telematics operations to complement the power from the main source but using it as the primary supply of power would make the supply unreliable in the long run.

Regenerative breaking is another alternative. However, for this we would need to handle high amounts of power, occurring only rarely in the use of the freight vehicle. Thus, the axle generator can be used an economical and reliable power source for charging the battery. It works even in a state where neither wind nor solar energy is available so it has an advantage over other power generation sources.

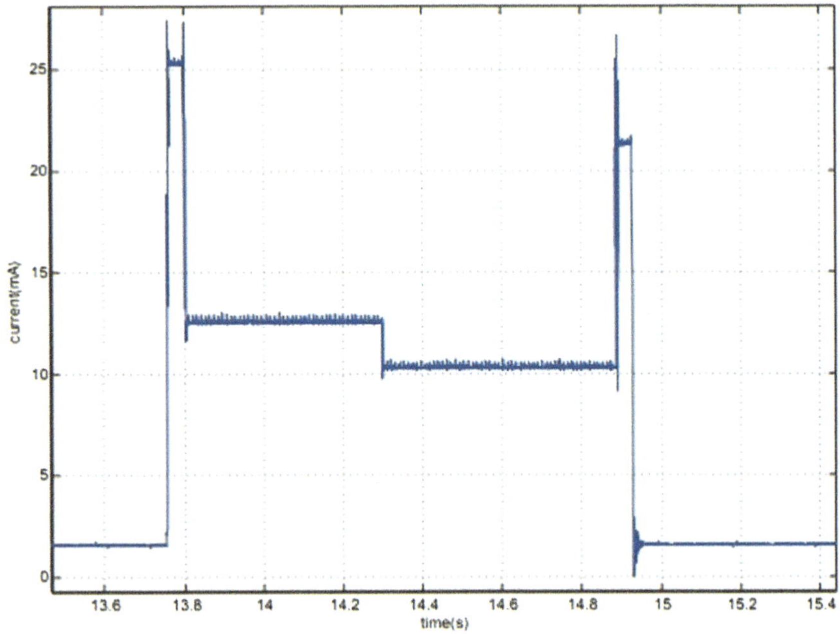

Fig. 3. Power consumption of zigbee modules [11]

6 Circuit Layout

The above circuit in Fig. 4, shows a circuit layout of a Permanent Magnet based Synchronous Generator to be used as an axle generator in the freight vehicle applications. The rotor shaft takes the speed of the axle as input to generate variable Sinusoidal Alternating (AC) voltage. This AC voltage is converted to nearly stable DC (Direct) voltage via a rectifier. The battery backup or DC link is placed in the middle. The DC voltage of the battery is converted to an appreciable AC voltage to power the electrical appliances in the freight vehicle. The "AC Link" is provided to have a flexible structure for joining other equipment in the future if needed.

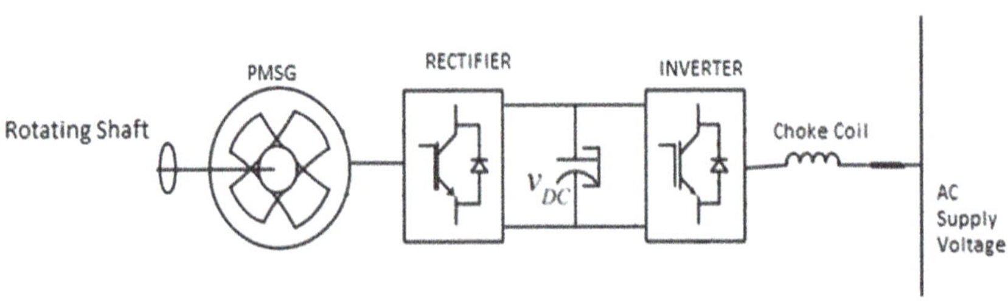

Fig. 4. Circuit layout of generator application

7 Axle Generator Design

Most leading manufacturers use synchronous generators in their products [12]. The design and modelling of the axle generator is dependent on the magnetic material, speed of rotor, number of poles of stator, induced voltage etc. A common problem in designing generators for such applications is the low speed angle of the vehicle. One solution for increasing the voltage output for low speeds can be an increase in number of poles in the design of the generator.

There is a choice between separately excited and permanent magnet type generators. Normally to avoid providing external power supply for excitation of windings it is better to use permanent magnet type.

An experimental setup of a generator having 86 mm diameter and 8 mm thickness is discussed here [13]. The generator is of the salient pole permanent magnet synchronous generator type. The generator poles are in the form of 6-pointed star with 6 windings. The design suggests to use electrical steel sheets 3414. The rotor is made up of Neodymium-iron-boron. There are 18 magnets of this type and all with the dimensions 7 mm × 7 mm × 7 mm. We now obtain a generator with 24 poles and 8 windings. The Fig. 5 shows the design for the generator.

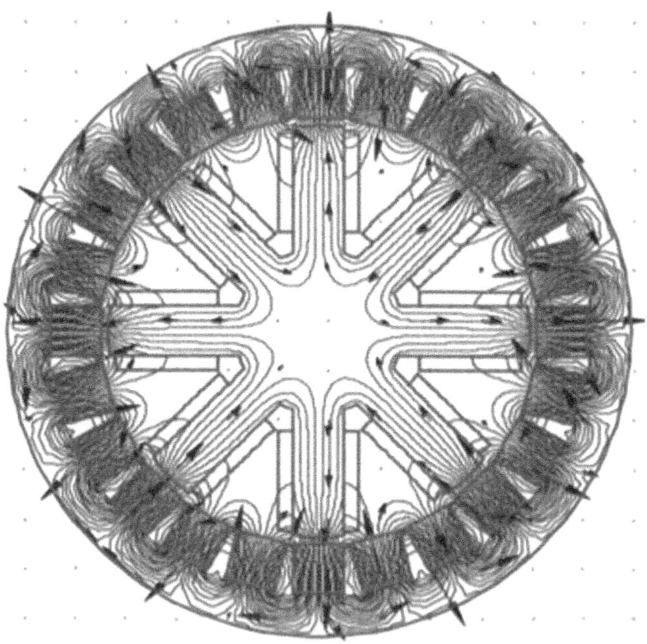

Fig. 5. Map of induced magnetic field [13]

The emf generated is found by the formula: -

$$E = nTPS\omega/\sqrt{2}$$

where

T = 200 is number of turns
P = 12 is the number of pole pairs
S = 0.000049 m^2 is the sectional area of magnetic pole
n = 8 is the number of coils
ω = 21 rad/s is the average shaft speed in a trip (this gives the average speed of the freight vehicle as 40 kmph).

The calculations show that at a voltage of 9.1 V the generated power output is nearly 1.204 W. This has been sized according to the low wagon speed and the power consumption requirements of the telematics.

8 Battery Backup

Since the axle generator only provides power, when the freight wagon is moving with a certain minimal speed, battery storage for powering the sensors and communication devices becomes necessary. So, in times when the system has an excess of power, it can be stored for later use. In principle the axle generator cannot run all the time to power the circuits and that is where the picture of a battery comes into play. The efficiency of the system as a whole depends on the battery backup as well. The study of standing time and running time of a freight vehicle is done and is used to size the battery. This is shown in Fig. 6 [2].

The charging time of the battery is 13% of the round-trip time and in this time the battery has to cover the energy consumption for sensor operations in the standing as well as running times.

Doing a study for a week, the time obtained for charging at 13% is 21.84 h while the generator power output is 1.204 W (at average speed of 40 KmpH) which means 26.23 Wh of energy is produced.

The total expected power consumption of all the sensors working at nominal voltage of 5 V comes out as 254.28 mW as discussed in Sect. 3. The communication channels which are operating at 3.3 V use up to 65 mW as shown in Sect. 4. The total power consumption results in 319.28 mW which is obtained after adding power consumption from both the channels. The total time in a week is 7 * 24 = 168 h. So, the watt hour estimation comes out as 319.26 mW * 168 h = 53.639 Whr. By doing so, we obtain the total speculated power consumption.

Comparing the figures obtained for generation and consumption of energy for the trip we have 26.23 Wh of energy production from a single axle generator against the total energy demand of 53.639 Wh. By this data we find the number of generators as 53.639 Wh/26.23 Wh = 2.0444 which can be rounded off to three taking into consideration the margin for losses. So, we have to use three such synchronous generators for meeting the energy demands of the vehicle.

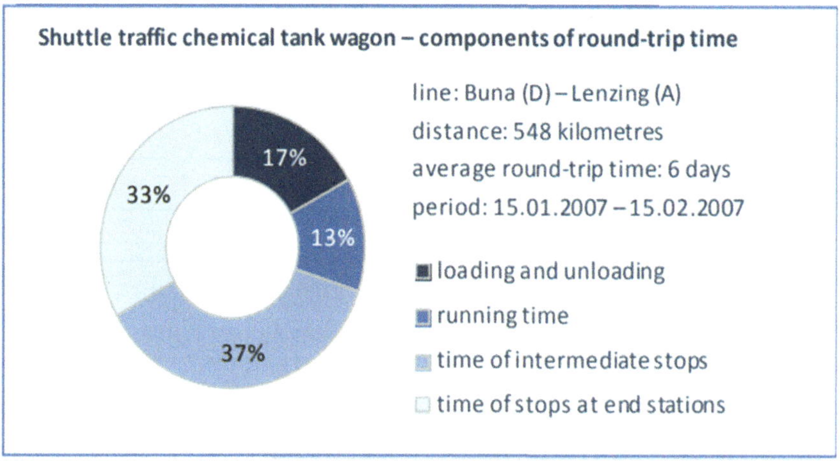

Fig. 6. Round trip of a freight wagon [2]

There are a number of working voltage levels. In this paper we choose a LiFeP battery which is described in [13]. This battery has nominal voltage of 6.0 V and is specified with a capacity of 5 Ah +− 5%. The charging voltage is at the start of charging 6.0 V and rises to 7.5 V [13]. 80% state of charge is achieved by charging below 7.0 V, <90% at voltages below 7.5 V at a constant current, as can be seen in Fig. 7. When 7.5 V is reached, further charging takes place at constant voltage. However, the synchronous generator generates a voltage of 9.12 V, so for charging the battery the voltage has to be adapted. For working with different operating voltages, we use in-built IC type voltage regulators for example IC 7806.

Each generator can charge with a mean charging current at 0.132 A, so per generator 2.89 Ah can be derived in 21.84 h. For three generators that would be 8.65 Ah, so two batteries could be charged by 4.32 Ahr. The power stored in the battery would be then 25.94 Whr (assuming the nominal voltage) which would not balance the power consumption of 53.68 Whr for the whole week. Assuming 6,4 V operation voltage, as the producer announces on their website [12] would lead to 55.30 Whr which would be sufficient under normal operational condition. Under adverse conditions (e.g. extended waiting periods in the sidings) an increase of the battery capacity could be helpful.

If reducing the voltage of the generators would lead to higher charging currents it could also work since three generators produce 78.89 Whr, charging two batteries would need around 75 Whr (assuming 80% charging efficiency).

The alternative would be the installation of four generators, producing 104.92 Whr. This power would be sufficient to charge the batteries and at the same time provide power to the consumers. Also, to be considered: Three batteries would have the advantage, that capacity loss of one or more batteries could be balanced. However, according to the specification provided in [14] the batteries should have more than 90% of their capacity during the first 400 cycles complete charging and discharging. Assuming 7 year with weekly charging and discharging the cycle, number would be 364 cycles.

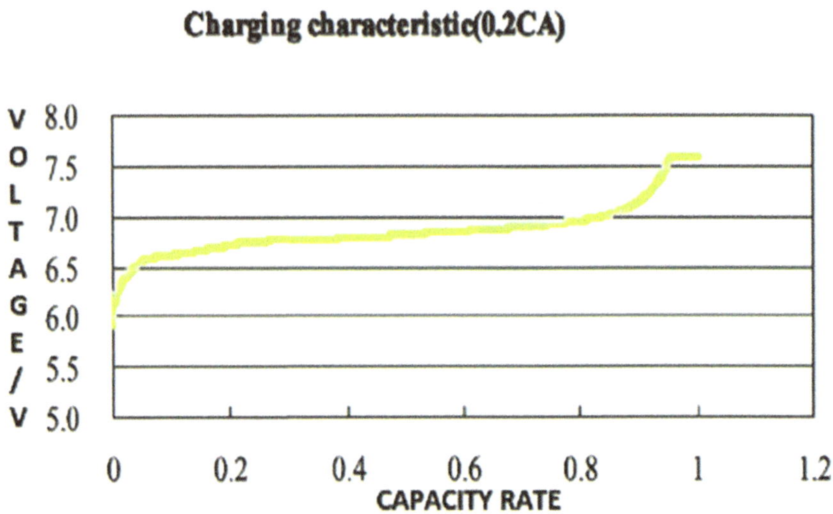

Fig. 7. Charging characteristics of (0.2CA) [14]

9 Reliability Analysis

There are a number of ways for increasing the reliability of the power supply system. Failures or downtimes due to mechanical wear and tear of one of the generators can be reduced by adopting the redundancy approach where we can provide one spare axle generator in parallel. This spare axle will be mechanically coupled but will get electrical connected to the electrical supply system as soon as the power from the main generator goes zero. But for going to this approach cost has to be within economical limits.

We can also check the signal strength from the sensors at regular intervals to monitor the status of the power supply using different software.

10 Conclusion

The paper suggests that the axle generator designed here coupled with the battery backup can serve as a possible reliable power supply for sensors as well as data communication protocols. The results of the voltage generated by axle generator at different running speeds can be successfully used to meet the electrical needs. Thus, a reliable energy supply for freight wagons which meets the service cycle requirements could be developed.

Acknowledgments. I would like to thank Prof. Manfred Enning, Chair of Railway Systems Engineering Aachen, Germany for his valuable technical inputs which made this work possible.

References

1. Jain, A., Bruckmann, D.: Application of the principles of energy exchanges to the rail freight sector. Transp. Res. Record **2609**(1), 28–35 (2017). https://doi.org/10.3141/2609-04
2. Galonske, N., Riebe, E., Toubol, A., et al.: The ViWaS project: future-proof solutions for wagonload transport. Trans. Res. Procedia **14**, 2850–2859 (2016). https://doi.org/10.1016/j.trpro.2016.05.402
3. Behrends, V., Haunschild, M., Galonske, N.: Smart telematics enabling efficient rail transport – development of the ViWaS research and development project. Transp. Res. Procedia **14**, 4430–4439 (2016). https://doi.org/10.1016/j.trpro.2016.05.365
4. Ballis, A., Dimitrou, L.: Issues on railway wagon asset management using advanced information systems. Transp. Res. Part C: Emerg. Technol. **18**(5), 807–820 (2010). https://doi.org/10.1016/j.trc.2009.09.003
5. Nagode, C., Ahmadian, M., Taheri, S.: Axle generator for freight car electric systems. In: 2012 Joint Rail Conference, pp. 761–765 (2012)
6. Lysenko, O.A.: The research of axel-box 24-pole generators for communications systems of freight railway wagons. In: Dynamics of Systems, Mechanisms and Machines (Dynamics), pp. 1–4 (2016)
7. Package, Thermoplastic PPS Surface Mount, Gauge, Patented Silicon Shear Stress Strain (2005) MPXV7002 Integrated Silicon Pressure Sensor On-Chip Signal Conditioned, Temperature Compensated and Calibrated. https://www.nxp.com/docs/en/data-sheet/MPXV7002.pdf. Accessed 30 Oct 2018
8. Texas Instruments (2018) LM35 Precision Centigrade Temperature Sensors. http://www.ti.com/lit/ds/symlink/lm35.pdf. Accessed 30 Oct 2018
9. Analog Devices ADXL335 Accelerometer: Datasheet. https://www.sparkfun.com/datasheets/Components/SMD/adxl335.pdf. Accessed 30 Oct 2018
10. Infineon Technologies AG TLE4941 plusC Differential Speed Sensor: Datasheet. https://www.infineon.com/dgdl/Infineon-TLE4941-DS-v01_01-en.pdf?fileId=db3a30433899edae0138c3cb0f870330. Accessed 30 Oct 2018
11. Sharp GP2Y0A21YK0F Distance Sensor: Datasheet. http://www.sharp-world.com/products/device/lineup/data/pdf/datasheet/gp2y0a21yk_e.pdf. Accessed 30 Oct 2018
12. Vossloh Kiepe Gmbh Axle Generator AG 40. https://www.batteryspace.com/lifepo4-battery-6v-5ah-30wh-10a-rate—replace-sla-with-5-times-longer-life-and-lighter-weight—un38-3-passed.aspx. Accessed 30 Oct 2018
13. Paller, G., Szármes, P., Élo, G.: Power consumption considerations of gsm-connected sensors in the agrodat. hu sensor network. Sens. Transducers **189**(6), 52 (2015)
14. Powerizer Batteries LiFePO4 battery: datasheet. http://www.batteryspace.com/prod-specs/6846.pdf. Accessed 30 Oct 2018
15. Asto Telematics. https://asto-telematics.com/en/. Accessed 30 Oct 2018

Liquid Petroleum Gas Supply Chain Challenges in Rural Medical Facilities in Zimbabwe

Gladys Makhanda, Oliver Pwaka, and Chengedzai Mafini(✉)

Vaal University of Technology, Vanderbijlpark, South Africa
chengedzaim@vut.ac.za

Abstract. Rural medical facilities in Zimbabwe have consistently faced energy-related challenges particularly in powering machines that are critical in their operations. The situation has been further worsened by the inconsistent supply of electricity from the national grid. This study aimed to investigate the supply chain challenges associated with the use of liquid petroleum gas (LPG) in rural medical facilities in Zimbabwe. The research followed a quantitative approach in which 97 respondents were surveyed. The respondents consisted of medical and maintenance staff drawn from selected provincial medical facilities in Zimbabwe, private LPG suppliers as well as non-governmental organisations involved in the procurement and supply of LPG. The collected data were analysed with the aid of the Statistical Package for Social Scientists (SPSS version 24.0).

The results of the study show that the performance of rural medical facilities improved with the use of LPG. The study further shows that various significant LPG supply-related challenges exist in rural medical facilities, which limit the performance of such institutions. Managerial implications are provided on how managers and employees in rural medical facilities can harness the supply and use of LPG as a viable alternative energy source in remote locations.

Keywords: Liquid-Petroleum Gas · Rural medical facilities · Quantitative modelling · Performance analysis · Sustainable logistics · Supply chain management

List of Abbreviations

ESMAP	Energy Sector Management Assistance Program
LPG	Liquid Petroleum Gas
MEER	Ministry of Electricity and Renewable Energy
NGO	Non-Governmental Organisation

1 Introduction and Background of the Study

There has been an increase in the demand for energy worldwide. This increase has often been met through the provision of electricity, particularly in developing nations. However, the demand for electricity and other forms of energy has been challenging to

U. Clausen et al. (Eds.): ICPLT 2019, LNLO, pp. 288–301, 2019.
https://doi.org/10.1007/978-3-030-13535-5_21

meet, especially given the rapid growth in the populations of these countries as well as the high levels of poverty which render the provision of power unaffordable. As noted by Hutton, Acharibasam and Apatinga (2014) as much as three billion people on a global scale are still using traditional fuels such as coal, dung, charcoal and wood for cooking. Most the people without access to reliable energy sources are concentrated in Africa, as shown in Fig. 1.

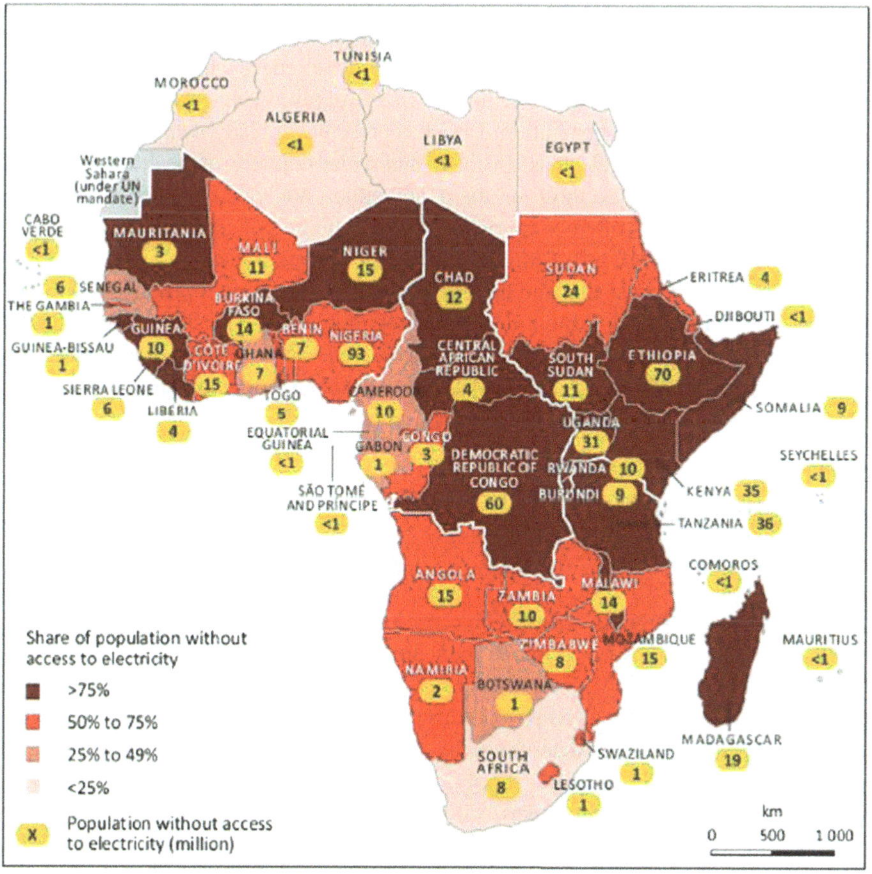

Fig. 1. Access to electricity in Africa Source: Avila, Carvallo and Shaw (2017)

Figure 1 presents the levels of access to electricity in Africa, which indicates that most people in that continent remain in areas where electricity is either inaccessible or in short supply. Because of poverty and rapid growth in populations, most African households, particularly those in the rural areas, continue to depend on traditional fuels which are unreliable and destructive to the environment. Osoro (2015) states that various petroleum products can be used to meet the energy demands of African communities. Within homes and industries, these sources can be used for various purposes such as cooking, heating and lighting (Osoro 2015). Liquid Petroleum Gas (LPG), which is a combination of various gases such as propylene, butane and propane, has emerged as an essential energy source that is widely viewed as a sustainable alternative to electricity (Miniaci, Scarpa and Valbonesi 2014).

This study aimed to investigate the supply chain challenges associated with the use of LPG in rural medical facilities in Zimbabwe. Even though LPG has been widely adopted as a sustainable and effective energy solution particularly in the remote areas of Zimbabwe where conventional energy sources such as electricity are scarce, there have been challenges negating the effectiveness of this novel energy source in the country. With electricity becoming expensive and in short supply owing to increased national demand against depressed generation, LPG has presented a ready solution, particularly for the less developed rural areas. In these areas, the delivery of health services has remained compromised as most of them are either not electrified or have erratic electricity supplies. This often compromises the functioning of equipment used to store vaccines in health institutions. Additionally, it has been suggested that LPG monitoring is not systematic nor consistent and that its use for sterilisation in reproductive health is not adequately monitored. Above this, end-user monitoring of LPG usage is neither consistent nor systematic. These challenges necessitate an impetus to study the supply chain issues and challenges leading to LPG shortages or unavailability within rural medical facilities in Zimbabwe, where it is a necessity for vaccine storage.

2 Literature Review

2.1 The Extent of Liquid Petroleum Gas Use in Rural Medical Facilities

Academic literature suggests that LPG has been adopted to a great extent in developing countries such as Zimbabwe, although there are unfilled gaps on the extent of its use in public health institutions. Kojima (2011) suggests in 30% of the countries worldwide, or more than one-half of all households use LPG as their primary energy source for cooking. The use of LPG is determined by income and a host of other factors such as accessibility, consistency of LPG supply, other fuel prices, the cost of purchasing or acquiring LPG cylinders and stoves, safety issues, unfamiliarity with making use of LPG as cooking fuel, poor education about LPG and cultural preferences (Ibid). The leading factors affecting the demand for LPG include marketing, distribution, financial constraints as well as consumer preferences. Factors enhancing the demand for LPG include increases in the prices of alternative energy sources such as charcoal, and those limiting its supply include the lack of infrastructure, price caps and supply constraints (Falzon et al. 2013).

Gujba, Mulugetta and Azapagic (2015) studied the life cycle, environmental impacts and costs of the household cooking sector in Nigeria and established that LPG is used to a vast extent in the form of cooking utensils which are alternatives to the highly expensive electrical and wood stoves. Abubakar (2014), in his examination of the influences of sustainability implementation on the competitive performance of oil and gas companies, relates the increased use of LPG to sustainable supply chains in the industry. The same author adds that LPG has increasingly become a dominant energy source owing to such business drivers as the aspiration to safeguard energy, proliferate market shares and improve competitiveness. However, a close examination of these and other related studies reveals that there is a critical empirical gap on the extent to which LPG has been adopted in the medical sector, mainly rural healthcare facilities.

The use of LPG leads to various benefits such as affordability, efficiency and friendliness to the environment (Asamoah, Amoakohene and Adiwokor 2012; Kojima 2013). However, Kaburia (2016) points to price, availability, distribution, infrastructure and safety issues as some of the challenges facing the LPG industry. Still, some literature (Energy Sector Management Assistance Program [ESMAP] 2007; Lascurain 2016) mention that in most settings, the available regulatory frameworks are vague with regard to the availability and use of LPG, which complicates its adoption as an alternative form of energy. However, extant literature has primarily disregarded the use of LPG in the medical sector, where LPG may be used for refrigeration and other purposes. Still, there is limited evidence of studies that have captured the use of LPG in rural areas and other remote settings. Thus, there is a need to address numerous critical literature gaps regarding LPG supply chain challenges that are specific to the context of rural-based medical facilities.

2.2 Strategies to Enhance the Effectiveness of Using Liquid Petroleum Gas in Vaccine Programs in Rural Medical Facilities

Because buying LPG cylinders can be a barrier to the use of this fuel type, the use of smaller cylinders which are less costly as is the common practice in Indonesia, or to lower the refilling costs of this source of energy, is recommended (Toft et al. 2016). However, some scholars (Ekouevi and Tuntivate 2012; Kojima 2013) argue against the use of small gas cylinders on the basis that such practices lead to increases in the unit prices of the gas, as has been the case in Senegal. According to the Ministry of Electricity and Renewable Energy (MEER 2015), it may also be useful to assist targeted LPG users with start-up implements such as free stoves, cylinders and installation, to lower the purchase price of such implements, or to offer the option of payment in instalments. The same source indicates that such a strategy has been implemented in Ecuador where the government introduced induction stoves and the payment is made through the electricity bill in 12 monthly instalments. Such implements can be purchased from any retail store, and the import of induction stoves is not subject to tariffs (MEER 2015). However, other scholars (Kojima 2013; Lascurain 2016) advise that this strategy is generally unhelpful to the poor who do not have the financial means to make payments for the even use of LPG throughout. To improve the effectiveness of such strategies, it is necessary that these grants are coupled with fuel subsidies, as in the case of Haiti and some Guatemalan urban settings.

In addition to the above, Toft et al. (2016) recommend the use of cost reduction measures while Chege (2013) advocates for proactive planning. However, Kojima (2013) argues that such strategies are ineffective as they can lead to increases in the unit prices of the gas. Lascurain (2016) places the responsibility of managing LPG supply chain challenges on governments, donors and international agencies while Kojima (2013) puts it on LPG firms that have to use new technology, among other internal measures. Moreover, Jahre et al. (2012) propose such internal measures as controlling the quantity of LPG ordered, the stock level and the costs of the order. These positions depict that academic literature remains inconclusive on whether LPG supply chain management processes are proactive or reactive. Also, there is no unanimity on the effectiveness of cost reduction measures in managing LPG supply chains. Still, it is unclear if the responsibility of ensuring LPG supply chain effectiveness lies on LPG firms or other external stakeholders such as governments and other non-governmental organisations.

3 Research Methodology

3.1 Assumptions

The primary assumption of the study was that rural medical facilities in Zimbabwe are similar, and hence the results obtained in this study can be applied to all rural medical facilities in the country. The study was also premised on an assumption of the homogeneity between medical and maintenance staff members of rural medical facilities throughout Zimbabwe. Hence the views obtained from the sample used in this study can be generalised to other medical and maintenance staff members in other rural medical facilities elsewhere in the country.

3.2 Design and Sample

The study followed a quantitative approach using the survey design. The research population was composed of 320 medical and maintenance staff drawn from selected provincial medical facilities in Zimbabwe, private LPG suppliers as well as non-governmental organisations involved in the procurement and supply of LPG. A research sample of 97 respondents was drawn using stratified sampling.

3.3 Procedures for Data Collection

Quantitative data were collected using a self-administered, five-section survey questionnaire. Section A elicited the demographic information of respondents. Section B focused on the extent of LPG use in rural medical facilities in Zimbabwe while Section C elicited information on the significant benefits and challenges accrued from the use of LPG in rural settings. Section D focused on the effect of LPG supply-chain challenges on the total vaccine losses in the rural medical facilities and Section E covered the possible strategies that can be used to enhance the effectiveness of using LPG in vaccine programs in rural medical facilities. Questionnaires were administered using the drop and collect method.

4 Research Results

4.1 Demographic Profile of Respondents

Out of the 97 respondents that participated in the study, 30 were females, and 67 were males. In terms of their age groups, the majority of respondents (41%; n = 40) were aged between 36 and 50 years. With respect to academic qualifications, respondents with first degrees were the largest number, being 35% (n = 34), followed by those who possessed academic, being a total of 28% (n = 27) and those with postgraduate degrees at 24% (n = 23). With regards to employment period, 35% (n = 43) of the respondents had been employed for periods ranging between two and five years in their organisations, followed by those who had been with their organisations for periods ranging between six and 10 years (22%; n = 21).

4.2 Scale Reliability and Validity

The reliability of the measurement scales was tested using Cronbach's alpha coefficient. The test indicated an overall alpha value of 0.714, which is above the 0.7 minimum cut off value recommended by Pallant (2013). To ensure face and content validity of the measurement scales, a pretest was conducted using 15 conveniently selected respondents, and the questionnaire was reviewed by a panel of two faculty experts in supply chain management. Feedback from the pretest sample and the faculty experts was used to improve the questionnaire.

4.3 Level of Adoption of Liquid Petroleum Gas

The study first tested for the degree of adoption of LPG in rural medical facilities. The results are indicated in Table 1.

Table 1. Level of adoption of LPG

Level of use	Frequency	Percentage (%)
High	80	83
Medium	15	16
Low	2	1
Total	97	100

As Table 1 shows, 83% (n = 82) of the respondents in the study highlighted that their organisations had adopted the use of LPG to a great extent. Those who suggested that their organisations had adopted LPG use to a medium extent were 16% (n = 15) of the total respondents in the study while 1% (n = 1) indicated a low adoption of LPG their organisation. These results indicate that rural medical facilities, private LPG suppliers and Non-governmental organisations (NGOs) were making use of LPG. These results may be attributed to the view that LPG has increasingly become a dominant energy source due to such business drivers as the desire to conserve energy, increase market share and improve competitiveness, as suggested by Abubakar (2014).

4.4 The Extent to Which the Procurement of Liquid Petroleum Gas Leads to the Achievement of Strategic Objectives

The study also tested for the extent to which the procurement of LPG leads to the achievements of strategic objectives in rural medical facilities. Strategic objectives of procurement such as reductions in threats to drug storage (threats), the enhancement of the medical facility's daily operations (operations), enhancement to the hospital's service delivery (service) and improvements to the overall hospital performance (performance) were considered. The results are presented in Table 2.

Table 2. Extent to which the procurement of liquid petroleum gas facilitates the achievement of strategic objectives

Statistic		Threats	Operations	Service	Performance
N	Valid	97	97	97	97
	Missing	0	0	0	0
Mean		3.3505	2.9175	2.7320	2.8351
Median		4.0000	3.0000	3.0000	3.0000
Std. deviation		1.25865	1.25557	1.17717	1.23052

As presented in Table 2 the mean score for the extent to which the LPG procurement reduces drug storage threats was 3.4 while that for the extent to which it enhances medical facilities' day to day operations was 2.9. The mean score for the extent to which procurement enhances service delivery in the medical facilities was 2.7 while that for the extent to which it enhances hospital performance was 2.8. Accordingly, the results show that sound LPG procurement reasonably enables rural medical facilities to manage drug storage threats and also leads to the enhancement of their daily operations, service delivery and performance.

The results of the present study may be linked to the fact that most strategic objectives in rural medical facilities are dependent on the availability of sources of energy. Objectives such as storage of drugs, daily operations, service delivery and overall hospital performance are to varying degrees driven by the availability of energy. In this case, this energy is LPG which is used for purposes such as cooking, central heating, water heating and refrigeration. Hence, the availability of LPG serves as an enabler for the achievement of strategic objectives in these specific areas within rural medical facilities.

4.5 Major Supply Chain Challenges in the Liquid Petroleum Gas Industry in Zimbabwe

The study further examined the leading supply chain challenges in the LPG industry in Zimbabwe. The results are presented in Table 3.

Table 3. Major liquid petroleum gas supply-chain challenges

Statistic		Supply	Demand	Operational	Security
N	Valid	97	97	97	97
	Missing	0	0	0	0
Mean		4.1203	1.4143	1.5310	1.9542
Median		3.0000	3.0000	3.0000	3.0000
Std. deviation		0.60225	0.36451	1.53757	1.92121

As shown in Table 3, the highest mean score of 4.12 was realised for the 'supply' factor which represents the dominance of supply chain challenges in the LPG industry. Mean scores for demand-side, operational and security issues ranged between 1.41 and 1.95. These results demonstrate that supply-side challenges exist significantly in the LPG industry in rural medical facilities, while those of demand, operation and security nature are insignificant in occurrence. As mentioned by Asamoah, Amoakohene and Adiwokor (2012) the majority of such challenges are critical in the LPG supply chain as because they relate to the existence of various stakeholders such as customers, retailers, distributors and producers as well as service providers. These challenges complicate the major functions of the LPG supply chain which are to gather, transmit, and distribute this product (Ríos-Mercado and Borraz-Sánchez 2014).

The dominance of supply chain related challenges is not surprising, given the long-standing economic problems experienced in Zimbabwe. The country has for some time been facing severe foreign currency deficiencies, leading to shortages of all imported products. Consequently, petroleum products such as petrol, diesel and LPG, among others are in short supply, which accounts for the supply chain challenges experienced by rural medical facilities.

4.6 Significance of the Knowledge of Liquid Petroleum Gas Supply Chain Challenges

The descriptive statistics of the extent to which the knowledge of LPG supply chain challenges is significant are presented in Table 4.

Table 4. Extent of the knowledge of liquid petroleum gas

Extent	Frequency	Percentage (%)
Very significant	0	0
Somewhat significant	55	57
Slightly significant	38	39
Not very significant	3	3
Not at all significant	1	1
Total	97	100

The results in Table 4 show that 57% (n = 55) of the respondents perceived that their LPG knowledge was somewhat significant, while 39% (n = 38) suggested their knowledge to be slightly significant. Those who suggested that their LPG knowledge was not very significant were 3% (n = 3) of the total respondents in the study while 1% (n = 1) of the respondents indicated that the level of their knowledge was not at all significant. None of the respondents in the study suggested that their level of LPG knowledge was very significant. The results, therefore, show that there was a significantly high number of respondents with knowledge of LPG.

The use of LPG in the country increased significantly over the previous years, due to shortages of electrical energy in Zimbabwe in the past decade. The increased use of LPG meant that many people in rural areas were exposed to this source of energy, and in the process, acquired some working knowledge of it. Hence the majority of respondents indicated that they had a somewhat significant knowledge of LPG.

4.7 The Significance of Current Risks to Medical Facilities' Supply Chain

The descriptive statistics of the significance of the current risks to medical facilities' LPG supply chains are presented in Table 5.

Table 5. Significance of current liquid petroleum gas risks

Extent	Frequency	Percentage (%)
Very significant	66	68
Somewhat significant	20	21
Slightly significant	5	5
Not very significant	4	4
Not at all significant	2	2
Total	97	100

The results in Table 5 show that 68% (n = 66) of the respondents in the study perceived that LPG risks are very significant. Those who suggested the risks to be somewhat significant were 21% (n = 20) of the respondents while 5% (n = 5), 4% (n = 4) and 2% (n = 2) suggested that the risks were slightly significant, not very significant and not at all significant, respectively. The results, therefore, suggest that supply chain risks are very significant in terms of the delivery of LPG to the medical facilities considered in this study.

4.8 The Extent to Which Medical Facilities Are Affected by Supply Chain Problems

The descriptive statistics of the extent to which medical facilities are affected by supply chain problems are presented in Table 6.

Table 6. Extent of supply chain problems

Problem	Mean	Median	Std. deviation	Variance	Skewness
Delays	2.9175	3.0000	1.23887	1.535	−.277
Disruptions	2.8247	3.0000	1.23331	1.521	−.100
Loss of key suppliers	3.1649	3.0000	1.34382	1.806	−.466
Customer dependence	3.6289	4.0000	1.20173	1.444	−.752
Forecasting errors	3.5773	4.0000	.96647	.934	−.471
Existence of low-priced competitor products	3.4124	4.0000	1.16147	1.349	−.659
Capacity	3.5567	4.0000	1.14536	1.312	−.333
Technical skills	3.5361	4.0000	1.15526	1.335	−.586
Counterfeit products	3.4330	4.0000	1.24923	1.561	−.581
System failure	4.2371	5.0000	1.24811	1.558	−1.284

It is vital to discover the mean of the extent of supply chain problems as this assists to determine the degree to which the results are accurately representative of the collected data. As presented in Table 6, the mean score for the existence of LPG delays was 2.9 while that for LPG delivery disruptions was 2.8. The mean response to the loss of key LPG suppliers was 3.2 while that for the problem of customer dependence was 3.6. The mean scores for the existence of such supply-chain problems as forecasting errors, low-priced products, poor delivery capacity, poor technical skills, counterfeit products and system failure were 3.6, 3.4, 3.6, 3.6, 3.4 and 4.2, respectively. These results confirm the position that there are some considerable challenges met in the use of LPG due to the distribution model of the gas.

System failure, with a mean score of 4.2, emerged as the most significant supply chain problem, followed by customer dependence (mean = 3.62) and forecasting errors (mean = 3.6). This result demonstrates that the most significant impact of the supply chain problems is felt in these areas.

The results of this study did not at all suggest that delays (mean = 2.9) influence the use of LPG and as such differ from the suggestions made by ESMAP (2007). Likewise, the results of the present study neither confirmed nor denied the existence of disruptions due to distribution monopolies (mean = 2.8) that can complicate the provision of all LPG-related services as suggested by Lascurain (2016).

The results signify that any shortages of LPG would lead to the failure of all systems that are dependent on this source of energy for their operation. Likewise, the quality of service delivery to customers, who depend on the rural medical facility for their health-related issues, would then deteriorate. It would also become very difficult to plan for future requirements such as drugs, due to inadequate storage facilities such as refrigeration space, which cannot function at all without LPG.

4.9 Effects of Liquid Petroleum Gas Supply Chain Disruptions

The descriptive statistics regarding the effects of LPG supply chain disruptions are presented in Table 7.

Table 7. Effects of LPG supply-chain disruptions

Effect	Mean	Median	Std. deviation	Variance	Skewness
Product shortages	4.6112	4.0000	1.23939	1.299	−.566
Late deliveries	4.6322	4.0000	1.74658	1.113	−.238
Product recalls	1.1491	4.0000	1.34643	1.468	−.119
Customer dissatisfaction	4.3872	4.0000	1.67801	1.912	−.642
Contractual disagreements	4.2371	5.0000	1.24811	1.558	−1.284
Lawsuits and other legal complications	4.2268	5.0000	1.01556	1.031	−1.203
Bad organisational image	4.2990	5.0000	.93725	.878	−1.179
Demurrage charges	3.7423	4.0000	1.33290	1.777	−.537
Product counterfeiting	1.0330	5.0000	1.11717	1.248	−1.980

The above results reveal that LPG supply chain disruptions lead to product shortages, late gas deliveries, customer dissatisfaction, contractual disagreements, lawsuits and legal problems as well as a poor organisational image. To a lesser extent, however, supply chain disruptions lead to LPG product recalls, demurrage charges and product counterfeiting. As presented in Table 7, a mean score of 4.6 was computed for the effect of LPG supply chain disruptions on LPG shortages, while a mean score of 4.6 was obtained for its effect on LPG late deliveries. The mean scores for the effect of supply chain disruptions on LPG recalls, customer dissatisfaction, contractual disagreements, lawsuits and legal complications as well as on bad organisational image were 1.1, 4.4, 4.2, and 4.2, respectively. For the effect of LPG supply chain disruptions on demurrage charges and product counterfeiting, mean scores of 3.7 and 1.0 were realised in the study.

The results of this study testify that late gas deliveries, customer dissatisfaction, contractual disagreements, lawsuits and legal problems as well poor organisational image are some of the bottlenecks that have to be managed if the use of LPG in rural medical facilities is to be effective. The results of the study are in line with Surajit and Dhalla's (2010) suggestion that the success of a supply chain is dictated by how well a company controls its supply base and mitigates supply bottlenecks and liabilities.

As shown by the results, the most significant effects of the LPG supply chain disruptions are late deliveries (mean = 4.63), followed by product shortages (mean = 4.61) and customer dissatisfaction (mean = 4.4). The least significant effects of LPG supply chain disruptions emerged as product counterfeiting. The latter result is not surprising, given that LPG only has alternatives such as electricity, solar and biogas, which are also either in short supply in Zimbabwe or are beyond the reach of most rural medical facilities in the country.

5 Conclusions and Managerial Implications

This study aimed to investigate the supply chain challenges associated with the use of LPG in rural medical facilities in Zimbabwe. The study revealed that most rural medical facilities in Zimbabwe are making use of LPG and are actively involved in its procurement. The use of LPG extends to private LPG suppliers and the NGOs, which shows that a more substantial proportion of organisations have adopted the use of LPG in recent years. The use of the LPG has been beneficial to rural medical facilities and enables them to manage drug storage and to enhance their daily operations, service delivery and performance. The use of LPG enhances service delivery by providing a reliable, efficient, cheap and readily available energy source to use in their refrigerators as well as for other energy needs. The study further showed that LPG supply challenges exist significantly in rural medical facilities. Some of these challenges include supply chain disruptions that greatly lead to product shortages, late gas deliveries, customer dissatisfaction, contractual disagreements, lawsuits and legal problems as well as poor organisational image.

The current study has several managerial implications. Since the use of LPG is beneficial to rural medical facilities, its adoption is encouraged throughout the healthcare industry, especially in those institutions located in remote areas. Stakeholders in the LPG industry, such as government, NGOs and private LPG suppliers should improve their coordination through proper and routine consultations, since dealing with existing supply chain challenges requires holistic stakeholder efforts. The use of LPG for vaccine programs in rural medical facilities can be enhanced by both employing trained technicians who can manage the LPG systems, as well as by training staff on the use and maintenance of the gas systems. Although NGOs and private companies are supporting most rural medical facilities, the support is often offered once-off, which creates a need to have in-house technicians who can attend to the gas systems when challenges arise.

In future, similar studies may be conducted in other essential rural facilities such as shopping centres, schools, police stations and any amenities intended to benefit the community. Given the persistent shortage of electrical energy in Zimbabwe, it would be interesting to extend the scope of the study to major towns and cities where LPG is also in regular use. Similar studies can also be conducted using the qualitative approach in order to obtain more in-depth insights from the users of LPG.

References

Abubakar, T.: A study of sustainability in the oil and gas supply chain. Doctoral thesis. University of Central Lancashire, United Kingdom (2014)

Acharibasam, J.B., Apatinga, G.A.: Ghana and the liguidified petroleum gas dilemma critical analysis of Ghana's LPG policy. Int. J. Energy Environ. Res. **2**(2), 1–8 (2014)

Asamoah, D., Amoakohene, R., Adiwokor, E.: Analysis of Liquefied Petroleum Gas (LPG) shortage in Ghana: a case of the Ashanti region. Int. J. Bus. Adm. **3**(5), 89–98 (2012)

Avila, N., Carvallo, J.P., Shaw, B., Kammen, D.M.: The energy challenge in sub-Saharan Africa: a guide for advocates and policymakers. Part 1: generating energy for sustainable and equitable development. Oxfam Research Backgrounder (2017)

Chege, S.M.: Strategies adopted by Liquified Petroleum Gas (LPG) companies to deal with the challenge of cross-filling activities in Kenya. University of Nairobi (2013)

Ekouevi, K., Tuntivate, V.: Energy Access for Cooking and Heating: Lessons Learned and the Way Forward. World Bank, Washington D.C (2012)

Energy Sector Management Assistance Program (ESMAP): Haiti: strategy to alleviate the pressure of fuel demand on national wood fuel resources. World Bank (2007)

Falzon, J., Vignati, F., Halstead, M., van der Linden, N., Pols, D.: Accelerating uptake of LPG in Maputo for lower-income households: study to support scoping of an intervention (2013). https://energypedia.info/ipages/b/bf/LPG_Maputo_Report_FINAL.pdf. Accessed 26 Oct 2018

Gujba, H., Mulugetta, Y., Azapagic, A.: The household cooking sector in Nigeria: environmental and economic sustainability assessment. Resources **4**, 412–433 (2015)

Hutton, G., Rehfuess, E., Tediosi, F.: Evaluation of the costs and benefits of interventions to reduce indoor air pollution. Energy. Sustain. Dev. **11**(4), 34–43 (2007)

Jahre, M., Dumoulin, L., Greenhalgh, L.B., Hudspeth, C., Limlim, P., Spindler, A.: Improving health in developing countries: reducing complexity of drug supply chains. J. Hum. Logist. Supply Chain. Manag. **2**(1), 54–84 (2012)

Kaburia, R.: Challenges of liquid petroleum gas penetration in Kenya. Int. J. Econ. Financ. **2**(3), 161–180 (2016)

Kojima, M.: The role of liquefied petroleum gas in reducing energy poverty. Extractive Industries For Development Series, No. 25, December 2011, 2nd edn. World Bank (2013)

Kojima, M.: The role of liquefied petroleum gas in reducing energy poverty. Extractive Industries for Development Series, No. 25, December 2011. World Bank (2011)

Lascurain, J.: Market Segmentation: Improved Cookstoves and Clean Fuels in Guatemala. Fast-Track Carbon. Global Alliance for Clean Cookstoves, Washington, D.C. (2016)

Ministry of Electricity and Renewable Energy (MEER): Plan Nacional de Cocción Eficiente. Ministerio de Electricidad y Energía Renovable, Ecuador (2015)

Miniaci, R., Scarpa, C., Valbonesi, P.: Fuel poverty and the energy benefits system: the Italian case. The Center for Research on Energy and Environmental Economics and Policy, Bocconi University, Milano (2014)

Osoro, A.: Challenges affecting performance of supply chain systems in the petroleum industries in Kenya. Doctoral thesis. Jomo Kenyatta University of Agriculture and Technology, Kenya (2015)

Ríos-Mercado, R.Z., Borraz-Sánchez, C.: Optimization problems in natural gas transportation systems: a state-of-the-art review. Los Alamos National Laboratory, Los Alamos, USA (2014)

Surajit, R., Dhalla, R.S.: Management of supply chain in petroleum corporations in India. In: Proceedings of the 2010 International Conference on Industrial Engineering and Operations Management, Dhaka, Bangladesh (2010)

Toft, L. Beaton, C., Lontoh, L.: International experiences with LPG subsidy reform: options for Indonesia. International Institute for Sustainable Development (2016). https://www.iisd.org/sites/default/files/publications/international-experiences-with-LPG-subsidy-reform.pdf. Accessed 26 Oct 2018

Author Index

U. Clausen et al. (Eds.): ICPLT 2019, LNLO, pp. 303–304, 2019.
https://doi.org/10.1007/978-3-030-13535-5

The manufacturer's authorised representative in the EU is Springer Nature Customer Service Centre GmbH, Europaplatz 3, 69115 Heidelberg, Germany. If you have any concerns regarding our products, please contact ProductSafety@springernature.com

Printed and bound by CPI Group (UK) Ltd, Croydon, CR0 4YY

30/09/2026

02225229-0008